Lecture Notes in Mathematics

Edited by J.-M. Morel, F. Takens and B. Teissier

Editorial Policy for Multi-Author Publications: Summer Schools / Intensive Courses

1. Lecture Notes aim to report new developments in all areas of mathematics and their applica-
    tions – quickly, informally and at a high level. Mathematical texts analysing new developments
    in modelling and numerical simulation are welcome. Manuscripts should be reasonably self-
    contained and rounded off. Thus they may, and often will, present not only results of the author
    but also related work by other people. They should provide sufficient motivation, examples and
    applications. There should also be an introduction making the text comprehensible to a wider
    audience. This clearly distinguishes Lecture Notes from journal articles or technical reports
    which normally are very concise. Articles intended for a journal but too long to be accepted by
    most journals, usually do not have this „lecture notes" character.

2. In general SUMMER SCHOOLS and other similar INTENSIVE COURSES are held to present
    mathematical topics that are close to the frontiers of recent research to an audience at the
    beginning or intermediate graduate level, who may want to continue with this area of work, for
    a thesis or later. This makes demands on the didactic aspects of the presentation. Because the
    subjects of such schools are advanced, there often exists no textbook, and so ideally, the
    publication resulting from such a school could be a first approximation to such a textbook.

    Usually several authors are involved in the writing, so it is not always simple to obtain a unified
    approach to the presentation.

    For prospective publication in LNM, the resulting manuscript should not be just a collection of
    course notes, each of which has been developed by an individual author with little or no co-
    ordination with the others, and with little or no common concept. The subject matter should
    dictate the structure of the book, and the authorship of each part or chapter should take
    secondary importance. Of course the choice of authors is crucial to the quality of the material
    at the school and in the book, and the intention here is not to belittle their impact, but simply
    to say that the book should be planned to be written by these authors jointly, and not just
    assembled as a result of what these authors happen to submit.

    This represents considerable preparatory work (as it is imperative to ensure that the authors
    know these criteria before they invest work on a manuscript), and also considerable editing
    work afterwards, to get the book into final shape. Still it is the form that holds the most
    promise of a successful book that will be used by its intended audience, rather than yet another
    volume of proceedings for the library shelf.

3. Manuscripts should be submitted (preferably in duplicate) either to Springer's mathematics
    editorial in Heidelberg, or to one of the series editors (with a copy to Springer). Volume editors
    are expected to arrange for the refereeing, to the usual scientific standards, of the individual
    contributions. If the resulting reports can be forwarded to us (series editors or Springer) this is
    very helpful. If no reports are forwarded or if other questions remain unclear in respect of
    homogeneity etc, the series editors may wish to consult external referees for an overall
    evaluation of the volume. A final decision to publish can be made only on the basis of the
    complete manuscript; however a preliminary decision can be based on a pre-final or
    incomplete manuscript. The strict minimum amount of material that will be considered should
    include a detailed outline describing the planned contents of each chapter.

    Volume editors and authors should be aware that incomplete or insufficiently close to final
    manuscripts almost always result in longer evaluation times. They should also be aware that
    parallel submission of their manuscript to another publisher while under consideration for
    LNM will in general lead to immediate rejection.

Continued on inside back-cover

Lecture Notes in Mathematics 1869

A. Dembo · T. Funaki

Lectures on Probability Theory and Statistics

Ecole d'Eté de Probabilités
de Saint-Flour XXXIII - 2003

Editor: Jean Picard

 Springer

Authors and Editors

Amir Dembo

Mathematics Department
Stanford University
Stanford, CA 94305
USA
e-mail: amir@math.stanford.edu

Tadahisa Funaki

Graduate School of Mathematical Sciences
University of Tokyo
Komaba
Tokyo 153-8914
Japan
e-mail: funaki@ms.u-tokyo.ac.jp

Jean Picard

Laboratoire de Mathématiques Appliquées
UMR CNRS 6620
Université Blaise Pascal (Clermont-Ferrand)
63177 Aubière Cedex
France
e-mail: jean.picard@math.univ-bpclermont.fr

Cover: Blaise Pascal (1623-1662)

Library of Congress Control Number: 2005931754

Mathematics Subject Classification (2000): 60J10, 60J65, 60K35, 28A80, 28A78, 31C15, 82C41
82B24, 82B31, 82B41, 82C24, 82C31, 82C41, 35J20, 35K55, 35R35

ISSN print edition: 0075-8434
ISSN electronic edition: 1617-9692
ISSN Ecole d'Eté de Probabilités de St. Flour, print edition: 0721-5363
ISBN-10 3-540-26069-2 Springer Berlin Heidelberg New York
ISBN-13 978-3-540-26069-1 Springer Berlin Heidelberg New York
DOI 10.1007/b136622

Springer is a part of Springer Science+Business Media
springeronline.com
© Springer-Verlag Berlin Heidelberg 2005
Printed in The Netherlands

Typesetting: by the authors and TechBooks using a Springer LaTeX package

Cover design: *design & production* GmbH, Heidelberg

Printed on acid-free paper   SPIN: 11429579   41/TechBooks   5 4 3 2 1 0

Preface

Three series of lectures were given at the 33rd Probability Summer School in Saint-Flour (July 6–23, 2003), by the Professors Dembo, Funaki and Massart. This volume contains the courses of Professors Dembo and Funaki. The course of Professor Massart, entitled "Concentration inequalities and model selection", will appear in another volume. We are grateful to the authors for their important contribution.

64 participants have attended this school. 31 of them have given a short lecture. The lists of participants and of short lectures are enclosed at the end of the volume.

The Saint-Flour Probability Summer School was founded in 1971. Here are the references of Springer volumes where lectures of previous years were published. All numbers refer to the *Lecture Notes in Mathematics* series, except S-50 which refers to volume 50 of the *Lecture Notes in Statistics* series.

1971: vol 307	1980: vol 929	1990: vol 1527	1997: vol 1717
1973: vol 390	1981: vol 976	1991: vol 1541	1998: vol 1738
1974: vol 480	1982: vol 1097	1992: vol 1581	1999: vol 1781
1975: vol 539	1983: vol 1117	1993: vol 1608	2000: vol 1816
1976: vol 598	1984: vol 1180	1994: vol 1648	2001: vol 1837 & 1851
1977: vol 678	1985/86/87: vol 1362 & S-50		2002: vol 1840
1978: vol 774	1988: vol 1427	1995: vol 1690	
1979: vol 876	1989: vol 1464	1996: vol 1665	

Further details can be found on the summer school web site
http://math.univ-bpclermont.fr/stflour/

Jean Picard
Clermont-Ferrand, September 2005

Contents

Part I

Amir Dembo: Favorite Points, Cover Times
and Fractals

Favorite Points, Cover Times and Fractals

Amir Dembo

Department of Mathematics, Stanford University, Stanford, CA 94305,
amir@math.stanford.edu

1 Overview

In this course we follow recent advances in the study of the fractal nature of certain random sets, emphasizing the methods used to obtain such results. We focus on some of the fine properties of the sample path of the most basic stochastic processes such as simple random walks, Brownian motion, and symmetric stable processes. As we shall see, probability on trees inspires many of our proofs, with trees used to model the relevant correlation structure. Along the way we also mention quite a few challenging open research problems. Among the methods that will be detailed here are

- Cover time for Markov chains (see Chap. 2).
- The dimension of discrete lim sup random fractals (see Chap. 3).
- The truncated second moment method (see Chap. 4).
- The KMT strong approximation construction (see Chap. 5).
- Ciesielski-Taylor identities (see Chap. 6).

The highly recommended survey of Taylor [Tay86] has many interesting examples of random fractals we cannot even mention in this course, as well as numerous references to earlier works in this field. It is also highly recommended to further study Le Gall's lecture notes [lG92] for a deep analysis of properties of the Brownian path that we only touch upon here.

1.1 Favorite Points and Cover Times

Most Visited Points

Let $S_n = \sum_{i=1}^n X_i$ denote a simple random walk in $\mathbb{Z}^2$. A natural question is: "How many times does the walk revisit the most frequently visited site in the first n steps?". This question was posed by Erdős and Taylor more than forty years ago in [ET60] and only recently resolved (in [DPRZ01]). More formally, let $T_n(x)$ denote the number of visits of S_n to x by time n, and let $T_n^* := \max_{x \in \mathbb{Z}^2} T_n(x)$. Then, almost surely,

$$\lim_{n \to \infty} \frac{T_n^*}{(\log n)^2} = \frac{1}{\pi} \qquad (1.1)$$

(in Chaps. 4 and 5 we outline the proof of (1.1)). For any $0 < \alpha < 1$ we call $x \in \mathbb{Z}^2$ an α-favorite point if $T_n(x) \geq (\alpha/\pi)(\log n)^2$ and let $\mathcal{F}_n(\alpha)$ denote the set of α-favorite points. It is also proved in [DPRZ01] that for $\alpha \in (0, 1]$,

$$\lim_{n \to \infty} \frac{\log |\mathcal{F}_n(\alpha)|}{\log n} = 1 - \alpha \qquad a.s. \qquad (1.2)$$

that is,

$$|\{x : T_n(x) \geq (\alpha/\pi)(\log n)^2\}| \sim n^{1-\alpha} .$$

In other words, the n^β-most visited point by time n is visited approximately

$$\frac{1-\beta}{\pi}(\log n)^2$$

times. Moreover, any random sequence $\{x_n\}$ in $\mathbb{Z}^2$ such that $T_n(x_n)/T_n^* \to 1$ must satisfy

$$\lim_{n\to\infty} \frac{\log |x_n|}{\log n} = \frac{1}{2} \qquad a.s.$$

This is a partial reply to a question of Révész (see Chap. 19 of [Rév90]): "What is the rate of convergence of the favorite point of the walk to infinity?" The analogous statement for the simple random walk on $\mathbb{Z}$ is contained in a well-known result of Bass and Griffin [BG85] (see also [LS04] and the references therein for recent developments in settling this question). The proof of these facts tells us also how the most visited site x_n^* is visited: typically the walk makes excursions of "all" time scales n^θ ($0 < \theta < 1$) between the visits to x_n^*.

Most of the problems we mention can be expressed in a Brownian setting. Indeed, for any Borel measurable function f from $0 \le t \le T$ to $\mathbb{R}^2$, let μ_T^f denote its *occupation measure*:

$$\mu_T^f(A) = \int_0^T \mathbf{1}_A(f_t)dt$$

for all Borel sets $A \subseteq \mathbb{R}^2$. Let $\{w_t\}_{t\ge 0}$ denote the planar Brownian motion started at the origin, and $\bar{\theta} = \inf\{t : |w_t| = 1\}$ the exit time of the unit disc $D(0,1)$ (where throughout $D(x,r)$ denotes the open disc in $\mathbb{R}^2$ of radius r centered at x). Since the path $\{w_t : 0 \le t \le \bar{\theta}\}$ is a compact set, it follows that $\mu_{\bar{\theta}}^w(D(x,r)) = 0$ for any x not in the path and all r small enough. It can be seen (using for example Lévy's uniform modulus of continuity for the upper bound, and techniques as those in [PT87] for the lower bound) that

$$\mathbb{P}\left(\frac{\log \mu_{\bar{\theta}}^w(D(x,r))}{\log r} \xrightarrow[r\to 0]{} 2 \quad \forall x \in \text{path}\right) = 1 .$$

Therefore, standard multi-fractal analysis must be refined in order to distinguish between highly visited and less visited points. This was done in [DPRZ01] where it is proved that for any $0 < a \le 2$,

$$\dim\{x : \lim_{r\to 0} \frac{\mu_{\bar{\theta}}^w(D(x,r))}{r^2(\log r)^2} = a\} = 2 - a \qquad a.s. \tag{1.3}$$

(throughout this chapter dim denotes the Hausdorff dimension of the set, see Sect. 3.1 for its definition and some of its properties). In Chap. 5 we outline the proof of (1.3) from a similar result on binary trees derived in Chap. 4. This, together with the appropriate upper bound yields

$$\limsup_{r\to 0} \sup_{x\in\mathbb{R}^2} \frac{\mu_{\bar{\theta}}^w(D(x,r))}{r^2(\log r)^2} = 2 \qquad a.s. \tag{1.4}$$

as conjectured by Perkins-Taylor in [PT87]. Note that for a typical x on the Brownian path,

$$\mu_{\bar{\theta}}^{w}(D(x,r)) \asymp r^2 |\log r|$$

(e.g. see [DPRZ01, Lemma 2.1]), so the *a-thick points*, i.e. those in the set considered in (1.3), correspond to unusually large occupation measure. The strong approximation of random walks by Brownian motion relates (1.1) and (1.2) of the discrete setting to (1.4) and (1.3). This derivation highlights the significance of the construction of Komlós-Major-Tsunády [KMT75] which asserts the existence of a Brownian motion w and a simple random walk S on the same probability space such that $|S_{[t]} - w_t| = O(\log t)$. Earlier approximations are not sharp enough for our task (for more details, see Chap. 5).

The proof of (1.3) and (1.4) relies on observing the Brownian motion upon hitting a sequence of concentric discs. When the radii of the discs are appropriately chosen, the observed process is approximately a simple random walk. The authors of [DPRZ01] study the probability of having numerous excursions at many scales (radii) around the same point. During each excursion, the Brownian motion "scores" some occupation measure around that point. Those excursions are independent and since their number is large the total occupation measure is highly concentrated around its mean. An alternative, simpler approach is to discretize the problem by taking a maximal collection of points in $D(0,1)$ such that $\inf_{l \neq j} |x_l - x_j| > \varepsilon$, and consider the random variable

$$Z = \sum_j \mathbf{1}_{\left\{ \frac{\mu(D(x_j,r))}{r^2(\log(r))^2} \geq a \right\}}$$

(where we use μ for $\mu_{\bar{\theta}}^{w}$). Indeed, for any point x there exists a j such that x_j is close to x so the occupation measure around x is approximately the same as the one around x_j. Thus, the event $\{Z \geq 1\}$, is approximately the same as

$$\sup_{x \in D(0,1)} \frac{\mu(D(x,r))}{r^2(\log r)^2} \geq a \ .$$

Since it is easily checked that

$$\mathbb{P}\left(\frac{\mu(D(0,r))}{r^2(\log r^2)} > \xi \right) \underset{r \to 0}{\to} e^{-\xi} \ ,$$

one can get upper bounds in (1.3) and (1.4) using the first moment method (i.e. bounding $\mathbb{P}(Z \geq 1)$ by $\mathbb{E}Z$). This is what Perkins and Taylor did in [PT87]. However, due to high correlations between the occupation measure at different points, the second moment of Z is too large, so applying the second moment method fails to produce a tight lower bound.

Late Points and Cover Time

Simulating a simple random walk on a 512×512 torus we observe that those points which are visited late appear as "islands" of various sizes in the simulation. A natural question is what are the geometric characteristics of such islands. Recent results on this topic, motivated by the physics paper [BH91] can be found in [DPRZ05]. More precisely, let $\tilde{X}_j$ be a simple random walk on the two dimensional torus $\mathbb{Z}_n^2 = \mathbb{Z}^2/n\mathbb{Z}^2$, with $\tau_x = \min\{j \geq 0 : \tilde{X}_j = x\}$ the first hitting time of x, and

$$C_n = \max_{x \in \mathbb{Z}_n^2}\{\tau_x\} ,$$

the cover time of $\mathbb{Z}_n^2$ by the simple random walk.

The simple random walk is a time-reversible Markov chain. There already exists a theory dealing with asymptotic for cover time of such processes to which the manuscript [AF01] is devoted. Whereas this theory provides the correct scaling in n it fails to provide the multiplying constant. Indeed, it was only recently shown in [DPRZ04] that

$$\lim_{n \to \infty} \frac{C_n}{(n \log n)^2} = \frac{4}{\pi} \quad \text{in probability} . \tag{1.5}$$

Previous work on this problem include the proof of the upper bound by D. Aldous [Ald89], as well as the proof of the lower bound $2/\pi$ by G. Lawler [Law92]. A nice informal description of this problem, due to H. Wilf [Wil89], is given in the introduction to [AF01]. This problem and those we described in the previous subsection, are much easier to handle in dimension $d \geq 3$. However, little is known beyond the limit of (1.5) or the corresponding limit for $d \geq 3$. For example,

Open problem 1. *Does* $\sqrt{C_n} - Med\{\sqrt{C_n}\}$ *multiplied by some appropriate normalization factor converge in distribution to a non-degenerate random variable, and if so what are the factor and the limit distribution?* Even the existence of a normalizing sequence that results with a tight, yet non-degenerate collection, is not obvious. For the corresponding problem for simple random walk on regular trees, a proof of tightness is contained in [BZ05].

As shown in [DPRZ04], the strong approximation theorems of [KMT75] allow one to obtain (1.5) from the corresponding problem for the Brownian motion on the unit torus which we describe next. Let $\{X_t\}$ denotes a Brownian motion on the two dimensional unit torus $\mathbb{T}^2$, with the corresponding hitting times,

$$\tau(x, \varepsilon) = \inf\{t > 0 : X_t \in D(x, \varepsilon)\} ,$$

and the ε-cover time,

$$C_\varepsilon = \sup_{x \in \mathbb{T}^2} \{\tau(x, \varepsilon)\} . \tag{1.6}$$

Equivalently, C_ε is the amount of time needed for the Wiener sausage of radius ε to completely cover $\mathbb{T}^2$. Then, it is shown in [DPRZ04] that

$$\lim_{\varepsilon \to 0} \frac{C_\varepsilon}{(\log \varepsilon)^2} = \frac{2}{\pi} \quad \text{a.s.} \tag{1.7}$$

The cover time problem (1.7) is, in a sense, "dual" to the Perkins-Taylor conjecture (1.4), in that it replaces "extremely large" occupation measure by "extremely small" occupation measure.

The general theory of cover times for Markov chains gives universal bounds. For example (see [AF01]), there exist constants k and K such that for any graph G with $|V|$ vertices

$$k|V| \log |V| \leq C_G \leq K|V|^3 \ ,$$

where C_G stands for the expected cover time of G by a simple random walk on this graph. In [JS00], Jonasson and Schramm proved that if G is a planar graph of maximal degree d then there are constants k_d and K_d depending only on d such that

$$k_d|V|(\log |V|)^2 \leq C_G \leq K_d|V|^2 \ . \tag{1.8}$$

Open problem 2. *Is the square lattice asymptotically the easiest to cover when $d = 4$? That is, does $C_G \geq (1/\pi)|V|(\log |V|)^2(1+o(1))$ as $|V| \to \infty$, for any collection of planar graphs of maximal degree $d = 4$?*

1.2 Fractal Geometry of Late and Favorite Points

Returning to the Brownian motion on the two dimensional unit torus $\mathbb{T}^2$, en route to (1.7) it is shown in [DPRZ04] that

$$\sup_{x \in \mathbb{T}^2} \limsup_{\varepsilon \to 0} \frac{\tau(x,\varepsilon)}{(\log \varepsilon)^2} = \frac{2}{\pi} \quad a.s. \ ,$$

and that for $a \leq 2$,

$$\dim \left\{ x \in \mathbb{T}^2 : \limsup_{\varepsilon \to 0} \frac{\tau(x,\varepsilon)}{(\log \varepsilon)^2} = \frac{a}{\pi} \right\} = 2 - a \quad a.s. \tag{1.9}$$

We call $x \in \mathbb{T}^2$ a *late point* if it is in the set considered in (1.9) for some $a > 0$.

Open problem 3. *Study the consistently late points, where the* $\limsup$ *in (1.9) is replaced by a limit or a* $\liminf$. The difficulty in doing so lies in the fact that the behaviors for different scales (i.e. ε's) are highly dependent when the scales are too close to each other.

Moving back to the discrete setting of a simple random walk (SRW) on the lattice torus $\mathbb{Z}_n^2$ which starts at the origin, the corresponding set of α-late points is now

$$\mathcal{L}_n(\alpha) = \{ x \in \mathbb{Z}_n^2 : \tau_x \geq \alpha(4/\pi)(n \log n)^2 \} \ .$$

It is shown in [DPRZ05] that for $\alpha \in (0, 1]$

$$\limsup_{n \to \infty} \frac{\log |\mathcal{L}_n(\alpha)|}{\log n} = 2(1 - \alpha) \quad \text{in probability} \,.$$

Thus, the size of $\mathcal{L}_n(\alpha)$ is typically $n^{2(1-\alpha)+o(1)}$. Several theorems in [DPRZ05] characterize the geometry of these sets. For example, for any fixed x in $\mathbb{Z}_n^2$ and $0 < \beta < 1$,

$$\lim_{n \to \infty} \frac{\log |\mathcal{L}_n(\alpha) \cap D(x, n^\beta)|}{\log n} = 2\beta - 2\alpha/\beta \quad \text{in probability} \,. \tag{1.10}$$

If the points of $\mathcal{L}_n(\alpha)$ were spread out uniformly in $\mathbb{Z}_n^2$, the limit in (1.10) would be $n^{2\beta-2\alpha}$, whereas we get much less α-late points near each fixed point x (as $2\beta - 2\alpha/\beta < 2\beta - 2\alpha$). This suggests a clustering pattern for the points of $\mathcal{L}_n(\alpha)$. It is confirmed by the following result of [DPRZ05]: for any $0 < \alpha, \beta < 1$, if Y_n is chosen uniformly in $\mathcal{L}_n(\alpha)$ then,

$$\lim_{n \to \infty} \frac{\log |\mathcal{L}_n(\alpha) \cap D(Y_n, n^\beta)|}{\log n} = 2\beta(1 - \alpha) \quad \text{in probability} \,. \tag{1.11}$$

An example of wrong heuristic: When exploring the number of pairs of late points one is tempted to apply the following natural approximations:

$$\mathbb{E}[\# \text{ of } \alpha\text{-late points within distance } n^\beta \text{ of each other}]$$
$$= n^2 n^{2\beta} \mathbb{P}(x, y \text{ are } \alpha\text{-late when } |x - y| \simeq n^\beta)$$
$$\simeq (\text{Typical value of such number of pairs})$$
$$\simeq (\# \text{ discs in } n^\beta \text{ grid with } \alpha\text{-late points})$$
$$\times (\text{Typical value of } |\mathcal{L}_n(\alpha) \cap D(x, n^\beta)| \text{ when } x \text{ is } \alpha\text{-late})^2$$

Such approximations are indicated in [BH91], but as seen in [DPRZ05], they fail to hold, with each of the three quantities displayed here exhibiting a different power growth exponent.

Open problem 4. *In [DPRZ05] the power growth exponent of pairs of α-late points within distance n^β of each other is computed. Extend this to a "full multi-fractal analysis". For example, find the power growth exponent of triplets (x_1, x_2, x_3) of α-late points, such that x_i is within distance n^β of x_j for $i, j = 1, 2, 3$.*

The techniques developed in [DPRZ01] and [DPRZ04] can answer many other related geometrical questions. For example, let R_n denote the maximal radius r such that for some x every point of $D(x, r) \cap \mathbb{Z}^2$ is in the range $\{S_1, S_2, \ldots, S_n\}$ of the simple random walk on $\mathbb{Z}^2$. Equivalently, R_n is the radius of the largest discrete disc that is completely covered by the planar simple random walk by time n. In 1990, P. Révész asked what is the asymptotic growth of R_n (see [Rév90, p. 247])? Based on numerical simulations (by Arvind Singh), Zhan Shi conjectured that $R_n \simeq n^{1/4+o(1)}$. This has been recently proved by combining the methods used in proving (1.1) and (1.5) (see

[DPR05] for details). Similarly, [DPR05] show that the largest discrete disc in the intersection of two independent planar simple random walk path, each run by time n, has radius $R_{n,2} \simeq n^{1/(2+2\sqrt{2})+o(1)}$, and that for any $0 < \alpha < 1$, the largest discrete disc of α-favorite points has radius $R_n(\alpha) \simeq n^{(1-\sqrt{\alpha})/4+o(1)}$. However,

Open problem 5. *The growth rate of the diameter $D_{n,2}$ of the largest connected component of the intersection of two independent planar simple random walk path, each run by time n is not known (Singh's numerical simulations suggest that $D_{n,2} = \sqrt{n}/(\log n)^b$ for some $1 < b < 2$). A related problem is to determine whether the intersection of two independent planar Brownian motion path, each run for a unit time, is almost surely a totally disconnected set.*

Upon understanding [DPRZ05], one is likely able to derive the corresponding results for favorite points. However, little is known beyond that. In particular,

Open problem 6. *Nothing is known about the evolution of the most favorite point $n \mapsto x_n^*$ in any dimension $d \geq 2$.*

Open problem 7. *How is the growth of time between the first and last visit to x_n^* prior to n, affected by the dimension d?*

Similarly, nothing is known about finer properties of the very late points. For example,

Open problem 8. *What is the distribution of the distance between the last two points to be covered by the simple random walk in $\mathbb{Z}_n^2$? In particular, does the chance that they are adjacent go to zero as n grows?*

1.3 Thick Points and Cover Time in $d \geq 3$ Dimensions

Ciesielski-Taylor Identities and Thick Points

Let $\{w_t\}$ be a Brownian motion in $\mathbb{R}^d$, $d \geq 3$, started at the origin. Recall our notation $\mu_\infty^w(B(0,r))$ for the total occupation measure of the ball $B(0,r)$ by w_t, and let

$$\sigma_d(0,r) = \inf\{t \geq 0 : \ w_t \notin B(0,r)\}$$

denote the exit time of w_t from $B(0,r)$. It is shown in [CT62] by computing moments that $\mu_\infty^w(B(0,r))$ and $\sigma_{d-2}(0,r)$ have exactly the same law for any $d \geq 3$ and $r > 0$. In the same paper Ciesielski and Taylor use this identity to prove that

$$\limsup_{r \to 0} \frac{\mu_\infty^w(B(0,r))}{r^2 \log|\log r|} = \frac{2}{\lambda_{d-2}} \quad \text{a.s.} \tag{1.12}$$

where λ_{d-2} is the minimal eigenvalue of the Laplacian operator in the unit ball of $\mathbb{R}^{d-2}$. In Chap. 6 we explain Kac's moments method, used by [CT62] to derive their identities. We also review in Sect. 6.3 an alternative derivation by [Yor91] of these identities as a consequence of Ray-Knight theorems.

It is shown in [DPRZ00a] that for any $d \geq 3$,

$$\sup_{x \in \mathbb{R}^d} \limsup_{r \to 0} \frac{\mu_\infty^w(B(x,r))}{r^2 |\log r|} = \frac{4}{\lambda_{d-2}} \quad \text{a.s.} \tag{1.13}$$

and for any $0 < a < 2$,

$$\dim \left\{ x \in \mathbb{R}^d \mid \limsup_{r \to 0} \frac{\mu_\infty^w(B(x,r))}{r^2 |\log r|} = \frac{2a}{\lambda_{d-2}} \right\} = 2 - a \quad \text{a.s.} \tag{1.14}$$

The appearance of λ_{d-2} is due to the Ciesielski-Taylor identity, though only the asymptotic exponential rate of decay of $\mathbb{P}(\mu_\infty^w(B(0,1)) \geq t)$ is relevant here. As we see in Chap. 3, contrary to the two-dimensional case, where excursions of all time scales are responsible for high occupation measure, here we can localize the time interval contributing to high occupation measure at x, resulting with a very well behaved second moment. This is due to the transience of the Brownian motion in dimension $d \geq 3$, because of which the typical thick point x occurs when the Brownian path stays for an unusually long time in its neighborhood, to which it does not return thereafter. Another difference with the two-dimensional case is that we get a different scaling when we replace the lim sup by lim inf in (1.13) and (1.14). Indeed, it is shown in [DPRZ00a] that

$$\frac{1}{d} \leq \sup_{x \in \mathbb{R}^d} \liminf_{r \to 0} \frac{\mu_\infty^w(B(x,r))}{r^2} \leq C_d . \tag{1.15}$$

Open problem 9. *Find the exact value in (1.15) and more generally, provide some understanding of the corresponding mechanism for consistently thick points.*

Cover Time for Brownian Motion on Compact Manifolds

Consider a Brownian motion X_t on a "nice" d-dimensional compact manifold M. For example, if $M = S^{d-1}$ then the Brownian motion on M is, up to a time change, the projection of the Brownian motion in $\mathbb{R}^d$ to the unit sphere. Let $C_\varepsilon(M)$ denote the ε-cover time of M by the process, i.e. $C_\varepsilon(M) = \sup_{x \in M} \{\tau(x, \varepsilon)\}$ where

$$\tau(x, \varepsilon) = \inf\{t > 0 : X_t \in B(x, \varepsilon)\} ,$$

and the ball $B(x, \varepsilon)$ is defined with respect to a suitable Riemannian metric in M. It is shown in [DPR03] that when $d \geq 3$,

$$\lim_{\varepsilon \to 0} \frac{C_\varepsilon(M)}{h_d(\varepsilon)} = d\kappa_M, \quad \text{a.s.} \tag{1.16}$$

where $h_d(\varepsilon) = \varepsilon^{2-d}(\log(1/\varepsilon))$, and κ_M is an explicit constant which perhaps surprisingly depends only on the volume and the dimension of M.

Open problem 10. *Nothing is known about the non-degenerate fluctuations of $C_\varepsilon(M)$ on finer scales, nor about finer estimates for $\mathbb{E}[C_\varepsilon(M)]$. Of particular interest is to find information about $C_\varepsilon(M)$ that depends on the geometry of M beyond its volume and dimension.*

The result (1.16) may be considered the dual of (1.13). Indeed, as we show in Chap. 2, the limit of $\mathbb{E}[C_\varepsilon(M)]/h_d(\varepsilon)$ follows from the general theory of cover time for Markov processes. Specifically, we follow there the derivation of such general bounds by Matthews in [Mat88a] (and [Mat88b]), then apply them for $M = \mathbb{T}^d$, $d \geq 3$ (for which no differential geometry is needed). In [DPR03] the methods we derive in Chap. 3 give the almost sure result (1.16) as well as the analog of (1.14) and other fractal properties of the set of late points in M.

1.4 Extensions

The results we described before mostly depend on the local properties of the stochastic processes we considered. They are thus not limited to Brownian motion or to simple random walks. In particular, sample path continuity is not essential. For instance, [DPRZ99] derive the analog of (1.13) and (1.14) for transient symmetric stable processes (i.e. when the index $\beta < d$), and Daviaud [Dav05] deals with the harder problem of getting the analogs of (1.3) and (1.4) for the Cauchy process on $\mathbb{R}$ (where $\beta = d = 1$ and the process is recurrent, like the planar Brownian motion).

While the Markov property of all these processes is of much help, [DPRZ02] deals with thick points for intersections of planar sample path, whereby the Markov property is partially lost.

As we see in Chap. 4, (1.1) and (1.4) are the analogs of natural properties of simple random walks on regular (b-ary) trees. In the same manner, [BDG01] pursues the similarity between the two dimensional discrete Gaussian free field ϕ_x on a large square $B_N = \{1, \ldots, N\}^2$, and a branching random walk type model on regular trees. In this context they derive independently of [DPRZ01] a similar refinement of the second moment method, en-route to proving that $\max_{x \in B_N} \phi_x$ is of order $g_N = 2\sqrt{2/\pi}\log N$ and the study of the field subject to the non-negativity constraint $\Omega_+ = \{\phi_x \geq 0 : x \in B_N\}$. Daviaud [Dav04] pursues this approach further, providing the analogs of the results of [DPRZ05] and [DPR05] in the context of this Gaussian free field. For example, it is shown in [Dav04] that conditioned on Ω_+, the largest disc within B_N for which all values of ϕ_x are below $2\eta g_N$ is of radius $N^{\eta(1+o(1))}$. The results of [Dav04] suggest the possibility of simpler proofs using an isomorphism between $\phi_x^2/2$ and the local time at x of a continuous time planar simple random walk. However,

Open problem 11. *Find a simpler proof, for example, of* (1.1), *that uses such an isomorphism.*

Some of the open problems we mentioned are about random fluctuations and non-degenerate limit distributions. These are believed to be related to behavior such as that of branching Brownian motion, and hence to KPP-type partial differential equations (consider further, [Ald91a], [Bra83], [Bra86], [McK75]).

Acknowledgment

I thank my coauthors Yuval Peres, Jay Rosen and Ofer Zeitouni, without whom this course would never have happened, and the National Science Foundation for funding the research on which this material is based. I also thank my students in Stanford whose notes have been the source for this manuscript, most notably Olivier Daviaud (Chap. 1), Debashis Paul (Chap. 2), Hernan Awad (Chap. 3), Jose Blanchet (Chap. 4), Ery Arias-Castro and Dimitrios Cheliotis (Chap. 6), Sanatan Rai, Stephanie Pereira, and Yonatan Gutman (editing and reviewing).

I am grateful to Marc Yor for his comments which greatly improved the exposition of Sect. 6.3, and to Dimitrios Cheliotis for creating the bulk of the figures in this manuscript. I am also grateful to the participants in the St. Flour summer school of 2003 for their comments and to the organizer, Jean Picard, for his hospitality. It was a pleasure and an honor to share the podium with Tadahisa Funaki and Pascal Massart.

2 Cover Time for Markov Chains

Let $X(t)$ be a time-homogeneous strong Markov process on the probability space $(\Omega, \mathcal{F})$ with topological state space $\mathcal{A}$ and almost surely right continuous sample path. Denote by $\mathbb{P}_a$ the probability measure of $X(t)$ when starting at $X(0) = a$. Let $A_1, \ldots, A_N$ be closed subsets of $\mathcal{A}$ for which the first entry times of A_j by the process $X(t)$,

$$T(A_j) = \inf\{t \geq 0 : X(t) \in A_j\}, \quad j = 1, \ldots, N,$$

are measurable. For any collection of sets $\{A_1, \ldots, A_i\}$, let

$$T(A_1, \ldots, A_i) = \max_{1 \leq j \leq i} T(A_j).$$

In this chapter we provide bounds on $T(A_1, \ldots, A_N)$ in terms of information about $\{T(A_i)\}_{i=1}^N$. Following Matthews [Mat88a] we deal in Sect. 2.1 with the mean $\mathbb{E}_a(T(A_1, \ldots, A_N))$ (see also [AF01] for a similar derivation). Section 2.2 utilizes the same method to bound the moment generating function of $T(A_1, \ldots, A_N)$, following [Mat88b] (note also that [Ald91b] shows that under a natural hypothesis, the cover time for a finite state time-homogeneous Markov chain is well approximated by its expectation, as the size of the state space tends to infinity). In Sect. 2.3 we demonstrate how the results of Sect. 2.1 apply for the ϵ-cover time $C_\epsilon(\mathbb{T}^d)$ of the Brownian motion on the d-dimensional unit torus $\mathbb{T}^d$, for $d \geq 3$ (c.f. (1.6)). We prove there that

$$\lim_{\epsilon \to 0} \frac{\mathbb{E}C_\epsilon(\mathbb{T}^d)}{h_d(\epsilon)} = d\kappa_{\mathbb{T}^d} \ , \tag{2.1}$$

for $h_d(\epsilon) = \epsilon^{2-d}|\log \epsilon|$ and some explicit constant $\kappa_{\mathbb{T}^d}$. The upper bound in (2.1) holds also when $d = 2$ (now with $h_2(\epsilon) = |\log \epsilon|^2$) but as we shall see (in Remark 2.14), this general theory fails to provide the matching lower bound when $d = 2$. Indeed [DPRZ04] prove the lower bound in this case by a sophisticated refinement of the second moment method, along the lines we shall further explore in Chaps. 4 and 5. We remark that [DPR03] combine the tail bounds of Sect. 2.2 with the stochastic co-dimension approach we explore in Chap. 3 to go beyond (2.1), getting the almost sure convergence of (1.16), as well as the ϵ-cover time $C_\epsilon(E)$ for subsets $E \subset M$ and the Hausdorff dimension of the set of a-late points.

2.1 Expected Cover Time

We further assume that $T(A_j)$ are stopping times with respect to the filtration $\mathcal{G}_t = \sigma(X(s), s \leq t)$. For example, this applies for any discrete time Markov chain, or when $X(t)$ has continuous sample path with probability one (see Proposition I.4.5 of [RY99]). If $X(t)$ is only right continuous, the same proof we give below applies, just taking everywhere the filtration $\cap_{s>t}\sigma(X(u), u \leq s)$ as $\mathcal{G}_t$, to guarantee that $T(A_j)$ be stopping times (see Proposition I.4.6. of [RY99]).

Fixing $a_0 \in \mathcal{A}$, let $\widehat{A}_i = \{a_0\} \cup_{j \neq i} A_j$ for $i = 1, 2, \ldots, N$ and define

$$\mu_- = \min_{1 \leq i \leq N} \inf_{a \in \widehat{A}_i} \mathbb{E}_a(T(A_i)), \qquad \mu_+ = \max_{1 \leq i \leq N} \sup_{a \in \widehat{A}_i} \mathbb{E}_a(T(A_i)) \ .$$

Our main theorem is then:

Theorem 2.1. *Let $\mathcal{D}_n$ denote the event that $T(A_1), \ldots, T(A_N)$ are distinct and non-zero. Then,*

$$\mathbb{P}_{a_0}(\mathcal{D}_n)\mu_- \sum_{i=1}^N \frac{1}{i} \leq \mathbb{E}_{a_0}(T(A_1, \ldots, A_N)) \leq \mu_+ \sum_{i=1}^N \frac{1}{i} \ .$$

Example 2.2. *An illustrative simple example is the coupon collector's problem, where $X(t)$ is a sequence of i.i.d. uniform samples on $\{1, \ldots, N\}$ and $A_i = \{i\}, i = 1, \ldots, N$ (while $\mathcal{A} = \{0, 1, \ldots, N\}$ and $a_0 = 0$). Then, $\mathbb{E}_a(T(A_i)) = \mu$ does not depend on $a \notin A_i$. Moreover, μ is the mean of a Geometric distribution with parameter $\frac{1}{N}$, so $\mu = N$ and we have that*

$$\mathbb{E}_0(T(A_1,\dots,A_N)) = \mu \sum_{i=1}^{N} \frac{1}{i} \sim N \log N \ .$$

The main idea of the proof of Theorem 2.1 is to introduce auxiliary randomness thus creating some exchangeability in the extended probability space we construct. To this end, let (S_N, G_N, γ) be the probability space of a random permutation σ on $\{1, \dots, N\}$ having a uniform law, and consider the product probability space $(\Omega \times S_N, \mathcal{F} \times G_N, \mathbb{P}_{a_0} \times \gamma)$. We denote by $\mathbb{P}$ the product probability measure $\mathbb{P}_{a_0} \times \gamma$ and the expectation with respect to $\mathbb{P}$ by $\mathbb{E}$. Define the projections

$$X(\omega, \pi) = X(\omega), \qquad \sigma(\omega, \pi) = \sigma(\pi) = \pi, \qquad \text{for } \omega \in \Omega, \ \pi \in S_N$$

For any $\pi \in S_N$ let $A_i^\pi = A_{\pi_i}$ with $T(A_i^\sigma)$ and $T(A_1^\sigma, \dots, A_i^\sigma)$ defined as before. Notice that

$$\mathbb{E}_{a_0}(T(A_1, \dots, A_N)) = \mathbb{E}(T(A_1^\sigma, \dots, A_N^\sigma)) \ ,$$

and define,

$$R_1 = T(A_1^\sigma)$$
$$R_i = T(A_1^\sigma, \dots, A_i^\sigma) - T(A_1^\sigma, \dots, A_{i-1}^\sigma), \quad i = 2, \dots, N \ .$$

We use hereafter the notation T_i for $T(A_1^\sigma, \dots, A_i^\sigma)$, with $T_0 = 0$, so $R_i = T_i - T_{i-1} \geq 0$ but it is possible to have $R_i = 0$. Let $(A_1^\omega, \dots, A_N^\omega)$ be a listing of $(A_1, \dots, A_N)$ in the order they are entered by $X(t)$, with ties broken by the convention that if $T(A_i) = T(A_j)$ then A_i appears before A_j in our list if $i < j$. Let $r_i = 1$ if A_i^σ is to the right of $A_1^\sigma, \dots, A_{i-1}^\sigma$ in the list $(A_1^\omega, \dots, A_N^\omega)$, and $r_i = 0$, otherwise. The following lemma is the key to the proof of the theorem.

Lemma 2.3. T_i and r_i are independent and

$$\frac{1}{i}\mathbb{P}_{a_0}(\mathcal{D}_n) \leq \mathbb{P}(R_i \neq 0) \leq \mathbb{P}(r_i = 1) = \frac{1}{i} \ , \quad \text{for all } i = 1, \dots, N \ . \qquad (2.2)$$

Proof of Lemma 2.3. Let $\{\sigma\}$ be an unordered list containing $\{\sigma_1, \dots, \sigma_i\}$ and $\mathcal{G}$ the σ-field generated by $\mathcal{F}$ and $\{\sigma\}$. Note that the list $(A_1^\omega, \dots, A_N^\omega)$ is measurable with respect to $\mathcal{F} \subset \mathcal{G}$, and the event $\{T_i \leq t\}$ is in $\mathcal{G}$. Because of the independence of σ and $X(\cdot)$ and the uniform law of σ, even when we are given $\{\sigma\}$ and $(A_1^\omega, \dots, A_N^\omega)$, the conditional probability of $\{r_i = 1\}$ is $\frac{1}{i}$. Thus, $\mathbb{P}(r_i = 1|\mathcal{G}) = \frac{1}{i}$ implying that the unconditional probability of $\{r_i = 1\}$ is also $\frac{1}{i}$. Combining all this we have by the tower property of conditional expectation,

$$\mathbb{P}(T_i \leq t, r_i = 1) = \mathbb{E}\big(\mathbf{1}_{\{T_i \leq t\}} \mathbb{P}(r_i = 1|\mathcal{G})\big)$$
$$= \frac{1}{i}\mathbb{P}(T_i \leq t) = \mathbb{P}(r_i = 1)\mathbb{P}(T_i \leq t) \ ,$$

proving the independence of T_i and r_i. By definition, the event $\{r_i = 0\}$ implies that $R_i = 0$ as well. Though the converse in general may fail, it holds when $T(A_i)$ are distinct and non-zero, so we have that

$$\mathbb{P}(r_i = 1, \mathcal{D}_n) \leq \mathbb{P}(R_i \neq 0) \leq \mathbb{P}(r_i = 1) \, .$$

With $\mathcal{D}_n \in \mathcal{F} \subset \mathcal{G}$ and $\mathbb{P}(r_i = 1|\mathcal{G}) = \frac{1}{i}$ we have the bounds of (2.2) and the proof of the lemma is complete. □

Proof of Theorem 2.1. Since $T(A_1, \ldots, A_N) \geq T(A_i)$ for all i it is easy to see that if either μ_- or μ_+ is infinite, the corresponding bound trivially hold. Hence, we assume hereafter that both μ_+ and μ_- are finite. By our assumptions, T_i are stopping times with respect to the filtration $\mathcal{F}_t$ generated by σ and $\mathcal{G}_t$. Hence,

$$\mathbb{E}_{a_0}(T(A_1, \ldots, A_N)) = \mathbb{E}(T_N) = \sum_{i=1}^{N} \mathbb{E}(R_i) \;=\; \sum_{i=1}^{N} \mathbb{E}(\mathbb{E}(R_i|\mathcal{F}^{i-1})) \, , \quad (2.3)$$

where $\mathcal{F}^{i-1}$ denote the stopped σ-field $\mathcal{F}_{T_{i-1}}$. Note that the event $\{R_i = 0\}$ is in $\mathcal{F}^{i-1}$. Using the strong Markov property of the process $X(\cdot)$ at T_{i-1} we have that

$$\mathbb{E}(R_i|\mathcal{F}^{i-1}) = \mathbf{1}_{R_i \neq 0} \, \mathbb{E}_{X(T_{i-1})}(T(A_i^\sigma)) \, .$$

By the right continuity of $t \mapsto X(t)$ we have that $X(T_{i-1}) \in \widehat{A_i^\sigma}$ for any $i = 1, \ldots, N$ and σ. Hence,

$$\mu_- \, \mathbf{1}_{R_i \neq 0} \leq \mathbb{E}(R_i|\mathcal{F}^{i-1}) \leq \mu_+ \, \mathbf{1}_{R_i \neq 0} \, . \quad (2.4)$$

Thus, combining (2.3) with (2.4) we conclude that

$$\mu_- \sum_{i=1}^{N} \mathbb{P}(R_i \neq 0) \leq \mathbb{E}_{a_0}(T(A_1 \ldots, A_N)) \leq \mu_+ \sum_{i=1}^{N} \mathbb{P}(R_i \neq 0) \, .$$

Applying the bounds (2.2) of Lemma 2.3 completes the proof. □

Remark 2.4. *Theorem 2.1 is a key ingredient in the proof of the non-trivial part of (1.8), namely the universal lower bound on the expected cover time of a planar graph G with vertex set V and maximal degree d, by a simple random walk on G. Indeed, Koebe's circle packing theorem states that for a finite planar graph G there exists a packing $\{C_v : v \in V\}$ of closed discs in $\mathbb{R}^2$ of radii $r_v > 0$, indexed by the vertices of G and such that C_v intersects C_u if and only if $\{v, u\}$ is an edge of G. Using this theorem Jonasson and Schramm deduce the existence of a positive constant c_d such that $R(u, v) \geq c_d \log(\text{dist}(C_u, C_v)/r_v)$ for any two vertices u, v of any planar graph G of maximal degree d. Here, the effective resistance $R(u, v)$ is i^{-1} where i is the current flowing into $v \in V$ when the voltage of u is set to one, that of v is set to zero and a unit resistor is put along each edge of G. From [Tet91] one has the relation*

$$\mathbb{E}_a(T(\{b\})) = \frac{1}{2} \sum_{w \in V} d_w (R(a,b) + R(b,w) - R(a,w)) \, ,$$

where d_w is the degree of vertex $w \in V$ of G, allowing [JS00] to convert information about effective resistances to information about the first hitting times $\mathbb{E}_a(T(\{b\}))$. Finally, Theorem 2.1 relates the latter to the expected cover time C_G, eventually yielding (1.8).

2.2 Tail Estimates

In the setting of Sect. 2.1, following [Mat88b], a refinement of the method of proof of Theorem 2.1 gives sharp bounds on the moment generating function of $T(A_1, \ldots, A_N)$ in terms of the functions

$$f^-(s) = \min_{1 \le i \le N} \inf_{a \in \hat{A}_i} \mathbb{E}_a \left[e^{sT(A_i)} \right]$$

and

$$f^+(s) = \max_{1 \le i \le N} \sup_{a \in \hat{A}_i} \mathbb{E}_a \left[e^{sT(A_i)} \right] .$$

Theorem 2.5. *For any $s \in \mathbb{R}$,*

$$\prod_{i=1}^N \left(\frac{i}{i - 1 + \frac{1}{f^-(s)}} \right) \le \mathbb{E}_{a_0} \left[e^{sT(A_1, \ldots, A_N)} \right] \le \prod_{i=1}^N \left(\frac{i}{i - 1 + \frac{1}{f^+(s)}} \right) . \quad (2.5)$$

Proof. In case $f^+(s) = \infty$ or $f^-(s) = 0$ the term of the product in (2.5) corresponding to $i = 1$ is infinite or zero, respectively, and (2.5) trivially holds. Hence, we assume hereafter that $\infty > f^+(s) > f^-(s) > 0$. Constructing the product space $(\Omega \times S_N, \mathcal{F} \times G_N, \mathbb{P}_{a_0} \times \gamma)$ as in the proof of Theorem 2.1 we use the notations $T_i = T(A_1^\sigma, \ldots, A_i^\sigma)$ and

$$R_1 = T_1, \ R_i = T_i - T_{i-1}, \ i = 2, \ldots, N \, ,$$

as before. Then, with the same notations used earlier, we have

$$\mathbb{E}_{a_0} \left[e^{sT(A_1, \ldots, A_N)} \right] = \mathbb{E}(e^{sT_N}) \, .$$

We saw that T_i is a stopping time with respect to the filtration $\mathcal{F}_t$ and that the event $\{R_i \ne 0\}$ is measurable on the σ-field $\mathcal{F}^{i-1}$. By definition,

$$\frac{1}{1 - 1 + \frac{1}{f^-(s)}} = f^-(s) \le \mathbb{E} \left[e^{sT(A_1^\sigma)} \right] \le f^+(s) = \frac{1}{1 - 1 + \frac{1}{f^+(s)}} . \quad (2.6)$$

We proceed to show that for $i \ge 2$,

$$\frac{i}{i - 1 + \frac{1}{f^-(s)}}\mathbb{E}\left[e^{sT_{i-1}}\right] \leq \mathbb{E}\left[e^{sT_i}\right] \leq \frac{i}{i - 1 + \frac{1}{f^+(s)}}\mathbb{E}\left[e^{sT_{i-1}}\right] . \tag{2.7}$$

Then, multiplying both sides of the inequalities (2.6) and (2.7) over $i = 1, \ldots, N$ we get the result (2.5) of the theorem.

Since T_{i-1} is measurable on $\mathcal{F}^{i-1}$, we have the decomposition

$$\mathbb{E}\left[e^{sT_i}\right] = \mathbb{E}\left[e^{s(T_{i-1}+R_i)}\right] = \mathbb{E}\left[e^{sT_{i-1}}\mathbb{E}\left(e^{sR_i}|\mathcal{F}^{i-1}\right)\right] . \tag{2.8}$$

Recall that the event $\{r_i = 0\}$ implies that $R_i = 0$ (see Lemma 2.3). Hence, we also have the decomposition

$$\mathbb{E}\left(e^{sR_i}|\mathcal{F}^{i-1}\right) = (1 - r_i) + r_i\mathbb{E}\left(e^{sR_i}|\mathcal{F}^{i-1}\right) . \tag{2.9}$$

Note that if $r_i = 1$ then $T(A_i^\sigma) \geq T_{i-1}$ and R_i is the extra time needed to enter A_i^σ starting at $X(T_{i-1})$ which is in $\widehat{A_i^\sigma}$. Consequently, using (2.9) we get by the strong Markov property of $X(\cdot)$ at T_{i-1} and the definitions of $f^-(\cdot)$ and $f^+(\cdot)$ that

$$(1 - r_i) + r_i f^-(s) \leq \mathbb{E}\left(e^{sR_i}|\mathcal{F}^{i-1}\right)$$
$$= (1 - r_i) + r_i\mathbb{E}_{X(T_{i-1})}\left[e^{sT(A_i^\sigma)}\right] \leq (1 - r_i) + r_i f^+(s) .$$

Note that this applies regardless of possible ties in T_i's and for all possible values of s, both positive and negative. We can rearrange the bounds above as

$$f^-(s) + (1 - r_i)(1 - f^-(s)) \leq \mathbb{E}\left(e^{sR_i}|\mathcal{F}^{i-1}\right)$$
$$\leq f^+(s) + (1 - r_i)(1 - f^+(s)) . \tag{2.10}$$

By Lemma 2.3, $\mathbb{P}(r_i = 0) = 1 - \frac{1}{i}$ and T_i and r_i are independent. With $\{r_i = 0\}$ implying that $T_i = T_{i-1}$ we thus have that,

$$\mathbb{E}\left[e^{sT_{i-1}}(1 - r_i)\right] = \mathbb{E}\left[e^{sT_i}(1 - r_i)\right] = \mathbb{E}\left[e^{sT_i}\right]\left(1 - \frac{1}{i}\right) .$$

Combining this with (2.8) and (2.10) we get (2.7) after some algebraic manipulations.   □

Remark 2.6. *If the state space $\mathcal{A}$ is finite then we can derive Theorem 2.1 from Theorem 2.5 by considering the limit as $s \uparrow 0$ of $\frac{1}{s}(\mathbb{E}(e^{sT_N}) - 1)$. Specifically, taking limits of the lower and upper bounds in (2.5) and using the fact that $\lim_{s\uparrow 0} f^\pm(s) = 1$, and that $0 \leq (e^{st} - 1)/s \leq t$ for $s < 0$, we get that*

$$f_*^- \sum_{i=1}^{N}\frac{1}{i} \leq \mathbb{E}_{a_0}(T(A_1, \ldots, A_N)) = \mathbb{E}(T_N) \leq f_*^+ \sum_{i=1}^{N}\frac{1}{i} ,$$

where

$$f_*^- = \lim_{s\uparrow 0}\frac{1}{s}(f^+(s) - 1) \quad and \quad f_*^+ = \lim_{s\uparrow 0}\frac{1}{s}(f^-(s) - 1) ,$$

equal μ^- and μ^+, respectively (when $\mathcal{A}$ is finite).

2.3 Brownian Cover Time for the d-Dimensional Torus, $d \geq 3$

Let $X(t)$ be the Brownian Motion on the d-dimensional unit torus $\mathbb{T}^d$, $d \geq 3$. For $\epsilon > 0$, let $T(x, \epsilon) = \inf\{t \geq 0 : X(t) \in B(x, \epsilon)\}$ where $B(x, \epsilon)$ is the ball of radius ϵ in $\mathbb{T}^d$ centered at x and the ϵ-cover time

$$C_\epsilon(\mathbb{T}^d) := \max_{x \in \mathbb{T}^d} \{T(x, \epsilon)\}.$$

In this section we bound $\mathbb{E}\left[C_\epsilon(\mathbb{T}^d)\right]$ using Matthews' method (that is, Theorem 2.1). To this end, we have by Lemma 2.2 of [DPR03] that

Lemma 2.7. *Let M be a smooth, compact, connected, Riemannian manifold without boundary of dimension $d \geq 3$, and let $\{X(t)\}_{t \geq 0}$ denote the Brownian motion on M. Let $T(m, r) = \inf\{t \geq 0 : X(t) \in B(m, r)\}$. For any $0 < \eta < 1$ there exists $r_0 = r_0(\eta) > 0$ such that for all $r \leq r_0$ and $m \in M$,*

$$(1 - 3\eta)\kappa_M r^{2-d} \leq \inf_{x \notin B(m, r/\eta)} \mathbb{E}_x(T(m, r))$$

$$\leq \sup_{x \in M} \mathbb{E}_x(T(m, r)) \leq (1 + 3\eta)\kappa_M r^{2-d}.$$

We prove below Lemma 2.7 in case $M = \mathbb{T}^d$, where no differential geometry is needed, and deduce by Theorem 2.1 that:

Theorem 2.8. *For any $x \in \mathbb{T}^d$,*

$$\lim_{\epsilon \to 0} \frac{\mathbb{E}_x(C_\epsilon(\mathbb{T}^d))}{h_d(\epsilon)} = d\kappa_{\mathbb{T}^d}, \qquad \text{where } h_d(\epsilon) = \epsilon^{2-d} \log\left(\frac{1}{\epsilon}\right).$$

Remark 2.9. *The proof of Theorem 2.8 is similar to the one given in [Mat88a] for Brownian motion on the sphere S^d, $d \geq 3$. Theorem 2.5 is combined in [DPR03] with the approach of Chap. 3 to get in addition almost sure convergence and Hausdorff dimension results.*

Proof. Fix $1 > \delta, \eta > 0$ and let $\epsilon > 0$ be small enough that there exist $\{x_j : j = 1, \ldots, N\}$ in $\mathbb{T}^d \backslash B(x, \epsilon^{1-\delta})$ such that

$$\inf_{l \neq j} \rho(x_l, x_j) \geq \epsilon^{1-\delta} \qquad \text{and} \qquad N \geq \epsilon^{-d(1-2\delta)},$$

where $\rho(x, y)$ denotes hereafter the distance between $x \in \mathbb{T}^d$ and $y \in \mathbb{T}^d$. Consider the closed subsets $A_j = \{y : \rho(y, x_j) \leq \epsilon\}$, $j = 1, \ldots, N$, of $\mathbb{T}^d$. Since $X(t)$ has continuous sample path, clearly $T(A_j) = T(x_j, \epsilon)$, so by Lemma 2.7 and our choice of x_j, there exists ϵ_0 such that for any $0 < \epsilon < \epsilon_0$,

$$\mu_- = \min_{1 \leq j \leq N} \inf_{x \in \hat{A}_j} \mathbb{E}_x(T(A_j)) \geq \kappa_{\mathbb{T}^d}(1 - \eta)\epsilon^{2-d}. \qquad (2.11)$$

With A_j disjoint for $j = 1, \ldots, N$, we have that $T(A_j)$ are almost surely distinct and non-zero. Hence, by Theorem 2.1 and our choice of N,

$$\mathbb{E}_x(T(A_1,\dots,A_N)) \geq \mu_- \sum_{i=1}^{N} \frac{1}{i} \geq d(1-3\delta)\mu_- \log\left(\frac{1}{\epsilon}\right) . \qquad (2.12)$$

Further, by definition $C_\epsilon(\mathbb{T}^d) \geq T(A_1,\dots,A_N)$, hence combining (2.11) and (2.12) we conclude that

$$\mathbb{E}_x(C_\epsilon(\mathbb{T}^d)) \geq (1-\eta)(1-3\delta)d\kappa_{\mathbb{T}^d}h_d(\epsilon) . \qquad (2.13)$$

To prove the complementary upper bound let $\epsilon > 0$ be small enough that there exists a *maximal* collection of points $\{y_j : j = 1,\dots,\widetilde{N}\}$ in $\mathbb{T}^d$ such that

$$\inf_{l \neq j} \rho(y_l, y_j) \geq \delta\epsilon \qquad \text{and} \qquad \widetilde{N} \leq \epsilon^{-d(1+\delta)} .$$

Consider the closed sets $\widetilde{A}_j = \{y : \rho(y, y_j) \leq (1-\delta)\epsilon\}$, $j = 1,\dots,\widetilde{N}$, noting that for any $y \in \mathbb{T}^d$, there exists $1 \leq j \leq \widetilde{N}$ such that $\widetilde{A}_j \subset B(y, \epsilon)$. Hence, by definition $C_\epsilon(\mathbb{T}^d) \leq T(\widetilde{A}_1,\dots,\widetilde{A}_{\widetilde{N}})$, and applying Theorem 2.1 and Lemma 2.7 we have that

$$\mathbb{E}_x(C_\epsilon(\mathbb{T}^d)) \leq \mathbb{E}_x(T(\widetilde{A}_1,\dots,\widetilde{A}_{\widetilde{N}}))$$

$$\leq \mu_+ \sum_{i=1}^{\widetilde{N}} \frac{1}{i} \leq (1+\eta)(1+2\delta)d\kappa_{\mathbb{T}^d}(1-\delta)^{2-d}h_d(\epsilon) . \qquad (2.14)$$

Dividing (2.13) and (2.14) by $h_d(\epsilon)$, then taking $\epsilon \to 0$, followed by $\delta \to 0$ and $\eta \to 0$ we complete the proof of the theorem. $\qquad\square$

Recall Poisson's equation that expresses quantities related to the exit from a domain D in $\mathbb{R}^d$ via the solution of the appropriate elliptic Partial Differential Equation (PDE).

Lemma 2.10. *Let D be an open, bounded subset of $\mathbb{R}^d$ and $f : D \to \mathbb{R}$ and $\kappa : \partial D \to \mathbb{R}$ be bounded continuous functions. Given $u : \overline{D} \to \mathbb{R}$, which is in $C^2(D)$ and solves the Poisson equation, $\frac{1}{2}\Delta u = -f$ in D subject to the boundary condition $u = \kappa$ on ∂D, we have the representation*

$$u(x) = \mathbb{E}_x\left(\kappa(w_{\theta_D}) + \int_0^{\theta_D} f(w_s)\,ds\right) ,$$

where w_t is a Brownian motion on $\mathbb{R}^d$, $w_0 = x$ and $\theta_D = \inf\{t \geq 0 : w_t \notin D\}$.

For example, see Exercise 4.2.25 of [KS91] for outline of the proof of Lemma 2.10.

Following [DPR03] and [DPRZ04], the key to the proof of Lemma 2.7 for $\mathbb{T}^d = (-\frac{1}{2}, \frac{1}{2}]^d$ is the corresponding converse result on $\mathbb{T}^d$:

Lemma 2.11. *Let Δ denotes the Laplacian in $(-\frac{1}{2}, \frac{1}{2}]^d$ with periodic boundary conditions. Then, the function $u(x) = \mathbb{E}_x(T(y,r))$ solves the PDE $\frac{1}{2}\Delta u = -1$ on the open set $D_y = \mathbb{T}^d \backslash \overline{B}(y,r)$ with boundary conditions $u = 0$ on ∂D_y.*

Remark 2.12. *While this lemma is a special case of results from the theory of Brownian motion on manifolds, see [Aub82, p. 104], we provide below a direct simple proof of it.*

Proof. Obviously, u is periodic with $u = 0$ on ∂D_y. Considering $x \in D_y$ and $\delta \geq \epsilon > 0$ such that $B(x, \delta) \subset D_y$, let $\theta_\epsilon = \inf\{t \geq 0 : X(t) \notin B(x, \epsilon)\}$ and note that the path of $\{X(t), t \leq \theta_\epsilon\}$ has the same law as $\{w_t, t \leq \theta_{B(x,\epsilon)}\}$, hence $\mathbb{E}_x(\theta_\epsilon) = \frac{1}{d}\epsilon^2$ (for example, by Lemma 2.10), with $X(\theta_\epsilon)$ distributed uniformly on $\partial B(x, \epsilon)$. Consequently, by the strong Markov property of $X(t)$ at θ_ϵ and linearity of the expectation we have that

$$u(x) = \frac{1}{d}\epsilon^2 + \int_{\partial B(0,\epsilon)} u((x+z)_{\mathbb{T}^d})\mu_\epsilon(dz) ,$$

where $\mu_\epsilon(\cdot)$ denotes the uniform (surface) measure on $\partial B(0, \epsilon)$. Averaging according to the surface area $\epsilon^{d-1} v_{d-1}$ of $B(0, \epsilon)$ for $0 < \epsilon \leq \delta$, we get the integrated form of the PDE,

$$u(x) = \frac{1}{d+2}\delta^2 + \frac{1}{V_\delta} \int_{B(0,\epsilon)} u((x+z)_{\mathbb{T}^d})dz , \qquad (2.15)$$

with V_δ denoting the volume of $B(0, \delta)$.

With $\xi = \inf_{x \in \overline{D}_y} \mathbb{P}_x(T(y, r) \leq 1) > 0$, it follows that $u(x)$ is bounded above on $\overline{D}_y$ by the mean of a Geometric random variable of parameter $\xi > 0$. With u bounded and measurable, (2.15) implies that it is also continuous on D_y, and applying (2.15) repeatedly we get that u is C^∞ on D_y and solves the stated PDE $\frac{1}{2}\Delta u = -1$ in D_y (for example, this is a trivial adaptation of the proof of Proposition 4.2.5 of [KS91]).   □

Note that $u(\cdot)$ of Lemma 2.11 can be written as $u(x) = e((x - y)_{\mathbb{T}^d})$, where $e(\cdot)$ solves the PDE $\frac{1}{2}\Delta e = -1$ on $D = \mathbb{T}^d \backslash \overline{B}(0, r)$ with $e = 0$ on ∂D. Suppose that g is Green's function on $\mathbb{T}^d \backslash \{0\}$, that is, $\Delta g = 1$ on $\mathbb{T}^d \backslash \{0\}$. Then $v = g + \frac{1}{2}e$ is harmonic on D (i.e., $\Delta v = 0$ on D) and continuous on $\overline{D}$. Consequently, $v(X(t \wedge T(0, r)))$ is a uniformly integrable martingale, hence by the optional stopping theorem $v(x) = \mathbb{E}_x[v(T(0, r))]$ obtains its maximum and minimum over $\overline{D}$ at points in $\partial D = \partial B(0, r)$ (this is the maximum principle, also valid in much generality on manifolds, c.f. Theorem 3.74 of [Aub98]). Since $e = 0$ on ∂D, it follows that

$$2[\inf_{z \in \partial B(0,r)} g(z) - g(x)] \leq e(x) \leq 2[\sup_{z \in \partial B(0,r)} g(z) - g(x)] ,$$

so after some algebra, Lemma 2.7 (with $\kappa_{\mathbb{T}^d} = 2c_d$) is a direct consequence of

Lemma 2.13. *Fixing $d \geq 3$, there exists Green's function g on $\mathbb{T}^d \backslash \{0\}$ such that $g(x) - c_d|x|^{2-d}$ is uniformly bounded on $\mathbb{T}^d$, for $c_d = \frac{1}{(d-2)v_{d-1}}$.*

Remark 2.14. *Note that $\kappa_{\mathbb{T}^d}$ of Lemma 2.7 is merely $2c_d$ of Lemma 2.13. As Lemma 2.13 holds for $d = 2$ with $c_d|x|^{2-d}$ replaced by $c_2 \log\left(\frac{1}{|x|}\right)$ and $c_2 = \frac{1}{2\pi}$, we have the analog of Lemma 2.7 for $d = 2$. That is,*

$$\left| \mathbb{E}_x(T(m,r)) - \kappa_{\mathbb{T}^2} \log\left(\frac{\rho(x,m)}{r}\right) \right| ,$$

is uniformly bounded with respect to m, r and $x \in \mathbb{T}^2 \backslash \overline{B}(m,r)$. As in the derivation of (2.14) this yields the upper bound for Theorem 2.8, now with $h_2(\epsilon) = |\log \epsilon|^2$. However, trying to prove the corresponding lower bound by an application of Theorem 2.1 for a finite collection of discs with ϵ^b-separated centers $\{x_j : j = 1,\ldots,N\}$ (for some fixed $0 < b < 1$), shall result with $N = \epsilon^{-2b(1+o(1))}$ and $\mu_- = (1 - b)\kappa_{\mathbb{T}^2}(1 + o(1))|\log \epsilon|$. Taking the optimal value of $b = 1/2$ gives a lower bound on $\mathbb{E}_x[C_\epsilon(\mathbb{T}^2)]/h_2(\epsilon)$ that is one fourth of our upper bound. Indeed, to close this gap and get the correct limiting constant $(= 2\kappa_{\mathbb{T}^2})$, one has to apply the more sophisticated method of Chaps. 4 and 5.

Proof of Lemma 2.13. Let $\phi \in C^\infty$ be such that $\phi(x) = 1$ for $x \in [0, \delta_1)$ and $\phi(x) = 0$ for $x \in (\delta_2, \frac{1}{2}]$ and some $\frac{1}{2} > \delta_2 > \delta_1 > 0$ (see Fig. 2.1). Define $h(z) = c_d\phi(|z|)|z|^{2-d}$. Since $\hat{h}(|z|) = h(z)$ is a radial function,

$$\Delta h = \hat{h}'' + \frac{d-1}{r}\,\hat{h}' ,$$

and with $r = |z|$,

$$|z|^{d-1}(\Delta h)(z) = \frac{d}{dr}(r^{d-1}\hat{h}'(r)) . \tag{2.16}$$

Since $\hat{h}'(r) = 0$ for $r > \delta_2$ we have that

$$\int_{\mathbb{T}^d} (\Delta h)(z)dz = \int_{B(0,\delta_2)} (\Delta h)(z)dz = v_{d-1} \int_0^{\delta_2} \frac{d}{dr}(r^{d-1}\hat{h}'(r))dr$$

$$= -v_{d-1} \lim_{r\downarrow 0}\{r^{d-1}\hat{h}'(r)\} .$$

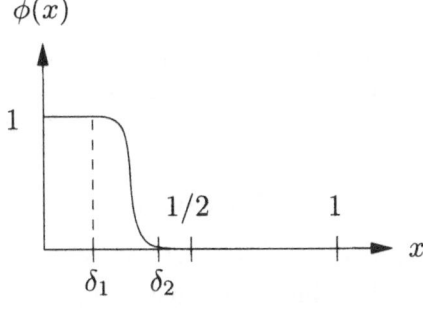

Fig. 2.1. A mollifying function $\phi(\cdot)$

With $\phi(x) = 1$ and $\phi'(x) = 0$ for $x \in [0, \delta_1)$, it follows that $\hat{h}'(r) = (2 - d)c_d r^{1-d}$ for $r < \delta_1$, so our choice of c_d results with

$$\int_{\mathbb{T}^d} (\Delta h)(z)dz = 1 . \tag{2.17}$$

Note that by (2.16) we have that $\Delta h(z) = 0$ for all $|z| < \delta_1$ or $|z| > \delta_2$, while in the annulus $\delta_1 \leq |z| \leq \delta_2$ the function $\Delta h(z)$ is C^∞. Thus, $H(z) = \Delta h(z) - 1$ is C^∞ on $\mathbb{T}^d$ and consequently has a Fourier series expansion,

$$H(z) = \sum_{k_1,\ldots,k_d} a_{k_1,\ldots,k_d} \prod_{i=1}^d \cos(2\pi k_i z_i) ,$$

with $a_{k_1,\ldots,k_d}$ rapidly decreasing. As a consequence of (2.17) we have $a_{0,\ldots,0} = 0$, so

$$F(z) = \sum_{k_1,\ldots,k_d} \frac{a_{k_1,\ldots,k_d}}{4\pi^2(\sum_{i=1}^d k_i^2)} \prod_{i=1}^d \cos(2\pi k_i z_i) \tag{2.18}$$

is also C^∞ on $\mathbb{T}^d$. It is easy to check that $\Delta F = -H$, so $g(z) = h(z) + F(z)$ has for $|z| > 0$,

$$\Delta g(z) = \Delta h(z) - H(z) = 1 .$$

Note that

$$g(z) - c_d|z|^{2-d} = F(z) - c_d|z|^{2-d}(1 - \phi(z))\mathbf{1}_{|z|\geq\delta_1} ,$$

is bounded on $\mathbb{T}^d$, as stated in the lemma. $\square$

3 Discrete Limsup Random Fractals

In this chapter we consider the elements of fractal geometry relevant in our context. Section 3.1 contains the necessary background on Hausdorff, Minkowski and packing dimensions. For related results and many interesting examples consider further the textbooks [Fal90] and [Mat95] on the subject of fractal geometry. Section 3.2 follows [KPX00] in defining the concept of discrete limsup random fractals and studying some of their properties. In Sect. 3.3 we demonstrate how such sets are used to provide the dimension of the set of Brownian thick points in dimension $d \geq 3$.

3.1 Background on Fractal Dimensions

Throughout this section A is a subset of a metric space (X, ρ), with $|A| = \sup\{\rho(x,y) : x, y \in A\}$ denoting the diameter of A. We start with the definition of Hausdorff measure and Hausdorff and Minkowski dimensions.

Hausdorff and Minkowski Dimensions

For $\alpha > 0$, and $\varepsilon > 0$ possibly infinite, let

$$\mathcal{H}^{\alpha}_{(\varepsilon)}(A) = \inf\left\{\sum_j |A_j|^{\alpha} : A \subset \bigcup_j A_j, |A_j| < \varepsilon\right\} .$$

Since $\mathcal{H}^{\alpha}_{(\varepsilon)}(A)$ is decreasing in ε, we have the existence of

$$\mathcal{H}^{\alpha}(A) = \lim_{\varepsilon \downarrow 0} \mathcal{H}^{\alpha}_{(\varepsilon)}(A) = \sup_{\varepsilon > 0} \mathcal{H}^{\alpha}_{(\varepsilon)}(A) ,$$

which is called the α-*Hausdorff measure* of A. Indeed, it can be shown that $\mathcal{H}^{\alpha}(\cdot)$ is a measure on the Borel subsets of (X, ρ).

It is easy to see that $\mathcal{H}^{\alpha}_{(\varepsilon)}(A)$ is decreasing in α for all $\varepsilon < 1$, hence so does $\alpha \mapsto \mathcal{H}^{\alpha}(A)$, prompting the following definition (c.f. Chap. 2 of [Fal90], and Fig. 3.1).

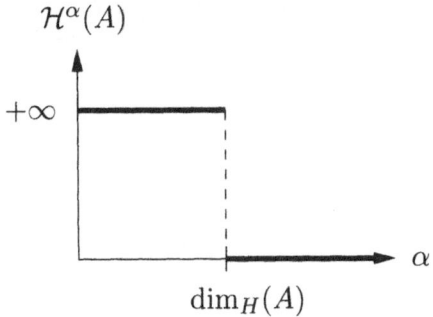

Fig. 3.1. The Hausdorff measure and dimension of a set A

Definition 3.1. *The Hausdorff dimension of A, denoted* $\dim_{\mathrm{H}}(A)$, *is*

$$\dim_{\mathrm{H}}(A) = \sup\left\{\alpha > 0 : \mathcal{H}^{\alpha}(A) = \infty\right\} .$$

For any $\beta > \alpha > 0$, and $\varepsilon < 1$ we have that

$$\mathcal{H}^{\beta}_{(\varepsilon)}(A) \leq \varepsilon^{\beta - \alpha} \mathcal{H}^{\alpha}_{(\varepsilon)}(A) .$$

Letting $\varepsilon \downarrow 0$ we see that if $\mathcal{H}^{\alpha}_{(\varepsilon)}(A) < \infty$ then necessarily $\mathcal{H}^{\beta}_{(\varepsilon)}(A) = 0$. Consequently, we have the alternative definition

$$\dim_{\mathrm{H}}(A) = \inf\left\{\alpha > 0 : \mathcal{H}^{\alpha}(A) = 0\right\} .$$

If $\dim_{\mathrm{H}}(A) = s$ then $\mathcal{H}^s(A)$ may be zero or infinite or anything in between. We note in passing that $\dim_{\mathrm{H}}(B) \leq \dim_{\mathrm{H}}(A)$ for any $B \subset A$. Chapter 2 of [Fal90] provides many other properties of $\dim_H(A)$ and examples where $\dim_H(A)$ is computed.

Considering a subset A of Euclidean space $\mathbb{R}^d$, let $N_\varepsilon(A)$ denote the minimum number of closed balls of radius ε needed to cover the set A. The growth of $N_\varepsilon(A)$ as $\varepsilon \downarrow 0$ provides an alternative definition of the dimension of the set A.

Definition 3.2. *The upper and lower Minkowski dimensions, denoted* $\overline{\dim}_M(A)$ *and* $\underline{\dim}_M(A)$, *are,*

$$\overline{\dim}_M(A) = \limsup_{\varepsilon \downarrow 0} \frac{\log N_\varepsilon(A)}{\log \frac{1}{\varepsilon}} \, ,$$

$$\underline{\dim}_M(A) = \liminf_{\varepsilon \downarrow 0} \frac{\log N_\varepsilon(A)}{\log \frac{1}{\varepsilon}} \, .$$

When $\overline{\dim}_M(A) = \underline{\dim}_M(A)$ we use the notation $\dim_M(A)$, called the Minkowski dimension of A. However, for many sets $\overline{\dim}_M(A) > \underline{\dim}_M(A)$.

The Minkowski dimension is often referred to as the box dimension since in Euclidean space, one can replace the balls with boxes (i.e. cubes). One may also restrict the cover considered to cubes on ε-mesh without changing the dimension (c.f. Proposition 3.1 of [Fal90]).

The Minkowski dimension serves as a simple upper bound of the Hausdorff dimension, that is,

Lemma 3.3. *For any set A,*

$$\underline{\dim}_M(A) \geq \dim_H(A)$$

Proof. Obviously, using the cover by closed balls we have that

$$\mathcal{H}_{2\varepsilon}^\alpha(A) \leq (2\varepsilon)^\alpha N_\varepsilon(A) \, .$$

If $\alpha > \underline{\dim}_M(A)$, then $\liminf_{\varepsilon \downarrow 0}(2\varepsilon)^\alpha N_\varepsilon(A) = 0$, implying that $\mathcal{H}^\alpha(A) = 0$, and by definition of the Hausdorff dimension, also $\alpha > \dim_H(A)$. $\square$

Out of the definition of Minkowski dimensions, one can easily verify that

$$\underline{\dim}_M(A) = \underline{\dim}_M(\overline{A}) \qquad \text{and} \qquad \overline{\dim}_M(A) = \overline{\dim}_M(\overline{A}) \, ,$$

for any set A, which is not the case for the Hausdorff dimension. The next example demonstrates some of the differences between these two dimensions.

Example 3.4. *Let $D = \mathbb{Q} \cap [0,1]$. Then,*

$$\dim_M(D) = \dim_M(\overline{D}) = \dim_M([0,1]) = 1 \, .$$

However, the Hausdorff dimension of any countable set is zero, so $\dim_H(D) = 0$ (indeed, enumerating the points of a countable set A and covering the i-th point by a ball of radius $2^{-i}\varepsilon$, we see that

$$\mathcal{H}^{\alpha}_{(\varepsilon)}(A) \leq \varepsilon^{\alpha} \sum_{i=1}^{\infty} \left(2^{-i}\right)^{\alpha} \leq c\varepsilon^{\alpha} ,$$

so letting $\varepsilon \downarrow 0$, we get $\mathcal{H}^{\alpha}(A) = 0$ for any $\alpha > 0$).

See Sects. 3.1–3.2 of [Fal90] for additional properties of Minkowski dimension.

Potential Theoretic Dimension Lower Bounds

Upper bounds on Hausdorff dimension are typically obtained by constructing efficient covers of the set, resulting with finite α-Hausdorff measures. Here we tackle the harder task of providing lower bounds. We start with the *Mass Distribution Principle*

Lemma 3.5. *If A supports a positive Borel measure μ, such that $\mu(D) \leq C|D|^{\alpha}$ for all Borel sets D, then*

$$\mathcal{H}^{\alpha}_{\infty}(A) \geq \frac{\mu(A)}{C} > 0 , \tag{3.1}$$

implying in particular that $\mathcal{H}^{\alpha}(A) \geq \frac{\mu(A)}{C}$ and therefore also $\dim_{\mathrm{H}}(A) \geq \alpha$.

Proof. If $A \subset \bigcup_j A_j$, then by sub-additivity of the measure μ,

$$\sum_j |A_j|^{\alpha} \geq \frac{1}{C} \sum_j \mu(A_j) \geq \frac{\mu(A)}{C} .$$

Since this bound holds for any cover of A, we are done.   □

The next proposition, due to [Fro35], replaces the infinitely many constraints on the measure μ of the mass distribution principle by a single energy constraint. It is thus easier to apply for specific sets.

Proposition 3.6. *Given a metric space, (X, ρ), suppose μ is a finite positive Borel measure supported on $A \subset X$, having finite α-energy, that is,*

$$\mathcal{E}_{\alpha}(\mu) \triangleq \int \frac{d\mu(x)d\mu(y)}{\rho(x,y)^{\alpha}} < \infty .$$

Then, $\mathcal{H}^{\alpha}_{\infty}(A) > 0$ and hence $\dim_{\mathrm{H}}(A) \geq \alpha$.

Remark 3.7. *A stronger result is stated and proved as Theorem 4.13 of [Fal90]. It is shown there that $\mathcal{H}^{\alpha}(A) = \infty$ and that conversely, $\mathcal{H}^{\beta}(A) > 0$ implies the existence of a positive Borel measure μ on A with $\mathcal{E}_{\alpha}(\mu) < \infty$ for all $\alpha < \beta$.*

Proof. Let $\phi_\alpha(\mu, x) := \int \frac{d\mu(y)}{\rho(x,y)^\alpha}$ and $A[M] = \{x \in A : \phi_\alpha(\mu, x) \leq M\}$.

Since $\int \phi_\alpha(\mu, x) d\mu(x) < \infty$ and μ is a positive measure it follows that $\mu(A[M]) > 0$ for some $M \in (0, \infty)$. Let $\nu := \mu|_{A[M]}$. Note that $\phi_\alpha(\nu, x) \leq \phi_\alpha(\mu, x) \leq M$ for all $x \in A[M]$. We will check that

$$\nu(D) \leq M\,(2|D|)^\alpha \qquad (3.2)$$

for any Borel set $D \subset X$. This, together with the mass distribution principle (Lemma 3.5) imply that $\mathcal{H}_\infty^\alpha(A) \geq \mathcal{H}_\infty^\alpha(A[M]) > 0$.

Thus, all that remains to be done is to verify (3.2). The latter trivially holds if either $|D| = \infty$ or $D \cap A[M] = \varnothing$, so we assume in what follows that $D \cap A[M] \neq \varnothing$ and that D is bounded. Let $x \in D \cap A[M]$ and $m := \max\{n \in \mathbb{Z} : D \subset B(x, 2^{-n})\}$. Note that $2^{-(m+1)} \leq |D|$ and

$$M \geq \phi_\alpha(\nu, x) = \int \frac{d\nu(y)}{\rho(x,y)^\alpha} \geq \int_D \frac{d\nu(y)}{\rho(x,y)^\alpha} \geq 2^{m\alpha}\nu(D)\,,$$

whence $\nu(D) \leq M\,2^{-m\alpha} \leq M(2|D|)^\alpha$, as desired. $\qquad\qquad\square$

Packing and Modified Box Dimensions

In Example 3.4 we saw one of the disadvantages of the box dimension, that possibly $\underline{\dim}_M(F) > 0$ for a countable set F. The modified box dimension eliminates this deficiency.

Definition 3.8. *The modified box dimension of a set $F \subseteq \mathbb{R}^d$ is*

$$\overline{\dim}_{MB} F := \inf\left\{\sup_i\{\overline{\dim}_M F_i\} : F \subset \bigcup_{i=1}^\infty F_i\right\},$$

with the infimum taken over all possible countable covers of F.

While the modified box dimension is harder to compute than the box dimension, it coincides for each Borel set $F \subset \mathbb{R}^d$ with the *packing dimension* of F, denoted $\dim_P F$. For the general definition of packing measure and dimension, see Sect. 3.4 of [Fal90].

Our next lemma summarizes a few useful facts regarding the packing and Hausdorff dimensions.

Lemma 3.9. *The following hold for any $F, F_i \subseteq \mathbb{R}^d$:*

(i) $\dim_H F \leq \dim_P F \leq \overline{\dim}_M F$.
(ii) $\dim_H F = \inf\{\sup_i\{\dim_H F_i\} : F \subset \cup_i F_i\}$.
(iii) $\dim_P(\cup_i F_i) = \sup_i \dim_P(F_i)$.
(iv) $\dim_H(\cup_i F_i) = \sup_i \dim_H(F_i)$.

Proof. (ii). Trivially, $\dim_H F \geq \inf \{\sup_i \{\dim_H F_i\} : F \subset \cup_i F_i\}$. For the converse, let $\{F_i : i \geq 1\}$ be a cover of F. Fix arbitrary $\delta > 0$, $\varepsilon > 0$ and $\gamma > 0$, setting $\alpha := \delta + \sup_i \dim_H F_i$. Since $\dim_H F_i \leq \alpha - \delta$, we have $\mathcal{H}^{\alpha}_{(\varepsilon)}(F_i) = 0$. Hence, for each i we can find a cover $\{E_{ij} : j \geq 1\}$ of F_i with $|E_{ij}| \leq \varepsilon$ and $\sum_j |E_{ij}|^{\alpha} < \gamma 2^{-i}$. Thus $\{E_{ij} : i, j \geq 1\}$ is a cover of F with sets of diameter at most ε such that $\sum_{i,j} |E_{ij}|^{\alpha} \leq \gamma$, whence $\mathcal{H}^{\alpha}_{(\varepsilon)}(F) \leq \gamma$. Since γ is arbitrary, $\mathcal{H}^{\alpha}_{(\varepsilon)}(F) = 0$, and since ε is arbitrary also $\mathcal{H}^{\alpha}(F) = 0$. Thus, $\dim_H(F) \leq \alpha$ and taking $\delta \downarrow 0$ completes the proof.

For (i), the second inequality is trivial, while the first one follows from (ii), Definition 3.8 and Lemma 3.3.

For (iii), it is immediate that $\sup_j \dim_P(F_j) \leq \dim_P(\cup_i F_i)$. For the converse, suppose $\dim_P F_i < \alpha$ for all i. Then, for each i there exists a cover $\{E_{ij} : j \geq 1\}$ of F_i such that $\sup_j \overline{\dim}_M(E_{ij}) < \alpha$. But then $\{E_{ij} : i, j \geq 1\}$ is a cover of $\cup_i F_i$ such that $\sup_{i,j} \overline{\dim}_M(E_{ij}) \leq \alpha$, hence $\dim_P(\cup_i F_i) \leq \alpha$ by Definition 3.8.

The proof of (iv) is almost identical to that of (iii) once we use (ii). □

3.2 Discrete Limsup Random Fractals

In this section we follow [KPX00] in introducing a family of random fractals defined by limsup operations. As we show in Theorem 3.11 such fractals almost surely intersect a set whose packing dimension exceeds a threshold, while not intersecting sets whose dimension is below this threshold. As a result, Theorem 3.18 provides a definite value for the non-random Hausdorff dimension of these discrete limsup random fractals. We start with the definition of these sets.

Definition 3.10. *Let $\mathcal{D}_n$ denote the collection of all cubes of the form*

$$\prod_{i=1}^{d} \left[k_i 2^{-n}, (k_i + 1)2^{-n} \right] \subseteq [0, 1]^d,$$

with $k = (k_1, \ldots, k_d) \in \mathbb{Z}^d$. Given collections $\{Z_n(I); I \in \mathcal{D}_n\}$ of $\{0, 1\}$-valued random variables indexed by $\mathcal{D}_n$, $n \geq 1$, they induce the discrete limsup random fractal $A := \limsup_n A(n)$, where,

$$A(n) := \bigcup_{I \in \mathcal{D}_n : Z_n(I) = 1} I^{\circ},$$

and I° denotes the interior of I (see Fig. 3.2).

In the remainder of this chapter we use the metric $\rho(x, y) = \max_{i=1}^{d} |x_i - y_i|$ on $\mathbb{R}^d$ when computing the diameter of sets and the corresponding Hausdorff measure (so $|I| = 2^{-n}$ for $I \in \mathcal{D}_n$, $n \geq 1$).

The first condition we impose on such set A is about the growth of the "average size" of its n-th scale, $\mathbb{E}[\mathcal{L}eb(A(n))] = 2^{-n\gamma(1+o(1))}$, or more precisely.

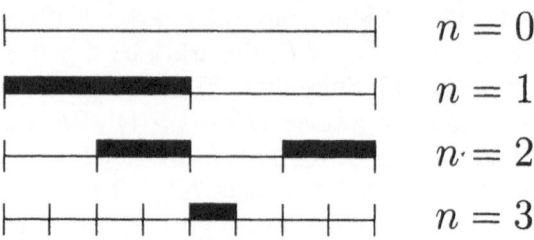

Fig. 3.2. Illustration of $A(n)$ for $d = 1$

The index assumption. *We say that the discrete limsup random fractal A has index $\gamma > 0$ if $p_n := \mathbb{E}\big[Z_n(I)\big]$ is independent of $I \in \mathcal{D}_n$ for all $n \geq 1$ and*

$$\lim_{n\to\infty} \frac{1}{n} \log_2 p_n = -\gamma,$$

for some $\gamma > 0$, where $\log_2$ denotes logarithm in base two.

Our second condition provides some control on the second moment of the size of the n-th scale of A, that is

The variance condition. *Suppose that*

$$\limsup_{n\to\infty} \frac{1}{n} \log_2 f_K(n) = 0 \qquad \text{for all } K > 1 , \tag{3.3}$$

where

$$f_K(n) := \max_{I \in \mathcal{D}_n} \#\Big\{ J \in \mathcal{D}_n :\ \mathbb{E}\big[Z_n(I)Z_n(J)\big] > K \mathbb{E}\big[Z_n(I)\big]\mathbb{E}\big[Z_n(J)\big]\Big\}$$

(throughout $\#\Gamma$ denotes the size of a finite set Γ, using $|\Gamma|$ for the same purpose only when there is no risk of confusion with the diameter).

In particular, both conditions apply when for each $n \geq 1$, the $Z_n(I)$, $I \in \mathcal{D}_n$ are i.i.d. random variables, but are applicable, of course, also to sets that show much more dependency than that.

Note that our conditions do not involve the dependency between the different scales, as the limsup in the definition of A allows us to ignore it.

Our next result shows that a discrete limsup random fractal satisfying the index and variance conditions introduces a dichotomy between "big" and "small" sets.

Theorem 3.11. *Suppose A is a discrete limsup random fractal with index γ, that satisfies the variance condition (3.3). Then, for any closed set $E \subset [0, 1]^d$,*

$$\mathbb{P}(A \cap E \neq \varnothing) = \begin{cases} 1, & \text{if } \dim_{\mathrm{p}}(E) > \gamma , \\ 0, & \text{if } \dim_{\mathrm{p}}(E) < \gamma . \end{cases} \tag{3.4}$$

Remark 3.12. *Theorem 3.1 of [KPX00] shows that (3.4) holds if E is a Borel set, and even E analytic suffices (see [BBT97, p. 484] for three equivalent definitions of analytic sets). To prove Theorem 3.11 in this generality one just replaces the simple proof we give of Lemma 3.16 with the elaborate proof of [JP95] for the more general result needed.*

Remark 3.13. *We can allow the set E in the theorem to be random, as long as it is independent of the probability space in which A is defined (just take the product space and condition on E).*

Remark 3.14. *Theorem 3.11 is strengthened in [DPRZ00b], where relaxing the variance condition to*

$$\lim_{K \to \infty} \limsup_{n \to \infty} \frac{1}{n} \log_2 f_K(n) = 0 , \qquad (3.5)$$

it is shown that $\mathbb{P}(A \cap E \neq \varnothing) > 0$ whenever $\dim_P(E) > \gamma$, and the full conclusion of Theorem 3.11 applies in the presence of a quasi-local approximation $\{Y_n(I) : I \in \mathcal{D}_n\}_{n \geq 1}$ of $\{Z_n(I) : I \in \mathcal{D}_n\}_{n \geq 1}$. For example, [DPRZ00b] apply the conclusion of Theorem 3.11 (see our Theorem 3.18), to determine the Hausdorff dimension of the set of Brownian thin points in $\mathbb{R}^d$, $d \geq 2$. In this context, the variance condition (3.3) applies only when the Brownian motion is transient, i.e. $d \geq 3$, but (3.5) allows [DPRZ00b] to handle successfully also the harder case of $d = 2$.

Proof. We start with the easier part of the proof, showing that $\dim_P(E) < \gamma \implies \mathbb{P}(A \cap E \neq \varnothing) = 0$. Assume first that $\overline{\dim}_M(E) < \gamma - \eta$ for some small $\eta > 0$. Then, by Definition 3.2 of the upper Minkowski dimension, we can find $\theta \in (0, \gamma - \eta)$, and $n_0 > 0$ such that for all $n \geq n_0$,

$$\#\left\{I \in \mathcal{D}_n : \ I \cap E \neq \varnothing\right\} \leq 2^{n\theta} \qquad (3.6)$$

(recall that the Minkowski dimension does not change even if we take the boxes in the cover on a fixed ε-mesh, here being $\mathcal{D}_n$ for $\varepsilon = 2^{-n}$). Since A has index γ it follows that there exists $n_1 > 0$ such that, for all $n > n_1$ and all $I \in \mathcal{D}_n$,

$$\mathbb{P}(A(n) \cap I \neq \varnothing) = p_n \leq 2^{-n(\gamma - \eta)} . \qquad (3.7)$$

Combining (3.6) and (3.7) we see that for each $n > n_2 := n_0 \vee n_1$,

$$\mathbb{P}(A(n) \cap E \neq \varnothing) \leq \sum_{I : I \cap E \neq \varnothing} \mathbb{P}(A(n) \cap I \neq \varnothing) \qquad (3.8)$$

$$\leq p_n \cdot \#\{I : I \cap E \neq \varnothing\} \leq p_n 2^{n\theta} \leq 2^{-n(\gamma - \eta - \theta)}$$

(where in the first inequality we used the fact that E is non-random).

Since $\theta < \gamma - \eta$, the Borel-Cantelli lemma implies that there exists a random variable n_3 such that almost surely, $E \cap A(n) = \varnothing$ for all $n \geq n_3$, implying that also $A \cap E = \varnothing$, almost surely.

Suppose now that $\dim_P(E) < \gamma$. Then, by Definition 3.8 of the packing dimension, there exists a cover $\{E_i : i \geq 1\}$ of E such that $\overline{\dim}_M(E_i) < \gamma$ for all i. As we have seen already, this implies $\mathbb{P}(A \cap E_i \neq \varnothing) = 0$ for all i, and since $\{E_i\}$ is countable, $\mathbb{P}(A \cap E \neq \varnothing) = 0$ as well.

The harder part of the proof, showing that $\dim_P(E) > \gamma \implies \mathbb{P}(A \cap E \neq \varnothing) = 1$, relies on the next two lemmas, whose proofs are deferred. It is in Lemma 3.15 that the variance condition is used.

Lemma 3.15. *Let F be a non-random set such that $\overline{\dim}_M(F) > \gamma$. Then, almost surely, $A(k) \cap F \neq \varnothing$ for infinitely many values of k.*

Lemma 3.16. *If E is closed with $\dim_P(E) > \alpha$ then there exists a non-empty, closed $E_\star \subset E$ such that $\overline{\dim}_M(E_\star \cap V) > \alpha$ for any V open such that $E_\star \cap V \neq \varnothing$.*

Equipped with Lemmas 3.15 and 3.16 we now complete the proof of the theorem. To this end, fix a closed set E such that $\dim_P(E) > \gamma$. Let $E_\star$ be the non-empty closed subset of E whose existence is guaranteed in Lemma 3.16 and V an open set such that $F = V \cap E_\star \neq \varnothing$. Note that $A = \bigcap_{n=1}^{\infty} B(n)$, where $B(n) = \bigcup_{k=n}^{\infty} A(k)$ are open sets (as $A(k)$ is the union of interiors of cubes). Further, by Lemma 3.16, $\overline{\dim}_M(F) > \gamma$, so by Lemma 3.15, almost surely $V \cap (B(n) \cap E_\star) \neq \varnothing$ for all $n \geq 1$. Letting V run over a countable base of open sets intersecting $E_\star$ we conclude that, almost surely, $B(n) \cap E_\star$ are dense in $E_\star$ for all $n \geq 1$.

With the relatively open sets $B(n) \cap E_\star$ being dense in the complete metric space $E_\star$, applying Baire's theorem in $E_\star$ (see [Roy88, p. 158]) we have that their intersection, namely $A \cap E_\star$, is also dense in $E_\star$. In particular, $A \cap E_\star$ is then non-empty. Since this applies almost surely and $E_\star \subset E$, we conclude that $\mathbb{P}(A \cap E \neq \varnothing) = 1$ as stated in the theorem. $\square$

We turn to provide the proofs of Lemmas 3.15 and 3.16.

Proof of Lemma 3.15. Fix a non-random set F such that $\overline{\dim}_M(F) > \gamma + \delta$ for some $\delta > 0$. Define $T_n := \sum_{\mathcal{I}_n} Z_n(I)$, where the sum is taken over $\mathcal{I}_n = \{I \in \mathcal{D}_n : I^\circ \cap F \neq \varnothing\}$. In words, T_n is the total number of cubes $I \in \mathcal{D}_n$ such that $I^\circ \cap F \cap A(n) \neq \varnothing$ (see Fig. 3.3). Our goal is to show that $\mathbb{P}\{T_n > 0 \text{ i.o. in } n\} = 1$. To this end, let $\mathcal{N}_n$ denote the size of the set $\mathcal{I}_n$. With $\overline{\dim}_M(F) > \gamma + \delta$, by Definition 3.2, the set

$$\mathfrak{N} := \left\{ n \geq 1 : \mathcal{N}_n \geq 2^{n(\gamma+\delta)} \right\} \tag{3.9}$$

is infinite. We know that $\mathbb{E}(T_n) = \mathcal{N}_n p_n$ and wish to bound $\mathbb{E}(T_n^2)$. To this end, fix $K > 1$ and for each $I \in \mathcal{D}_n$, let $\mathcal{B}_n(I)$ denote the collection of all $J \in \mathcal{D}_n$ such that $\mathbb{E}(Z_n(I)Z_n(J)) > K p_n^2$. Recall that $\#\mathcal{B}_n(I) \leq f_K(n)$ and $Z_n(I) \in \{0, 1\}$ so $\mathbb{E}(Z_n(I)Z_n(J)) \leq \mathbb{E}(Z_n(I)) = p_n$ for all $I, J \in \mathcal{D}_n$. Therefore,

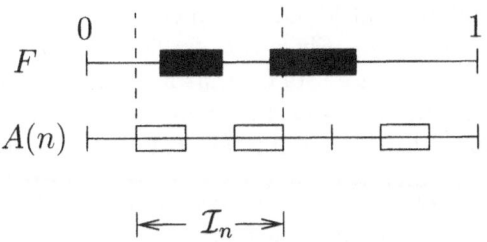

Fig. 3.3. Illustration of F and $\mathcal{I}_n$, where $d = 1$ and $T_n = 2$

$$\mathbb{E}(T_n^2) = \sum_{I \in \mathcal{I}_n} \sum_{J \in \mathcal{I}_n} \mathbb{E}\big(Z_n(I)Z_n(J)\big)$$

$$\leq K\mathcal{N}_n^2 p_n^2 + \sum_{I \in \mathcal{I}_n, J \in \mathcal{B}_n(I)} \mathbb{E}\big(Z_n(I)Z_n(J)\big)$$

$$\leq K\mathcal{N}_n^2 p_n^2 + \mathcal{N}_n p_n f_K(n) \,.$$

Consequently, by Chebyshev's inequality, we obtain that

$$\mathbb{P}(T_n = 0) \leq \frac{\mathrm{Var}(T_n)}{\mathbb{E}(T_n)^2} \leq K - 1 + \frac{f_K(n)}{\mathcal{N}_n p_n} \,.$$

By our index assumption, $\mathcal{N}_n p_n \geq 2^{n(\delta + o(1))}$ for $n \in \mathfrak{N}$ (see (3.9)), whereas by our variance condition $f_K(n) = 2^{o(n)}$. Thus,

$$\limsup_{n \to \infty : n \in \mathfrak{N}} \mathbb{P}(T_n = 0) \leq K - 1 \,.$$

Since $K > 1$ is arbitrary, we see that $\mathbb{P}(T_n = 0) \to 0$ as $n \to \infty$ in $\mathfrak{N}$ and hence

$$\mathbb{P}(T_n > 0 \text{ i.o.}) \geq \limsup_{n \to \infty} \mathbb{P}(T_n > 0) = 1 \,,$$

as claimed.   □

Proof of Lemma 3.16. Let $\{O_i\}$ be a countable base of open sets for $\mathbb{R}^d$ (for example, all open balls with rational center and radii). We split the set E as $E = E_0 \cup E_\star$ where

$$E_0 = \bigcup_{i : \dim_{\mathrm{p}}(E \cap O_i) \leq \alpha} E \cap O_i \,,$$

and $E_\star = E \setminus E_0$. Obviously, $E_\star$ is a closed set. By part (iii) of Lemma 3.9 we know that $\dim_{\mathrm{p}}(E_0) \leq \alpha$, and since $\dim_{\mathrm{p}}(E) > \alpha$, necessarily $E_\star$ is non-empty. Suppose that $\dim_{\mathrm{p}}(E_\star \cap V) \leq \alpha$ for some open set V. Then, for any $O_i \subseteq V$ we have that $\dim_{\mathrm{p}}(E_\star \cap O_i) \leq \alpha$. With $\dim_{\mathrm{p}}(E_0 \cap O_i) \leq$

$\dim_{\mathrm{P}}(E_0) \leq \alpha$, it follows that for each such O_i also $\dim_{\mathrm{P}}(E \cap O_i) \leq \alpha$, hence by construction $E \cap O_i \subseteq E_0$. Since this applies to all $O_i \subseteq V$ and any open set V is the union of the base sets O_i that are contained in it, we have that $E \cap V \subseteq E_0$, that is, $E_\star \cap V = \varnothing$. In conclusion, we see that if $E_\star \cap V \neq \varnothing$ then necessarily $\overline{\dim}_{\mathrm{M}}(E_\star \cap V) \geq \dim_{\mathrm{P}}(E_\star \cap V) > \alpha$, as stated.   □

Remark 3.17. *As mentioned earlier, Lemma 3.16 is the only part where the assumption that E is closed plays a role in our proof of Theorem 3.11. The same construction always produces a non-empty $E_\star$ with the desired property, but in general it might not be a closed set (which we needed in order to apply Baire's category theorem). However, Lemma 3.16 holds even for E an analytic set, but its proof is then much more delicate (see [JP95] for details).*

The next result shows that our discrete limsup random fractals have non-random Hausdorff dimension $d - \gamma$, where γ is the index of the fractal. This is what we shall use in applications (for example, see Sect. 3.3).

Theorem 3.18. *Suppose A is a discrete limsup random fractal with index γ that satisfies the variance condition. Then, almost surely $\dim_{\mathrm{H}}(A) = d - \gamma$.*

Theorem 3.18 does not rely directly on the variance condition but rather on the index assumption for the upper bound, and on the conclusion of Theorem 3.11 for the lower bound. We do not use the full power of latter theorem here. For example, Theorem 3.18 is merely a special case of Corollary 3.3 of [KPX00] which shows that for any closed (or even any analytic) set $E \subset [0,1]^d$,

$$\dim_{\mathrm{H}}(E) - \gamma \leq \dim_{\mathrm{H}}(A \cap E) \leq \dim_{\mathrm{P}}(E) - \gamma \quad \text{a.s.} \tag{3.10}$$

(clearly, taking $E = [0,1]^d$ in (3.10) results with Theorem 3.18).

The lower bound is derived here by a simplified version of the stochastic co-dimension argument of Lemma 3.4 of [KPX00]. The general idea, going back to [Tay66], is that if a set A intersects many independent random sets of suitable dimension, then a lower bound on the dimension of A follows. In doing so here, we use the random closed *fractal percolation* sets $\varUpsilon_\alpha$ constructed independently of A as follows. Consider the natural tiling of the unit cube $[0,1]^d$ by 2^d closed cubes of side $1/2$. Let $\varXi_1$ be a random sub-collection of these cubes, where each cube has probability $2^{-\alpha}$ of belonging to $\varXi_1$, and these events are mutually independent. At the k-th stage, if $\varXi_k$ is nonempty, tile each cube $Q \in \varXi_k$ by 2^d closed sub-cubes of side 2^{-k-1} (with disjoint interiors) and include each of these sub-cubes in $\varXi_{k+1}$ with probability $2^{-\alpha}$ (independently). Finally, let

$$\varUpsilon_\alpha = \bigcap_{k=1}^{\infty} \bigcup_{Q \in \varXi_k} Q \,,$$

(see Fig. 3.4), and denote by $\mathbb{P}_\alpha$ the law of $\varUpsilon_\alpha$. Our next lemma, due to Hawkes [Haw81], provides the intersection properties of $\varUpsilon_\alpha$.

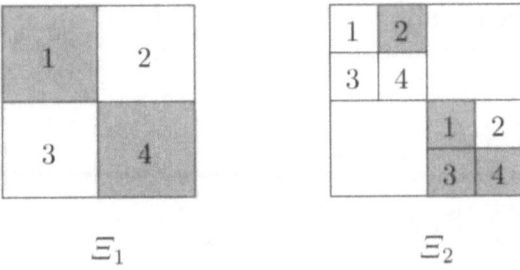

Fig. 3.4. First two levels in construction of Υ_α, $d = 2$

Lemma 3.19. *(a) If a Borel set $F \subset [0,1]^d$ intersects the random set Υ_α with positive probability, then $\dim_{\mathrm{H}}(F) \geq \alpha$.*
(b) Conversely, if Borel set $F \subset [0,1]^d$ has $\dim_{\mathrm{H}}(F) > \alpha$ then Υ_α intersects F with positive probability.

There is a tree associated with each realization of fractal percolation. The vertices at level k correspond to the cubes $Q \in \Xi_k$ and a vertex v at level k is the parent of a vertex u at level $k+1$ if the cube corresponding to v contains the cube corresponding to u (see Fig. 3.5). Since for any $0 < \alpha < d$ the tree that corresponds to Υ_α is a super-critical Galton-Watson tree of Binomial$(2^d, 2^{-\alpha})$ offspring law, that survives with positive probability, it follows that $\mathbb{P}_\alpha(\Upsilon_\alpha \neq \varnothing) > 0$. See also the lecture notes [Per99] and the papers [Per96a, Per96b] for more on fractal percolation and its relation to probability on trees and to intersections of random sets.

Proof. We prove here only the easy part (a) of the lemma which is all we use in proving Theorem 3.18. Part (b) is used by [KPX00] to derive (3.10). See [Lyo90] for a modern proof of this part when F is closed and note that the general case follows since any Borel (or even analytic) set F with $\dim_{\mathrm{H}}(F) > \alpha$ contains a compact subset with the same property (see [Dav52]).

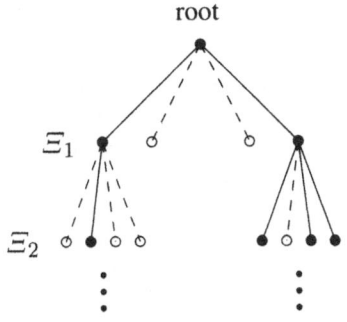

Fig. 3.5. The tree of the fractal percolation of Fig. 3.4

Turning to the proof of (a), fix Borel $F \subset [0,1]^d$ with $\mathbb{P}_\alpha(F \cap \Upsilon_\alpha \neq \varnothing) = \eta > 0$. Then, for any countable collection of cubes $I_j \in \mathcal{D}_{n_j}$ that cover F we have

$$
\begin{aligned}
\eta &\leq \sum_j \mathbb{P}_\alpha(I_j \cap \Upsilon_\alpha \neq \varnothing) \\
&\leq \sum_j \mathbb{P}_\alpha(I_j \cap \Xi_{n_j} \neq \varnothing) = \sum_j 2^{-\alpha n_j} = \sum_j |I_j|^\alpha .
\end{aligned}
\tag{3.11}
$$

Recall that any $I \subset [0,1]^d$ with $|I| \in [2^{-(n+1)}, 2^{-n}]$ can be covered by at most 2^d cubes from $\mathcal{D}_n$. Consequently, the restriction to covering F by countable collections of cubes $I_j \in \mathcal{D}_{n_j}$ at most increases $\mathcal{H}^\alpha(F)$ by a multiplicative factor of $2^{d+\alpha}$. Thus, by (3.11) the α-Hausdorff measure of F is strictly positive, implying that $\dim_{\mathrm{H}}(F) \geq \alpha$. □

Proof of Theorem 3.18. We start by proving that for any $\theta > d - \gamma$ the Hausdorff measure $\mathcal{H}^\theta(A)$ is almost surely zero, and hence $\dim_{\mathrm{H}}(A) \leq d - \gamma$ almost surely. Recall that $B(m) := \cup_{n \geq m} A(n)$ is a cover of A by cubes of maximal diameter 2^{-m}, with exactly $S_n = \sum_{I \in \mathcal{D}_n} Z_n(I)$ cubes of diameter 2^{-n} for $n = m, m+1, \ldots$. Hence, by definition,

$$
\mathcal{H}^\theta_{(2^{-m})}(A) \leq \sum_{n=m}^{\infty} S_n 2^{-n\theta} .
\tag{3.12}
$$

Recall that $\mathbb{E}(S_n) = 2^{nd} p_n = 2^{n(d-\gamma+o(1))}$ by the index assumption, so with $d - \gamma - \theta < 0$ we have that

$$
\mathbb{E}\left[\sum_{n=1}^{\infty} S_n 2^{-n\theta} \right] = \sum_{n=1}^{\infty} \mathbb{E}(S_n) 2^{-n\theta} < \infty .
$$

Consequently, almost surely $\sum_n S_n 2^{-n\theta} < \infty$ implying that $\sum_{n \geq m} S_n 2^{-n\theta} \to 0$ as $m \to \infty$. Thus, also $\mathcal{H}^\theta_{(\delta)}(A) \to 0$ when $\delta \downarrow 0$ as needed to establish that $\mathcal{H}^\theta(A) = 0$.

Turning to the proof of the lower bound, fix $\beta < d - \gamma$ and $\alpha \in (\gamma, d - \beta)$. Note that the intersection of independent Υ_α and Υ_β has the same distribution as $\Upsilon_{\alpha+\beta}$, and in particular is nonempty with positive probability. Applying part (a) of Lemma 3.19 for $F = \Upsilon_\beta$ we deduce that $\mathbb{P}_\beta(\dim_{\mathrm{H}}(\Upsilon_\beta) \geq \alpha) = \xi > 0$. Fixing $m < \infty$ we consider the closed random fractal

$$
\Gamma_m = \bigcup_{i=1}^{m} \Upsilon_\beta^{(i)} ,
$$

where $\Upsilon_\beta^{(i)}$ are i.i.d. copies of Υ_β that are also independent of the given discrete limsup random fractal A of index γ. Since $\alpha > \gamma$ and

$$\dim_{\mathrm{P}}(\Gamma_m) \geq \dim_{\mathrm{H}}(\Gamma_m) \geq \max_{i=1}^{m}\{\dim_{\mathrm{H}}(\Upsilon_\beta^{(i)})\} ,$$

applying Theorem 3.11 for $E = \Gamma_m$ we have that

$$\mathbb{P}_\beta^m \times \mathbb{P}(A \cap \Gamma_m \neq \varnothing) \geq \mathbb{P}_\beta^m(\dim_{\mathrm{P}}(\Gamma_m) \geq \alpha) \geq 1 - (1 - \xi)^m . \qquad (3.13)$$

However, $A \cap \Gamma_m$ is the finite union of $A \cap \Upsilon_\beta^{(i)}$ so $\dim_{\mathrm{H}}(A) < \beta$ implies that $\mathbb{P}_\beta^m(A \cap \Gamma_m \neq \varnothing) = 0$ by m applications of part (a) of Lemma 3.19. Consequently,

$$\mathbb{P}(\dim_{\mathrm{H}}(A) \geq \beta) \geq \mathbb{P} \times \mathbb{P}_\beta^m(A \cap \Gamma_m \neq \varnothing) . \qquad (3.14)$$

Combining (3.13) and (3.14) we have that $\mathbb{P}(\dim_{\mathrm{H}}(A) \geq \beta) \geq 1 - (1 - \xi)^m$. Taking $m \to \infty$ we see that almost surely $\dim_{\mathrm{H}}(A) \geq \beta$. Since this applies for all $\beta < d - \gamma$, the proof of the theorem is complete. $\qquad\qquad \square$

3.3 Thick Times for d-Dimensional Brownian Motion, $d \geq 3$

In this section we consider the Brownian motion w_t in $\mathbb{R}^d$, $d \geq 3$, and prove (1.13), that is,

$$\sup_{x \in \mathbb{R}^d} \limsup_{\varepsilon \to 0} \frac{\mu_\infty^w(B(x, \varepsilon))}{\varepsilon^2 |\log \varepsilon|} = \frac{4}{\lambda_{d-2}} \quad \text{a.s.} \qquad (3.15)$$

where λ_{d-2} is the minimal eigenvalue of the Laplacian operator in the unit ball of $\mathbb{R}^{d-2}$ and

$$\mu_T^w(A) = \int_0^T \mathbf{1}_A(w_t) dt$$

is the occupation measure of A by w_t, $0 \leq t \leq T$. This result is derived in [DPRZ00a] as a consequence of the fact that

$$\dim_{\mathrm{H}}\left\{ x \in \mathbb{R}^d \Big| \limsup_{\varepsilon \to 0} \frac{\mu_\infty^w(B(x, \varepsilon))}{\varepsilon^2 |\log \varepsilon|} = \frac{2a}{\lambda_{d-2}} \right\} = 2 - a \quad \text{a.s.} \qquad (3.16)$$

for each $0 < a < 2$, while the set considered in (3.16) is empty when $a > 2$. By continuity of the Brownian sample path it suffices to consider in (3.15) only x in the range of w_t, $t \geq 0$ (for otherwise $\mu_\infty^w(B(x, \varepsilon)) = 0$ for all $\varepsilon > 0$ small enough). Consequently, up to the precise value of the limit, (3.15) follows also from

Theorem 3.20. *There exist constants* $\theta_d > 0$, $d \geq 3$, *such that for* $0 < \alpha < 2/\theta_d$,

$$\dim_{\mathrm{H}}\left\{ t \geq 0 \Big| \limsup_{\varepsilon \to 0} \frac{\mu_\infty^w(B(w_t, \varepsilon))}{\varepsilon^2 |\log \varepsilon|} \geq \alpha \right\} = 1 - \frac{\theta_d}{2}\alpha \quad \text{a.s.} \qquad (3.17)$$

while the subset of thick times Thick$_{\geq \alpha}$ *considered in (3.17) is empty for* $\alpha > 2/\theta_d$.

We establish Theorem 3.20 by the approach of Sect. 3.2, specifically, applying Theorem 3.18 to get a sharp lower bound on $\dim_{\mathrm{H}}(\mathsf{Thick}_{\geq \alpha})$. In doing so, we rely on the tail estimate

$$\lim_{u \to \infty} e^{\theta_d u} \mathbb{P}(\mu_\infty^w(B(0,1)) > u) = \psi_d , \qquad (3.18)$$

with ψ_d and θ_d positive, finite constants. In Sect. 6.1 we prove (3.18) by applying Kac's moment formula, in a similar manner to that of [CT62]. This derivation of (3.18) follows [DPRZ99] where it is done also for all transient symmetric stable processes. Ciesielski and Taylor go further than that, identifying the law of $\mu_\infty^w(B(0,1))$ with that of the time it takes Brownian motion in $\mathbb{R}^{d-2}$ to exit the unit ball. This obviously identifies θ_d as half of λ_{d-2} and completes the proof of (3.15). In Sect. 6.3 we establish the latter identity by following the alternative derivation of [Yor91].

Recall Kaufman's uniform dimension doubling property

$$\mathbb{P}\big(\dim_{\mathrm{H}}(w(\mathcal{S})) = 2\dim_{\mathrm{H}}(\mathcal{S}) \text{ for all Borel } \mathcal{S} \in [0, \infty]\big) = 1 , \qquad (3.19)$$

where $w(\mathcal{S}) = \{w_t : t \in \mathcal{S}\} \subseteq \mathbb{R}^d$, and w_t is the Brownian motion in $\mathbb{R}^d$, $d \geq 2$ (c.f. [Kau69], or Sect. 22 of [Per01] for a more accessible proof).

While we shall not do so here, note that Corollary 3.3 of [DPRZ00b] is a refinement of Theorem 3.18 that allows for the more delicate task of bounding the dimension of sets such as $\mathsf{Thick}_\alpha = \mathsf{Thick}_{\geq \alpha} \setminus \cup_{\beta > \alpha} \mathsf{Thick}_{\geq \beta}$. Indeed, (3.16) is derived in [DPRZ00a] for $a = \alpha \theta_d$ by lower bounding $\dim_{\mathrm{H}}(\mathsf{Thick}_\alpha)$ using the potential theoretic method of Proposition 3.6 and then applying (3.19) to the *random* Borel set $\mathcal{S} = \mathsf{Thick}_\alpha$. As shown in [DPRZ00a], the sets $\mathsf{Thick}_{\geq \alpha}$ have different packing and Hausdorff dimensions, with $\dim_{\mathrm{P}}(\mathsf{Thick}_{\geq \alpha}) = 1$ for all $\alpha \leq 2/\theta_d$, so $\mathsf{Thick}_{2/\theta_d}$ is almost surely non-empty (using instead Corollary 3.3 of [DPRZ00b] one can check that also $\dim_{\mathrm{P}}(\mathsf{Thick}_\alpha) = 1$).

Upper Bounds

We prove the upper bound on $\dim_{\mathrm{H}}(\mathsf{Thick}_{\geq \alpha})$ by the construction of an efficient cover of $\mathsf{Thick}_{\geq \alpha} \cap [0, T]$, relying on Brownian scaling and the tail estimate (3.18). To this end, fixing $T \geq 1$, we start by discretizing the scales using the monotonicity of $\varepsilon \mapsto \mu_\infty^w(B(w_t, \varepsilon))$ and continuity of $h(\varepsilon) = \varepsilon^2 |\log \varepsilon|$. Specifically, fixing $\delta > 0$ and $\varepsilon_n = (1 - \delta)^n$ we see that for all n large enough,

$$\inf_{\varepsilon_n \geq \varepsilon \geq \varepsilon_{n+1}} h(\varepsilon) \geq (1 - \delta)^2 h(\varepsilon_n) ,$$

so for such n and ε, by the monotonicity of $\varepsilon \mapsto \mu_\infty^w(B(w_t, \varepsilon))$

$$(1 - \delta)^2 \frac{\mu_\infty^w(B(w_t, \varepsilon))}{h(\varepsilon)} \leq \frac{\mu_\infty^w(B(w_t, \varepsilon_n))}{h(\varepsilon_n)} .$$

Consequently,

$$\text{Thick}_{\geq \alpha} \cap [0,T] \subseteq \{0 \leq t \leq T : \limsup_{n \to \infty} \frac{\mu_\infty^w(B(w_t, \varepsilon_n))}{h(\varepsilon_n)} \geq (1-\delta)^2 \alpha\} := D_\alpha$$

Upon discretizing the scales we next discretize the space $[0,T]$ using Lévy's modulus of continuity theorem, which states that almost surely

$$\limsup_{\zeta \downarrow 0} \frac{\rho(\zeta)}{\sqrt{2\zeta |\log \zeta|}} = 1 , \qquad (3.20)$$

where

$$\rho(\zeta) = \sup_{\substack{0 \leq t,s \leq T \\ |t-s| \leq \zeta}} |w_s - w_t| ,$$

is the modulus of continuity of a one dimensional Brownian motion evaluated at ζ (c.f. Theorem I.2.7 of [RY99] and Brownian scaling). Taking $\tilde{\varepsilon}_n = \varepsilon_n^{2/(1-\delta)}$, we thus have that $\rho(\tilde{\varepsilon}_n) \leq \delta \varepsilon_n / d$ for some finite random variable $n_0 = n_0(w, \delta)$ and all $n \geq n_0$. Consequently, for any $n \geq n_0$, if $t_{j,n} \leq t \leq t_{j+1,n}$ for some $t_{j,n} = j\tilde{\varepsilon}_n \in [0,T]$, then

$$B(w_t, \varepsilon_n) \subseteq B(w_{t_{j,n}}, (1+\delta)\varepsilon_n) ,$$

and hence also

$$\mu_\infty^w(B(w_t, \varepsilon_n)) \leq \mu_\infty^w(B(w_{t_{j,n}}, (1+\delta)\varepsilon_n)) . \qquad (3.21)$$

With both space and scale discrete, we now fix $\alpha > 0$ and construct our covers of D_α. To this end, let $K_n = \lceil T/\tilde{\varepsilon}_n \rceil$ and

$$\mathcal{A}_n = \{0 \leq j \leq K_n - 1 : \mu_\infty^w(B(w_{t_{j,n}}, (1+\delta)\varepsilon_n)) \geq (1-\delta)^3 \alpha h(\varepsilon_n)\} . \qquad (3.22)$$

Observe that if $t \in D_\alpha$ then necessarily,

$$\mu_\infty^w(B(w_t, \varepsilon_{n_k})) \geq (1-\delta)^3 \alpha h(\varepsilon_{n_k}) ,$$

for all k and some subsequence $n_k = n_k(w,t) \to \infty$. Hence, in view of (3.21) and (3.22), the union over all sufficiently small scales forms the cover of D_α

$$\mathcal{C}_m = \bigcup_{n \geq m} \bigcup_{j \in \mathcal{A}_n} [t_{j,n}, t_{j+1,n}] ,$$

which is composed of intervals of maximal diameter $\tilde{\varepsilon}_m$. Further, $\tilde{\varepsilon}_m \to 0$ as $m \to \infty$ and $m \mapsto \mathcal{C}_m$ is monotone decreasing, so the covers $\mathcal{C}_m$ provide the upper bound on the γ-Hausdorff measure of D_α,

$$\mathcal{H}^\gamma(D_\alpha) \leq \sum_{n=1}^{\infty} (\#\mathcal{A}_n) \tilde{\varepsilon}_n^\gamma := Z_\gamma$$

If $Z_\gamma < \infty$ almost surely for some $\gamma > 0$, then $\mathcal{H}^\gamma(D_\alpha) < \infty$ and hence

$$\dim_H(\mathsf{Thick}_{\geq\alpha} \cap [0,T]) \leq \dim_H(D_\alpha) \leq \gamma \,.$$

Considering the countable union over $T = 1, 2, \ldots$, we thus deduce by part (iv) of Lemma 3.9 that $\dim_H(\mathsf{Thick}_{\geq\alpha}) \leq \gamma$, almost surely.

The easiest way to control *random covers* is by computing expected values. To demonstrate this important principle, showing below that for all n large enough,

$$\mathbb{E}(\#\mathcal{A}_n) = \sum_{j=0}^{K_n-1} \mathbb{P}(j \in \mathcal{A}_n) \leq K_n \varepsilon_n^{(1-\delta)^6 \theta_d \alpha} \,, \tag{3.23}$$

we see that for $\alpha < 2/\theta_d$ and $\gamma = 1 - (1-\delta)^8 \theta_d \alpha / 2 > 0$,

$$\mathbb{E}(Z_\gamma) \leq 2T \sum_{n=1}^{\infty} \widetilde{\varepsilon}_n^{(1-\delta)^7 \theta_d \alpha / 2 - 1 + \gamma} < \infty$$

(recall that $\varepsilon_n = \widetilde{\varepsilon}_n^{(1-\delta)/2}$ and $K_n \leq 2T/\widetilde{\varepsilon}_n$). Thus, necessarily $Z_\gamma < \infty$ almost surely, and taking $\delta_l \downarrow 0$ gives the stated upper bound

$$\dim_H(\mathsf{Thick}_{\geq\alpha}) \leq 1 - \frac{\theta_d}{2}\alpha \qquad a.s.$$

Moreover, if $\alpha > 2/\theta_d$ and $\delta > 0$ is small enough for $(1-\delta)^7 \theta_d \alpha > 2$, then by (3.23)

$$\sum_{n=1}^{\infty} \mathbb{P}(\mathcal{A}_n \neq \varnothing) \leq \sum_{n=1}^{\infty} \mathbb{E}(\#\mathcal{A}_n) < \infty \,,$$

so by the Borel-Cantelli lemma, with probability one the sets $\mathcal{A}_n$ are empty for all n large enough. This implies of course that the cover $\mathcal{C}_m$ of D_α is empty when m is large enough, hence so is the set $\mathsf{Thick}_{\geq\alpha} \cap [0,T]$. Taking the countable union over $T = 1, 2, \ldots$, leads to the stated conclusion that $\mathsf{Thick}_{\geq\alpha}$ is almost surely empty for any fixed $\alpha > 2/\theta_d$.

Turning to the remaining task of establishing (3.23), observe that by Brownian scaling

$$Q(r,\xi) := \sup_{t\geq0} \mathbb{P}(\mu_\infty^w(B(w_t,r)) \geq \xi) = Q(1, r^{-2}\xi) \,.$$

Since $r^{-2}\xi \geq (1-\delta)^5 \alpha |\log\varepsilon|$ when $r = (1+\delta)\varepsilon$ and $\xi = (1-\delta)^3 \alpha h(\varepsilon)$, the bound (3.23) is an immediate consequence of the fact that

$$\lim_{u\to\infty} e^{(1-\delta)\theta_d u} Q(1,u) = 0 \,. \tag{3.24}$$

Turning to establish (3.24), let $\overline{w}$ denote the two-sided Brownian motion in $\mathbb{R}^d$ (i.e. $\overline{w}_t = w_t$ and $\overline{w}_{-t} = w'_t$ for any $t \geq 0$, with w and w' two independent copies of the Brownian motion). With a slight abuse of notations we let $\mu_\infty^{\overline{w}}(A)$ be the occupation time of the set A by $\{\overline{w}_t : -\infty < t < \infty\}$. Clearly,

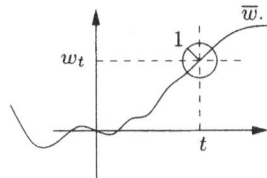

Fig. 3.6. Illustration of $\mu_\infty^{\overline{w}}(B(w_t,1))$

$\mu_\infty^w(B(w_t,1)) \leq \mu_\infty^{\overline{w}}(B(\overline{w}_t,1))$ for any $t \geq 0$. Further, the Gaussian process $\overline{w}_{t+\cdot} - \overline{w}_t$ has the same law as $\overline{w}_\cdot$. We thus see that

$$Q(1,u) \leq \sup_{t\geq 0} \mathbb{P}(\mu_\infty^{\overline{w}}(B(\overline{w}_t,1)) \geq u) = \mathbb{P}(\mu_\infty^{\overline{w}}(B(0,1)) \geq u) = \mathbb{P}(\mathcal{J} + \mathcal{J}' \geq u) \,,$$

with $\mathcal{J}$ and $\mathcal{J}'$ denoting two independent copies of $\mu_\infty^w(B(0,1))$. We have by (3.18) that $\mathbb{P}(\mathcal{J} \geq u) \leq Ce^{-\theta_d u}$ for some $C < \infty$ and all $u \geq 0$. Consequently, using integration by parts,

$$\mathbb{P}(\mathcal{J} + \mathcal{J}' \geq u) = \int_0^\infty d\mathbb{P}_{\mathcal{J}}(v)\mathbb{P}(\mathcal{J}' \geq u - v) \leq C\int_0^\infty d\mathbb{P}_{\mathcal{J}}(v)e^{-\theta_d \max(u-v,0)}$$

$$\leq Ce^{-\theta_d u}\left[1 + \theta_d \int_0^u e^{\theta_d v}\mathbb{P}(\mathcal{J} \geq v)dv\right] \leq C(1 + C\theta_d u)e^{-\theta_d u} \,. \qquad (3.25)$$

With (3.24) a direct consequence of (3.25), we are done with the proof of the upper bounds.

The Dimension Lower Bound

We wish to derive lower bounds on $\dim_{\mathrm{H}}(\mathsf{Thick}_{\geq\alpha})$ by the method of Sect. 3.2. To satisfy the variance condition assumed there, it is crucial to be able to consider the occupation measure of a ball of radius ε over a small time interval whose length tends to zero rapidly enough as $\varepsilon \to 0$. Such a *localization phenomenon* holds for transient Brownian motion: with a relatively small drop in probability, the balls of radius ε that have the exceptional occupation measure of order $h(\varepsilon)$, accumulate most of this measure in a surprisingly short time interval (of length at most $\varepsilon^2|\log\varepsilon|^b$ for some b, e.g. $b = 4$ works). Upon Brownian scaling, it is a direct consequence of the following lemma which we shall prove in page 43.

Lemma 3.21. *Let $\{w_t : t \geq 0\}$ be a Brownian motion in $\mathbb{R}^d$, $d \geq 3$. There exists $c = c_d > 0$ such that for all u sufficiently large,*

$$\mathbb{P}(\mu_{u^4}^w(B(0,1)) \geq u) \geq ce^{-\theta_d u} \,. \qquad (3.26)$$

Remark 3.22. *This localization phenomenon breaks down in dimension two, where the correct scaling of occupation measure, and the techniques needed to establish it, are quite different (c.f. Sect. 5.2).*

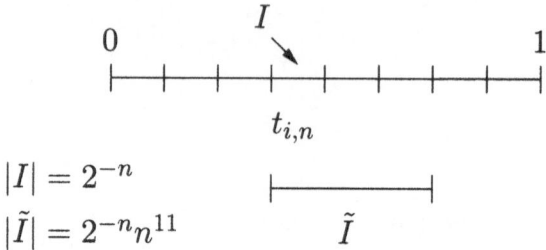

Fig. 3.7. Typical intervals I and $\tilde{I}$

Fixing $\alpha < 2/\theta_d$, our goal is to find a discrete limsup random fractal A of sufficiently large index which satisfies the variance condition and show that it is a subset of $\mathsf{Thick}_{\geq \alpha}$. To this end, let $t_{i,n} = i2^{-n}$ for $i = 0, \ldots, 2^n - 1$. Setting $\varepsilon_n = n^3 2^{-n/2}$ and $\delta_n = n^5 \varepsilon_n^2$ we associate with each $I = [t_{i,n}, t_{i,n} + 2^{-n}] \in \mathcal{D}_n$ the enlarged interval $\tilde{I} = [t_{i,n}, t_{i,n} + \delta_n)$, and the random variable $Z_n(I) \in \{0, 1\}$ such that

$$Z_n(I) = 1 \quad \Longleftrightarrow \quad \int_{\tilde{I}} \mathbf{1}_{B(w_{t_{i,n}}, \varepsilon_n)}(w_s) ds \geq \alpha h(\varepsilon_n) .$$

Let A denote the discrete limsup random fractal induced in this manner according to Definition 3.10. Noting that $\{w_{t+s} - w_t\}_{s \geq 0}$ is a Brownian motion for any fixed $t \geq 0$, it follows that

$$\mathbb{P}(Z_n(I) = 1) = \mathbb{P}(\mu_{\delta_n}^w(B(0, \varepsilon_n)) \geq \alpha h(\varepsilon_n)) := p_n ,$$

is independent of $I \in \mathcal{D}_n$. With $u_n = \alpha |\log \varepsilon_n| \to \infty$ and $\delta_n \varepsilon_n^{-2} = n^5 \geq u_n^4$ for all n large enough, by Brownian scaling and (3.26) we have that

$$p_n = \mathbb{P}(\mu_{n^5}^w(B(0, 1)) \geq \alpha |\log \varepsilon_n|) \geq c \varepsilon_n^{\theta_d \alpha} ,$$

for all n large enough. The corresponding upper bound on p_n follows directly from (3.18), so our set A has the index

$$\gamma = - \lim_{n \to \infty} \frac{1}{n} \log_2 p_n = \frac{\theta_d}{2} \alpha .$$

If j and i are such that $t_{j,n} - t_{i,n} \geq \delta_n$ then the corresponding enlarged intervals $\tilde{J}$ and $\tilde{I}$ are disjoint. The corresponding random variables $Z_n(J)$ and $Z_n(I)$ depend only on the increments of the Brownian motion in $\tilde{J}$ and $\tilde{I}$, respectively. By the independence of the Brownian increments it follows that then $\mathsf{Cov}(Z_n(I), Z_n(J)) = 0$. For each $I \in \mathcal{D}_n$ this applies to all but at most $f_K(n) = 2\delta_n 2^n = 2n^{11}$ intervals $J \in \mathcal{D}_n$, so obviously our set A satisfies the variance condition (3.3).

By Theorem 3.18 we have that almost surely, $\dim_{\mathrm{H}}(A) = 1 - \gamma$, so it remains only to show that $A \subset \mathsf{Thick}_{\geq \alpha}$. To this end, recall that Lévy's modulus continuity theorem guarantees the existence with probability one of $n_0(w)$ finite such that $|w_t - w_{t_{i,n}}| \leq n2^{-n/2} := \beta_n$ for all $n \geq n_0$, i and $t \in [t_{i,n}, t_{i,n} + 2^{-n}]$ (see (3.20), noting that $n2^{-n/2} \gg \rho(2^{-n})$ for all large n). Thus, if $t \in I$ with $Z_n(I) = 1$ and $n \geq n_0$, then necessarily

$$\mu_\infty^w(B(w_t, \varepsilon_n + \beta_n)) \geq \mu_\infty^w(B(w_{t_{i,n}}, \varepsilon_n)) \geq \alpha h(\varepsilon_n) . \qquad (3.27)$$

By Definition 3.10, each point t of A satisfies (3.27) for a subsequence $n_k = n_k(t, w) \to \infty$. Since $h(\varepsilon_n + \beta_n)/h(\varepsilon_n) \to 1$, it follows that any such t is in $\mathsf{Thick}_{\geq \alpha}$, completing the proof of the dimension lower bound.

Localization Phenomenon

We conclude by proving the localization Lemma 3.21 (see also [DPRZ99, Lemma 2.2] for a more general version, applicable to all symmetric transient stable processes).

Proof of Lemma 3.21. Let $\sigma_z := \inf\{s \geq 0 : |w_s| = z\}$, denote the exit time of the Brownian motion from $B(0, z)$. Solving Poisson's PDE as in Lemma 2.10 (for $D = B(0, z)$, $f = 1$ and $\kappa = 0$) it is easy to check that $\mathbb{E}^x(\sigma_z) \leq \mathbb{E}^0(\sigma_z) = z^2/d$ for all $x \in B(0, z)$. Hence, by Kha'sminskii's condition, for some $c > 0$, all $u > 1$ and $z = z(u) = u^{4/3}$,

$$\mathbb{P}(\sigma_z > u^4) \leq c^{-1} e^{-cu^4/z^2} = c^{-1} e^{-cu^{4/3}}$$

(apply Theorem 6.7 for part (ii) of Example 6.5). We use hereafter the notation $\mathcal{J}^s = \mu_s^w(B(0,1))$ with $\mathcal{J} = \mathcal{J}^\infty$. By the monotonicity of $s \mapsto \mathcal{J}^s$, clearly,

$$\mathbb{P}(\mathcal{J}^{u^4} > u) \geq \mathbb{P}(\mathcal{J}^{\sigma_z} > u) - \mathbb{P}(\sigma_z > u^4) ,$$

so we get (3.26) as soon as we show that $e^{\theta_d u} \mathbb{P}(\mathcal{J}^{\sigma_z} > u)$ is bounded away from zero. To this end, let $\tau_1 = \inf\{s \geq \sigma_z : |w_s| = 1\}$ denote the time of first return to $\partial B(0, 1)$ by the Brownian motion after it exits $B(0, z)$. Let $\mathcal{F}_s$ denote the σ-field generated by $\{w_\xi : 0 \leq \xi \leq s\}$ and $\mathcal{F}_{\sigma_z}$ the corresponding stopped σ-field. With the Brownian motion transient in $\mathbb{R}^d$, $d \geq 3$, we have that

$$\mathbb{P}(\tau_1 < \infty \,|\, \mathcal{F}_{\sigma_z}) = z^{2-d} < 1 . \qquad (3.28)$$

(Indeed, solving Poisson's PDE in the annulus $B(0, R) \setminus B(0, 1)$, taking $f = 0$ and $\kappa = \mathbf{1}_{\partial B(0,1)}$ in Lemma 2.10, one has that the probability of Brownian motion starting at x entering $B(0, 1)$ before exiting $B(0, R)$ is $(|x|^{2-d} - R^{2-d})/(1 - R^{2-d})$. Taking $R \to \infty$ we deduce that the probability of ever returning to $B(0, 1)$ is $|x|^{2-d}$, and (3.28) follows by the strong Markov property at σ_z, where almost surely $|w_{\sigma_z}| = z$). Since no Brownian occupation measure of $B(0, 1)$ is accumulated during the time interval $[\sigma_z, \tau_1]$, it follows that

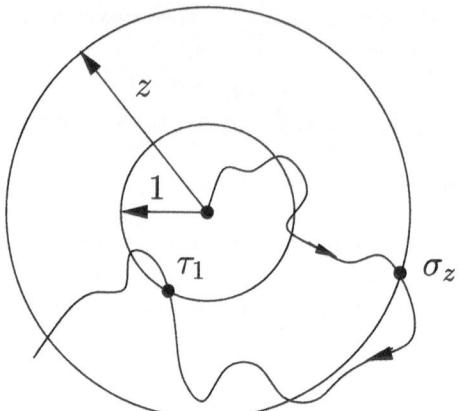

Fig. 3.8. A sketch of σ_z and τ_1 for a Brownian realization in $\mathbb{R}^2$

$$\mathcal{J} = \mathcal{J}^{\sigma_z} \mathbf{1}_{\tau_1=\infty} + (\mathcal{J}^{\sigma_z} + \mathcal{J}_{\tau_1})\mathbf{1}_{\tau_1<\infty} \, ,$$

where $\mathcal{J}_s = \int_s^\infty \mathbf{1}_{\{|w_\xi|<1\}} d\xi$. Consequently,

$$\mathbb{P}(\mathcal{J} > u) \le \mathbb{P}(\mathcal{J}^{\sigma_z} > u) + \mathbb{P}(\mathcal{J}^{\sigma_z} + \mathcal{J}_{\tau_1} > u, \tau_1 < \infty) \, ,$$

so in view of (3.18) it suffices to show that

$$\lim_{u \to \infty} e^{\theta_d u} \mathbb{P}(\mathcal{J}^{\sigma_z} + \mathcal{J}_{\tau_1} > u, \tau_1 < \infty) = 0 \, . \tag{3.29}$$

To this end, note that $\mathcal{J}^{\sigma_z}$ and τ_1 are both measurable on $\mathcal{F}_{\tau_1}$. Further, if $\tau_1 < \infty$, then by the strong Markov property at τ_1 and radial symmetry of the Brownian motion, $\mathcal{J}_{\tau_1}$ is independent of $\mathcal{F}_{\tau_1}$. So, the law of $\mathcal{J}_{\tau_1}$ is then that of an independent copy $\tilde{\mathcal{J}}$ of $\mathcal{J}$ for a Brownian motion that starts at $\partial B(0,1)$. Recall that $\mathcal{J}^{\sigma_z}$ is measurable on $\mathcal{F}_{\sigma_z}$ and by (3.28), the event $\{\tau_1 < \infty\}$ is independent of $\mathcal{F}_{\sigma_z}$. Combining the above observations and the monotonicity of $s \mapsto \mathcal{J}^s$,

$$\mathbb{P}(\mathcal{J}^{\sigma_z} + \mathcal{J}_{\tau_1} > u, \tau_1 < \infty) = \mathbb{P}(\mathcal{J}^{\sigma_z} + \tilde{\mathcal{J}} > u, \tau_1 < \infty)$$
$$= \mathbb{P}(\tau_1 < \infty)\mathbb{P}(\mathcal{J}^{\sigma_z} + \tilde{\mathcal{J}} > u) \le z^{2-d}\mathbb{P}(\mathcal{J} + \tilde{\mathcal{J}} > u) \, . \tag{3.30}$$

Since $\mathbb{P}(\tilde{\mathcal{J}} > u) \le \mathbb{P}(\mathcal{J} > u)$ for any $u > 0$, it follows from (3.25) that $\mathbb{P}(\mathcal{J} + \tilde{\mathcal{J}} > u) \le C(1 + C\theta_d u)e^{-\theta_d u}$ for some $C < \infty$ and all $u > 0$. If $u \to \infty$ then $z^{2-d}u \to 0$ for $d \ge 3$ and our choice of $z(u) = u^{4/3}$. Therefore, (3.30) implies that (3.29) holds, thus completing the proof of the lemma. $\qquad \square$

4 Multi-Scale Truncated Second Moment

As we have seen in Sect. 3.3, the "usual" methods for deriving lower bounds on the Hausdorff dimension of fractal sets fail to work for thick points of planar Brownian motion, i.e., the sets considered in (1.3). This is not merely a technical problem. Rather, it is a consequence of the fact that a typical thick point x is the result of many excursions, possibly reaching far away from x, thus creating greater dependence between the events "x is a thick point" and "y is a thick point" for far away points y. In this chapter, we develop an approach to solve this problem by a refinement of the second moment method. We do this in the context of a seemingly unrelated question: the rate of growth of the number of favorite leaves for a simple random walk (SRW) on a regular b-ary tree of depth N as $N \to \infty$.

As we shall see in Sect. 5.1, favorite points for SRW on $\mathbb{Z}^2$ are related to thick points of Brownian motion on $\mathbb{R}^2$ by the KMT strong approximation. As we further explain in Sect. 5.3, these problems are closely related to the one we treat here. Indeed, the growth asymptotic of number of favorite leaves for the SRW on a regular tree may be considered a toy model for computing the dimension of the set of thick points for the planar Brownian motion.

Recall that, in Sect. 2.3, we found the growth asymptotic for the ϵ-cover time for a Brownian motion in $\mathbb{T}^d$, $d \geq 3$, and noted that the "general" method we derive in Chap. 2 fails to provide the correct lower bound in case $d = 2$. Indeed, the correct lower bound is provided in [DPRZ04] based on a toy model with SRW on a regular tree; in this case, that of the time it takes the walk to visit all vertices of the tree. While the asymptotic of the latter has already been found in [Ald91a], this solution relies too much on the tree structure, hence does not help with solving the corresponding Brownian question (see also [Per03]). So, in Sect. 4.4 we outline the modification of our method which yields the growth asymptotic for the ϵ-cover time for a Brownian motion in $\mathbb{T}^2$ (and henceforth the cover time for a SRW on the two dimensional lattice torus $\mathbb{Z}_n^2$).

4.1 Favorite Leaves for SRW on b-ary Tree

Let T be a b-ary rooted regular tree of depth N. That is, the degree of each vertex v of T, denoted $\deg(v)$, is $b + 1$, except for the root, denoted 0, whose degree is b and the b^N leaves of the tree, each of whom has degree one. Let $u \leftrightarrow v$ denote the shortest path in T between u and v. The length of the path $u \leftrightarrow v$ is denoted $d(u, v)$, with $d(0, v) = |v|$ called the height, or level, of v. For any $u, v \in T$ let $u \wedge v$ denote the vertex at which the path $0 \leftrightarrow u$ and $0 \leftrightarrow v$ separate. We denote by ∂T the set of leaves of T and for $v \in \partial T$, we call $0 \leftrightarrow v$ the ray of v. The SRW on T is a time homogeneous Markov chain S_t whose state space is T and its transition kernel is

$$\mathbb{P}(S_{t+1} = u | S_t = v) = \begin{cases} \frac{1}{\deg(v)} & \text{if } d(u, v) = 1 \text{,} \\ 0 & \text{otherwise ,} \end{cases}$$

see also Fig. 4.1. Let x_0 denote the left most leaf of T. The SRW S_t on T starts at $S_0 = x_0$ and stops at time $\tau_0 = \inf\{t \geq 0 : S_t = 0\}$, that is, at the first visit

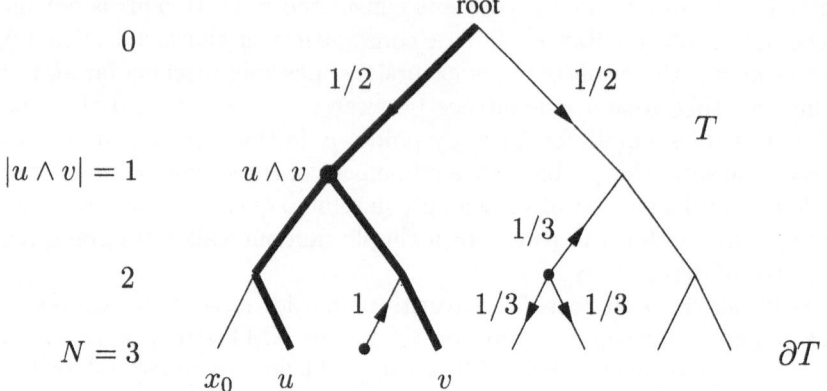

Fig. 4.1. Marked rays $0 \leftrightarrow u$, $0 \leftrightarrow v$ on T and SRW transitions

to the root. Let $\widetilde{T}_x$ denote the number of visits of $x \in T$ by $\{S_0, S_1, \ldots, S_{\tau_0}\}$ and

$$\mathcal{F}_N(\alpha) = \{x \in \partial T : \widetilde{T}_x \geq \alpha N^2 \log b\} \,,$$

be the set of α-favorite leaves of T. This chapter is devoted to the proof of

Theorem 4.1. *For any* $0 < \alpha < 1$

$$\lim_{N \to \infty} \frac{1}{N} \log |\mathcal{F}_N(\alpha)| = (1 - \alpha) \log b \quad \text{in probability} \,. \tag{4.1}$$

Moreover,

$$\lim_{N \to \infty} \frac{1}{N^2} \max_{x \in \partial T} \{\widetilde{T}_x\} = \log b \quad \text{in probability} \,. \tag{4.2}$$

We start with the easy part of the proof, whereby we prove the upper bounds in (4.1) and (4.2) using first moment estimates. To this end, since $|\partial T| = b^N$,

$$
\begin{aligned}
\mathbb{E}[|\mathcal{F}_N(\alpha)|] &= \sum_{x \in \partial T} \mathbb{P}(\widetilde{T}_x \geq \alpha N^2 \log b) \\
&= \frac{b^N}{N(b-1)} (1 + o(1)) \left(1 - \frac{1}{N}\right)^{\alpha N^2 \log b - 1} \\
&= b^{N(1-\alpha)(1+o(1))} \,,
\end{aligned}
\tag{4.3}
$$

where we have used the identity

$$\mathbb{P}(\widetilde{T}_x \geq t) = \frac{|x \wedge x_0|}{N} \left(1 - \frac{1}{N}\right)^{t-1} \,, \tag{4.4}$$

and hereafter $o(1)$ denote quantities that approach zero as $N \to \infty$. To derive (4.4), assume without loss of generality that the walk starts at $x \wedge x_0$ since before the first visit to $x \wedge x_0$ it neither visits x nor 0. Further, $\widetilde{T}_x$ is measurable on the restriction of S_t to the ray $0 \leftrightarrow x$. The latter is a SRW $\{Y_t\}$ on $\{0, 1, \ldots, N\}$ with the root represented by 0, $x \wedge x_0$ by $l = |x \wedge x_0|$ and x by N (see Fig. 4.2). The probability of Y_t reaching N before 0, starting at $Y_0 = l$, is $\frac{l}{N}$. Upon reaching N, the SRW Y_t next moves to $N - 1$ and thereafter with probability $1 - \frac{1}{N}$ returns to N before visiting 0, resulting with (4.4). Fixing $\delta > 0$, we have by (4.3) that

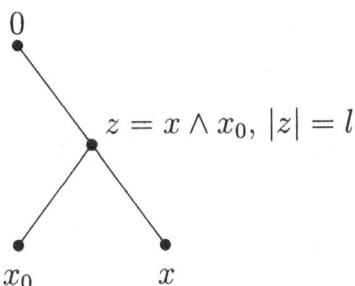

$$z = x \wedge x_0, \ |z| = l$$

Fig. 4.2. Reduced tree for deriving (4.4)

$$\mathbb{P}\left(|\mathcal{F}_N(\alpha)| \geq e^{N(1-\alpha+\delta)\log b}\right) \leq \mathbb{E}(|\mathcal{F}_N(\alpha)|)e^{-N(1-\alpha+\delta)\log b}$$
$$\leq b^{-N\delta(1+o(1))} \ . \qquad (4.5)$$

Taking $N \to \infty$, we obtain the upper bound for (4.1):

$$\limsup_{N\to\infty} \mathbb{P}\left(\frac{1}{N}\log|\mathcal{F}_N(\alpha)| \geq (1-\alpha+\delta)\log b\right) = 0 \ .$$

Considering $\alpha > 1$ and $\delta > 0$ small enough so that $\eta = 1 - \alpha + \delta < 0$, we get from (4.5) that

$$\mathbb{P}(|\mathcal{F}_N(\alpha)| \geq 1) \leq \mathbb{P}(|\mathcal{F}_N(\alpha)| \geq b^{N\eta}) \to 0 \ ,$$

as $N \to \infty$, implying that,

$$\lim_{N\to\infty} \mathbb{P}(\mathcal{F}_N(\alpha) = \varnothing) = 1 \ . \qquad (4.6)$$

Note that when the set $\mathcal{F}_N(\alpha)$ is empty, necessarily $\frac{1}{N^2}\max_{x\in\partial T}\{\widetilde{T}_x\} < \alpha \log b$. Hence, with (4.6) holding for all $\alpha > 1$, we have that for any $\eta > 0$,

$$\limsup_{N\to\infty} \mathbb{P}\left(\frac{1}{N^2}\max_{x\in\partial T}\{\widetilde{T}_x\} \geq \log b + \eta\right) = 0 \ ,$$

which is the upper bound for (4.2).

To complete the proof of (4.1) and (4.2), it suffices to show that for any $\delta > 0$ and $\alpha < 1$,

$$\lim_{N \to \infty} \mathbb{P}\left(|\mathcal{F}_N(\alpha)| \le e^{N(1-\alpha-\delta)\log b}\right) = 0 \,. \tag{4.7}$$

Typically, one attempts to do this by showing that

$$\mathrm{Var}(|\mathcal{F}_N(\alpha)|) = o(\mathbb{E}(|\mathcal{F}_N(\alpha)|)^2) \,,$$

for then $\frac{|\mathcal{F}_N(\alpha)|}{\mathbb{E}(|\mathcal{F}_N(\alpha)|)} \to 1$ and (4.7) follows by (4.3). However, this would not work for any $\alpha > \frac{1}{2}$ since there exists an $\eta = \eta(\alpha) > 0$ such that

$$\liminf_{N \to \infty} \frac{1}{N} \log_b \mathbb{E}(|\mathcal{F}_N(\alpha)|^2) \ge 2(1-\alpha) + \eta \,. \tag{4.8}$$

Indeed, with $t_N = \alpha N^2 \log b$, we have for any $1 > \xi > 0$ that,

$$\mathbb{E}(|\mathcal{F}_N(\alpha)|^2) \ge \sum_{(x,y) \in \partial T_\xi} \mathbb{P}(\widetilde{T}_x \ge t_N, \widetilde{T}_y \ge t_N) \,, \tag{4.9}$$

where

$$\partial T_\xi = \{(x,y) : x, y \in \partial T, |x \wedge y| = \xi N, |x \wedge x_0| = 1\} \,.$$

There are b^{l-1} vertices $z \in T$ with $|z| = l \ge 1$ and $|z \wedge x_0| = 1$, each having $b^{2(N-l-1)}b(b-1)/2$ pairs $x, y \in \partial T$ such that $z = x \wedge y$, so $|\partial T_\xi| = b^{(2-\xi)N(1+o(1))}$ and as

$$q_{\alpha,\xi,N} = \mathbb{P}(\widetilde{T}_x \ge t_N, \widetilde{T}_y \ge t_N) \,,$$

is independent of the specific choice of $(x,y) \in \partial T_\xi$, we establish (4.8) for $\eta = 2\alpha(1 - \frac{1}{1+\xi}) - \xi$, by (4.9) and the following lower bound on $q_{\alpha,\xi,N}$, noting that when $\alpha > 1/2$, such η is positive provided $\xi > 0$ is small enough.

Lemma 4.2.

$$q_{\alpha,\xi,N} \ge b^{-2\alpha(1+\xi)^{-1}N(1+o(1))} \,.$$

Proof. Let $z = x \wedge y$ for $x, y \in \partial T_\xi$. Using the same reduction technique we applied when proving (4.4), it suffices to consider a SRW, starting at 1, on the star shape of Fig. 4.3. With probability $\frac{1}{\xi N}$ we reach z before time τ_0. Upon departing from z, consider which of the following four vertices, $0, z, x, y$, is visited first. With probability $\frac{1}{3\xi N}$ it is 0, with probability $\frac{1}{3(1-\xi)N}$ it is x and with the same probability it is y. Otherwise it is z. Consequently, the number R of excursions away from z to $\{x\} \cup \{y\}$ before time τ_0 is a Geometric random variable with success probability

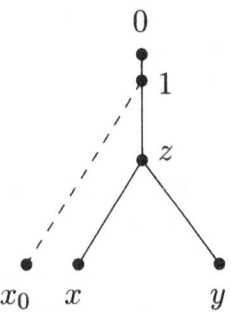

Fig. 4.3. Reduced tree for the joint law of $(\widetilde{T}_x, \widetilde{T}_y)$

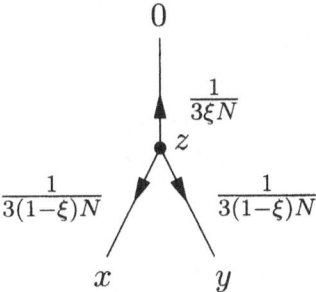

Fig. 4.4. Transition probabilities away from z for $|z| = \xi N$

$$p = \frac{\frac{1}{3\xi N}}{\frac{1}{3(1-\xi)N} + \frac{1}{3(1-\xi)N} + \frac{1}{3\xi N}} = \frac{1-\xi}{1+\xi} \qquad (4.10)$$

(see also Fig. 4.4). Let R_x denote the number of excursions from z to x and back to z prior to time τ_0, with R_y the number of excursions from z to y and back, prior to τ_0, so $R = R_x + R_y$.

By simple large deviations estimates we will see that

$$\mathbb{P}(R_x = R_y = r_N) \geq b^{-\alpha I_2(\rho)N(1+o(1))} , \qquad (4.11)$$

for $r_N = [\alpha N\rho \log b]$ and $I_2(\rho) = 2\rho \log((1+\xi)/(2\xi))$. Further, we will see that

$$\mathbb{P}(\widetilde{T}_x \geq t_N | R_x = r_N) \geq b^{-\alpha J_A(\rho)N(1+o(1))} , \qquad (4.12)$$

for $J_A(\rho) = \rho \log(\rho/A) - \rho + A$, with $A = 1/(1-\xi)$. Observing that conditional upon R_x and R_y the random variables $\widetilde{T}_x$ and $\widetilde{T}_y$ are independent of each other, with the law of $\widetilde{T}_x$ given $R_x = r$ being the same as that of $\widetilde{T}_y$ given $R_y = r$, it follows from (4.11) and (4.12) that

$$\begin{aligned} q_{\alpha,\xi,N} &\geq \mathbb{P}(R_x = R_y = r_N)\mathbb{P}(\widetilde{T}_x \geq t_N | R_x = r_N)^2 \\ &\geq b^{-\alpha N(1+o(1))\inf_{\rho>0}[I_2(\rho)+2J_A(\rho)]} . \end{aligned} \qquad (4.13)$$

The infimum over ρ in (4.13) is obtained at $\rho_2^* = 2\xi A/(1+\xi) > 0$, and considering this value of ρ in (4.13) provides the stated lower bound on $q_{\alpha,\xi,N}$.

Turning to derive (4.11) note that for any $r \geq 1$,

$$\mathbb{P}(R_x = R_y = r) \geq \frac{1}{\xi N}\mathbb{P}(R = 2r)\mathbb{P}(R_x = r|R = 2r) .$$

Recall that $\mathbb{P}(R = r) = p(1-p)^{r-1}$ for $p = \frac{1-\xi}{1+\xi}$ of (4.10). Further, as the law of R_x conditional upon $R = 2r$ is a Binomial $(2r, \frac{1}{2})$, it follows that $\mathbb{P}(R_x = r_N|R = 2r_N) \geq e^{-o(N)}$ when $r_N = O(N)$ and thus

$$\lim_{N\to\infty} \frac{1}{N}\log_b \mathbb{P}(R_x = R_y = r_N) \geq \lim_{N\to\infty} \frac{1}{N}\log_b \mathbb{P}(R = 2r_N) = -\alpha I_2(\rho) ,$$

which is (4.11).

Turning to derive (4.12), let $\widetilde{G}_i$ for $i = 1, \ldots, R_x$ denote the total number of visits to x during the i-th excursion away from z via x. Observe that $\widetilde{T}_x = \sum_{i=1}^{R_x} \widetilde{G}_i$, with the i.i.d. random variables $\widetilde{G}_i$ independent of the value of R_x. Hence,

$$\mathbb{P}(\widetilde{T}_x \geq t_N|R_x = r_N) = \mathbb{P}\left(\sum_{i=1}^{r_N} \widetilde{G}_i \geq t_N\right) .$$

Further, observe that each random variable $\widetilde{G}_i$ has a Geometric distribution with success probability A/N. That is,

$$\mathbb{P}(\widetilde{G}_i = k) = \frac{A}{N}\left(1 - \frac{A}{N}\right)^{k-1} , \quad k = 1, 2, \ldots \qquad (4.14)$$

(for $A = \frac{1}{1-\xi}$). Let E_i be i.i.d. Exponential random variables, each of parameter A. The probability of the event $\{\sum_{i=1}^{r_N} E_i \geq Ns\}$ is the probability that a Poisson random variable of parameter ANs is at most $r_N - 1$. Taking $s = \alpha \log b$ we note that $(r_N - 1)/(Ns) \to \rho$ as $N \to \infty$. Estimating the tails of the relevant Poisson distribution, we thus easily verify that

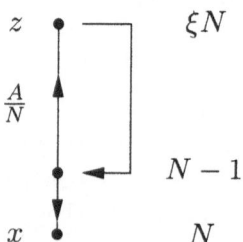

Fig. 4.5. Illustration for the Geometric law of (4.14)

$$\lim_{N\to\infty} \frac{1}{N} \log \mathbb{P}\left(\sum_{i=1}^{r_N} E_i \geq Ns\right) \geq -sJ_A(\rho) \,.$$

Recall that as $N \to \infty$ each of the variables $N^{-1}\widetilde{G}_i$ converges in distribution to the corresponding Exponential variable E_i. Though somewhat cumbersome, it is not hard to show that also

$$\lim_{N\to\infty} \frac{1}{N} \log \mathbb{P}\left(\sum_{i=1}^{r_N} \widetilde{G}_i \geq N^2 s\right) = \lim_{N\to\infty} \frac{1}{N} \log \mathbb{P}\left(\sum_{i=1}^{r_N} E_i \geq Ns\right) \,,$$

and hence to establish (4.12) upon noting that $t_N = N^2 s$.   □

The proof of Lemma 4.2 provides an insight to the reason for the excessive growth of $\mathrm{Var}(|\mathcal{F}_N(\alpha)|)$. It is indeed not hard to show that both (4.11) and (4.12) provide the correct rate of decay as $N \to \infty$. Further, by similar reasoning

$$\mathbb{P}(R_x = r_N) = b^{-\alpha I_1(\rho)N(1+o(1))} \,,$$

for $I_1(\rho) = \rho \log(1/\xi)$, leading to the estimate

$$\mathbb{P}(\widetilde{T}_x \geq t_N) = b^{-\alpha N(1+o(1)) \inf_{\rho>0} [I_1(\rho)+J_A(\rho)]} \,. \tag{4.15}$$

The infimum over ρ in (4.15) is obtained at $\rho_1^* = \xi A < \rho_2^*$. Thus, the main contribution to the variance of $|\mathcal{F}_N(\alpha)|$ is due to the rare events of excessive number of excursions (of order $\alpha N \rho_2^* \log b$) between a vertex z at level ξN and the leaves of the sub-tree rooted at z. Such events do not contribute much to the mean of $|\mathcal{F}_N(\alpha)|$, as the typical number of such excursions for α-favorite leaves of T is only of order $\alpha N \rho_1^* \log b$. However, the occurrence of such rare event yields too many α-favorite leaves at the sub-tree rooted at z, hence resulting with the excessive growth of the variance. This suggests the strategy that we pursue in Sect. 4.2. Namely, we shall replace the set $\mathcal{F}_N(\alpha)$ by a set $\mathcal{S}_N(\alpha)$ of leaves along the rays of which we control various excursion counts, keeping them within a relatively small distance from the typical excursion count profile of an α-favorite leaf. When correctly implemented it yields a huge variance reduction without much change in the mean.

4.2 n-Perfect Leaves: First Moment

With $\mathbb{E}(|\mathcal{F}_N(\alpha)|^2) \gg (\mathbb{E}(|\mathcal{F}_N(\alpha)|))^2$, we aim at finding a modified set of leaves $\mathcal{S}_N(\alpha)$ such that,

$$\lim_{N\to\infty} \mathbb{P}(\mathcal{S}_N(\alpha) \subseteq \mathcal{F}_N(\alpha - \delta)) = 1, \quad \text{for all } \delta > 0 \,, \tag{4.16}$$

$$\lim_{N\to\infty} \frac{1}{N} \log \mathbb{E}(|\mathcal{S}_N(\alpha)|) = (1 - \alpha) \log b \,, \tag{4.17}$$

$$\operatorname{Var}(|\mathcal{S}_N(\alpha)|) = o(\mathbb{E}(|\mathcal{F}_N(\alpha)|)^2) . \tag{4.18}$$

Clearly, the construction for each $0 < \alpha < 1$ of $\mathcal{S}_N(\alpha)$ with properties (4.16)–(4.18) shall complete the proof of Theorem 4.1.

Towards this end, we first find the typical number of steps the SRW makes from the ancestor of $x \in \mathcal{F}_N(\alpha)$ at level $k-1$ to the ancestor of same x at level k. As we have already seen, it suffices to consider the restriction of the SRW to $0 \leftrightarrow x$, represented by a SRW on $\{0, 1, \ldots, N\}$ and denoted $\{Y_t\}$. Ignoring the first and last part of the path Y_t, we can decompose it into independent excursions, each starting at level $(N-1)$ and run until the first return to level N without ever visiting 0. Let $\mathbb{P}_j(\cdot)$ denote the law of $\{Y_t\}_{t \geq 1}$ starting at $Y_0 = j$, and $\tau_l = \inf\{t \geq 1 : Y_t = l\}$ be the corresponding hitting time of $l \in \{0, 1, \ldots, N\}$. Our next lemma provides the typical value of the number of steps Z_k from $k-1$ to k in one excursion of Y_t starting at $Y_0 = N-1$ and ending at $\min(\tau_0, \tau_N)$.

Lemma 4.3. *When both $k \gg 1$ and $N - k \gg 1$,*

$$\mathbb{E}_{N-1}(Z_k \mathbf{1}_{\{\tau_0 > \tau_N\}}) = \frac{k^2}{N^2}(1 + o(1)) .$$

Proof. If $Z_k > 0$ then necessarily $\tau_N > \tau_k$, so by the strong Markov property of $\{Y_t\}$ at τ_k we have,

$$\mathbb{E}_{N-1}(Z_k \mathbf{1}_{\{\tau_0 > \tau_N\}}) = \mathbb{P}_{N-1}(\tau_N > \tau_k)\mathbb{E}_k(Z_k \mathbf{1}_{\{\tau_0 > \tau_N\}})$$

$$= \frac{1}{N-k} \sum_{j=1}^{\infty} \mathbb{P}_k(Z_k \geq j, \tau_0 > \tau_N)$$

Further, by the strong Markov property at consecutive visits to $(k-1)$ followed by returns to k (c.f. Fig. 4.6), we have,

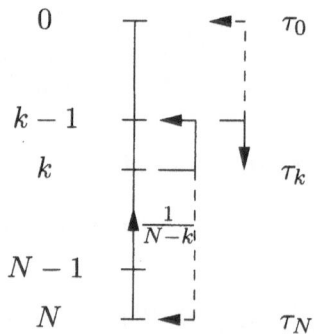

Fig. 4.6. Path decomposition for $Z_k \mathbf{1}_{\{\tau_0 > \tau_N\}}$ with $Y_0 = N - 1$

$$\mathbb{P}_k(Z_k \geq j, \tau_0 > \tau_N) = [\mathbb{P}_k(\tau_N > \tau_{k-1})\mathbb{P}_{k-1}(\tau_0 > \tau_k)]^j \, \mathbb{P}_k(\tau_0 > \tau_N)$$

$$= \left[\left(1 - \frac{1}{N-k+1}\right)\left(1 - \frac{1}{k}\right)\right]^j \frac{k}{N} \, .$$

With $Q = (1 - \frac{1}{N-k+1})(1 - \frac{1}{k})$, if $k \gg 1$ and $N - k \gg 1$, then

$$\mathbb{E}(Z_k \mathbf{1}_{\{\tau_0 > \tau_N\}}) = \frac{kQ}{(N-k)N(1-Q)} = \frac{k^2}{N^2}(1 + o(1)) \, ,$$

as stated.   □

Noting that $\mathbb{P}_{N-1}(\tau_0 > \tau_N) = 1 - \frac{1}{N}$, we deduce from Lemma 4.3 that the number of steps by the walk from $k-1$ to k during $\alpha N^2 \log b$ excursions from $N-1$ to N none of which visit 0, is $\alpha k^2(1+o(1)) \log b$. This suggests choosing the set $\mathcal{S}_N(\alpha)$ to consist of those leaves such that for any $k = 1, \ldots, N$,

$$\widetilde{L}_k = \sum_{i=1}^{\tau_0} \mathbf{1}_{\{Y_i = k-1, Y_{i+1} = k\}}$$

is near the typical value of $\alpha k^2 \log b$. Obviously, such a choice satisfies (4.16) and is likely to satisfy (4.18) as well. Unfortunately, the probability that $\widetilde{L}_k$ is near its typical value at *every* level k between 1 and N is too small, so the expected number of leaves with this property fails to grow at the rate $b^{N(1-\alpha)}$ that is needed for (4.17).

We thus follow the more sophisticated strategy of considering the birth-death process $\{X_l\}$, obtained by observing only the excursions of $\{Y_t\}$ between the discrete levels h_k, $k = 0, 1, \ldots, n$, where $h_0 = 0$, $h_1 = 1$ and for $k \geq 2$ let $h_k = [ck \log k]$ for some universal positive constant c (c.f. Fig. 4.7). Note that $\{X_l\}$ is a Markov chain on $\{0, 1, \ldots, n\}$ with

$$p_k = \mathbb{P}(X_l = k+1 | X_{l-1} = k) = 1 - \mathbb{P}(X_l = k-1 | X_{l-1} = k)$$

$$= \frac{h_k - h_{k-1}}{h_{k+1} - h_{k-1}}, \quad k = 1, \ldots, n-1, \tag{4.19}$$

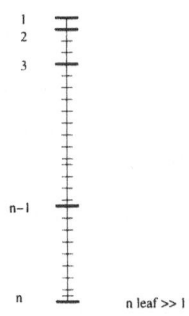

Fig. 4.7. The levels h_k, $k = 1, \ldots, n$, where $N = h_n$

depending on k. We make 0 an absorbing state by setting $p_0 = 0$ and impose the reflection at the leaf $(X_l = n)$ by setting $p_n = 0$. With h_k growing not much faster than linear, by Lemma 4.3, the likely occupation measure profile has about $n_k = (\frac{k}{n})^2 n_n$ steps of X_l from $k-1$ to k, provided both $h_n - h_k \gg 1$ and $h_k \gg 1$. We thus set the nominal values of

$$L_k = \sum_{l=1}^{\infty} \mathbf{1}_{\{X_l=k-1,X_{l+1}=k\}} \qquad (4.20)$$

to be $n_k = \zeta k^2 \log k$ for $k = 2, \ldots, n$ and ζ a fixed constant to be determined later. We say that a leaf $x \in \partial T$ is n-perfect if $|L_k - n_k| \le k$ for $k = 2, \ldots, n$.

In Sect. 4.3 we construct the set $\mathcal{S}_N(\alpha)$ having properties (4.16), (4.17), and the "moral equivalent" of (4.18). This is done by the appropriate union of sets of the form

$$\mathcal{S}_n^v(\alpha) = \{x \in \partial T^v : x \text{ is n-perfect between } \tau_v \text{ and } \sigma_{v^f}\}, \qquad (4.21)$$

where T^v is the subtree of v and all its descendents within T whose depth is $h_n - 1$, v^f denotes the parent of v in T (see Fig. 4.8), τ_v denotes the first time the SRW on T visits v and σ_{v^f} is the time of first hitting v^f after τ_v.

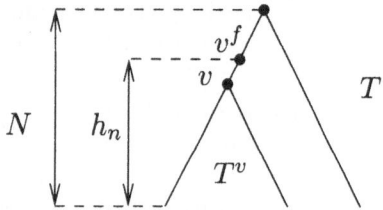

Fig. 4.8. The tree T and a typical sub-tree T^v

Our next proposition estimates the probability that a specific leaf of the tree is n-perfect.

Proposition 4.4. *For any choice of $c, \zeta > 0$ we have as $n \to \infty$ that*

$$q_n = \mathbb{P}(x \text{ is n-perfect}) = (n!)^{-\zeta(1+o(1))}.$$

Remark 4.5. *Our proof of Proposition 4.4 is an adaptation of the derivation of such an estimate in the context of thick points for planar Brownian motion, done in Lemma 7.1 of [DPRZ01].*

Proof. The key to the computation of q_n is the observation that $\{L_k\}$ is a non-homogeneous Markov chain, with

$$L_{k+1} = \sum_{i=1}^{L_k} G_i, \qquad \text{for } k \ge 2,$$

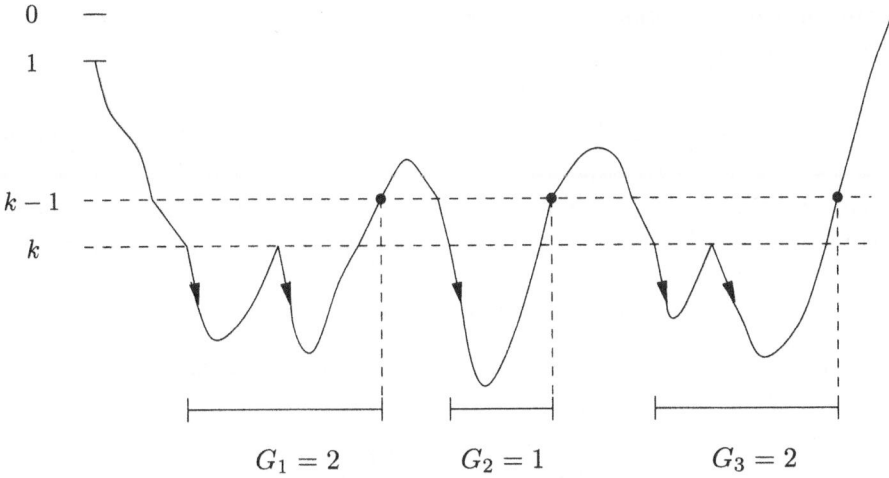

Fig. 4.9. Path decomposition for the $L_k = 3$ excursions of X_l

where the G_i are independent, identically distributed, Geometric random variables such that for p_k of (4.19),

$$\mathbb{P}(G_i = j) = (1 - p_k)p_k^j , \qquad j = 0, 1, 2, \ldots \qquad (4.22)$$

Indeed, referring to Fig. 4.9, the number of "down" segments is L_k and G_i denotes the number of transitions from k to $k+1$ during the i-th "down" segment. Thus, each G_i has the law of the number of transitions $k \to k+1$ before the first $k \to k-1$ transition in a birth-death Markov chain starting at k, given by (4.22) and they are i.i.d. random variables. In conclusion, the Markov chain $\{L_k, k \geq 1\}$ on the non-negative integers has transition probabilities

$$\mathbb{P}(L_{k+1} = m \mid L_k = l + 1) = \binom{m + l}{m} p_k^m (1 - p_k)^{l+1} , \qquad (4.23)$$

for $k \geq 1$, $m, l \geq 0$, while $\mathbb{P}(L_{k+1} = 0 | L_k = 0) = 1$ for all $k \geq 2$. We thus have that

$$q_n = \mathbb{P}(n_k - k \leq L_k \leq n_k + k \, ; \, 2 \leq k \leq n)$$

$$= \sum_{\substack{l_2, \ldots, l_n \\ |l_k - n_k| \leq k}} \prod_{k=1}^{n-1} \mathbb{P}(L_{k+1} = l_{k+1} | L_k = l_k) , \qquad (4.24)$$

where $l_1 = 1$. We next approximate the Markov transition kernel of (4.23) for the values of l, m relevant here.

Lemma 4.6. *For some $C = C(\zeta) < \infty$ and all $k \geq 2$, $|m - n_{k+1}| \leq k + 1$, $|l + 1 - n_k| \leq k$,*

$$C^{-1} \frac{k^{-\zeta}}{\sqrt{k^2 \log k}} \leq \mathbb{P}(L_{k+1} = m \mid L_k = l + 1) \leq C \frac{k^{-\zeta}}{\sqrt{k^2 \log k}} . \qquad (4.25)$$

Proof of Lemma 4.6.. It suffices to consider $k \gg 1$, and since $n_k - 2k \to \infty$, the binomial coefficient in (4.23) is well approximated by Stirling's formula

$$m! = \sqrt{2\pi} m^m e^{-m} \sqrt{m} (1 + o(1)) .$$

With $n_k = \zeta k^2 \log k$ it follows that for some $C_1 < \infty$ and all k large enough, if $|m - n_{k+1}| \leq 2k$, $|l - n_k| \leq 2k$ then

$$\left| \frac{m}{l} - 1 - \frac{2}{k} \right| \leq \frac{C_1}{k \log k} . \qquad (4.26)$$

Hereafter, we use the notation $f \sim g$ if f/g is bounded and bounded away from zero as $k \to \infty$, uniformly in $\{m : |m - n_{k+1}| \leq 2k\}$ and $\{l : |l - n_k| \leq 2k\}$. We then have by (4.23) and the preceding observations that

$$\mathbb{P}(L_{k+1} = m \mid L_k = l + 1) \sim \frac{(m + l)^{m+l}}{\sqrt{l} \, l^l m^m} p_k^m (1 - p_k)^l$$

$$\sim \frac{\exp(-lI(\frac{m}{l}, p_k))}{\sqrt{k^2 \log k}} , \qquad (4.27)$$

where

$$I(\lambda, p) = -(1 + \lambda) \log(1 + \lambda) + \lambda \log \lambda - \lambda \log p - \log(1 - p) .$$

Our choice of $h_k = [ck \log k]$ results with $h_k - h_{k-1} = c \log k (1 + o(1))$ and $2h_k - h_{k-1} - h_{k+1} = O(\frac{1}{k})$. Using (4.19) we thus have then that

$$p_k = \frac{1}{2} + O\left(\frac{1}{k \log k} \right) .$$

The function $I(\lambda, p)$ and its first order partial derivatives vanish at $(1, 1/2)$, with the second derivative $\frac{\partial^2 I}{\partial \lambda^2}(1, 1/2) = 1/2$ (see also Fig. 4.10). Thus, by a Taylor expansion to second order of $I(\lambda, p)$ at $(1, 1/2)$, the estimate (4.26) results in

$$\left| I\left(\frac{m}{l}, p_k \right) - \frac{1}{k^2} \right| \leq \frac{C_2}{k^2 \log k} \qquad (4.28)$$

for some $C_2 < \infty$, all k large enough and m, l in the range considered above. Since $|l - \zeta k^2 \log k| \leq 2k$, combining (4.27) and (4.28), we establish (4.25). $\square$

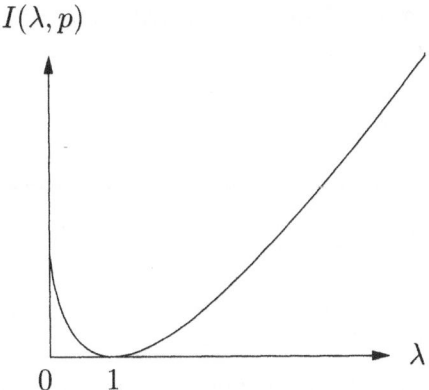

$I(\lambda, p)$

λ

$0 \quad 1$

Fig. 4.10. The function $I(\lambda, p)$ for $p = 1/2$

Equipped with Lemma 4.6 and (4.24) we see that

$$q_n = \sum_{l_2,\ldots,l_n} \prod_{k=1}^{n-1} \mathbb{P}(L_{k+1} = l_{k+1} | L_k = l_k) \sim \sum_{l_2,\ldots,l_n} \prod_{k=1}^{n-1} \frac{k^{-(\zeta+1)}}{\sqrt{\log k}} \ .$$

The number of vectors $(l_2, \ldots, l_n)$ considered here is at least $n!$ and at most $3^n n!$ (c.f. (4.24)). Since $n^{-1} \log(n!) \to \infty$ and for some $\eta_n \to 0$

$$\prod_{k=2}^{n} \log k = (n!)^{\eta_n} \ ,$$

We see that as $n \to \infty$,

$$q_n = (n!)^{-\zeta(1+o(1))} = e^{-\zeta(n \log n)(1+o(1))} \ ,$$

thus completing the proof of the proposition. $\qquad\qquad\qquad\qquad\square$

Recall that $h_n = [cn \log n]$ so setting $\zeta = c\alpha \log b$ leads to $q_n = e^{-h_n \alpha \log b(1+o(1))}$. If $\mathcal{S}_N(\alpha)$ involves checking which of a set of $b^{N(1+o(1))}$ leaves is n-perfect for some n such that $h_n = N(1 + o(1))$, then (4.17) would hold for this choice of ζ.

We turn to check if there is hope that (4.16) would hold as well. We start with the nominal value of L_n for our choice of h_n and ζ which is

$$n_n = \zeta n^2 \log n = \alpha h_n^2 \log b / (c \log n)$$

The random variable L_n records the number of excursions of the SRW starting at level h_{n-1} that reach the leaf x by time τ_0. The typical number of visits to x in each such excursion is rather large. Indeed, the situation here is very similar to the computation we had leading to (4.4). Let G_i denote now the number of visits to x by Y_t during the i-th excursion. Then, G_i are i.i.d. random variables, each having the Geometric distribution with success probability $1/(h_n - h_{n-1})$. Hence,

$$\mathbb{E}(G_i) = h_n - h_{n-1} = c\log n(1 + o(1)),$$

implying that

$$\mathbb{E}\left(\sum_{i=1}^{n_n - n} G_i\right) = \alpha h_n^2 \log b(1 + o(1)).$$

Thus, n-perfect leaves, for which $L_n \geq n_n - n$, are very likely to be $(\alpha - \delta)$-favorite, as formalized in the next lemma.

Lemma 4.7. *Let L denote the number of excursions from level h_{n-1} to a leaf x at level h_n prior to τ_0 and $\tilde{T}_x$ the number of visits to x during these L excursions. Then, for any $\delta > 0$, there exists $\theta(\delta) > 0$ such that,*

$$\mathbb{P}\left(\tilde{T}_x \leq (1 - 2\delta)n_n(h_n - h_{n-1})|L \geq n_n - n\right) \leq e^{-\theta(\delta)(n_n - n)}.$$

Proof. Recall that

$$\tilde{T}_x = \sum_{i=1}^{L} G_i$$

where G_i are i.i.d. each of Geometric distribution with success probability $p_n = 1/(h_n - h_{n-1})$. The random variable G_i represents the number of visits to x during the i-th excursion starting at the parent of x and ending when visiting level h_{n-1}. Since L is independent of the number of visits to x in each of these excursions and $\tilde{T}_x$ is increasing in L, we have that for each $c, t, \lambda > 0$

$$\mathbb{P}\left(\tilde{T}_x \leq t|L \geq c\right) \leq \mathbb{P}\left(\sum_{i=1}^{c} G_i \leq t\right) \leq e^{\lambda t}\mathbb{E}\left(e^{-\lambda G_1}\right)^c.$$

Taking $c = n_n - n$ and $t = (1 - 2\delta)n_n(h_n - h_{n-1})$ we have that $t \leq (1 - \delta)c/p_n$ for all large n (as $n_n \gg n$). Thus, it suffices to find a sequence $\lambda_n > 0$ such that for some $\theta > 0$ and all n large enough,

$$e^{\lambda_n(1-\delta)/p_n}\mathbb{E}\left(e^{-\lambda_n G_1}\right) \leq e^{-\theta}. \tag{4.29}$$

To establish (4.29), pick $\kappa > 0$ and $\lambda_n = \kappa p_n$. Recall that $p_n G_1$ converges weakly to an Exponential random variable with unit mean, yielding that

$$e^{\lambda_n(1-\delta)/p_n}\mathbb{E}\left(e^{-\lambda_n G_1}\right) \to \frac{e^{\kappa(1-\delta)}}{\kappa + 1}.$$

Choosing $\kappa > 0$ small enough, this proves (4.29) and thus completes the proof of the lemma. $\square$

4.3 Multi-Scale Truncated Second Moment: n-Perfect Leaves

We wish to estimate $\mathbb{E}(|\mathcal{S}_n^v(\alpha)|^2)$ en-route to verifying (4.18). To this end, we start by bounding the relevant correlation.

Lemma 4.8. *There exists $\gamma_n \to 0$ such that for any $l = 2, \ldots, n$,*

$$q_{n,l} \triangleq \sup_{\substack{h_{l-1} \le |x \wedge y| < h_l \\ |x \wedge x_0| = 1, |y \wedge x_0| = 1}} \mathbb{P}\big(x \text{ and } y \text{ are } n\text{-perfect}\big) \le q_n^2 \, (l!)^{\zeta + \gamma_l} . \qquad (4.30)$$

Proof. We may and shall assume without loss of generality that n is large enough for $n_{n-2} \ge n - 1$. By Proposition 4.4 we know that $q_n \ge (n!)^{-\zeta - \delta_n}$ for some $\delta_n \to 0$. Hence, assuming without loss of generality that $f(n) := ((n-1)!)^{\zeta + \gamma_{n-1}} \ge (n!)^{\zeta + \delta_n}$ is nondecreasing, it suffices to consider only $l \le n - 2$. Let L_k^x for $x \in \partial T$, $k = 1, 2, \ldots, n$, denote the number of excursions from the vertex at height h_{k-1} to the vertex at height h_k on the ray $0 \leftrightarrow x$ prior to first visit of 0 (the root of T) by the SRW on the b-ary tree. With $n_k = \zeta k^2 \log k$ we shall write $L \overset{k}{\sim} n_k$ if $|L - n_k| \le k$. Let $\mathcal{G}_l^y$ denote the σ-field generated by the excursions of the SRW from level h_{l-1} to level h_l on the ray $0 \leftrightarrow y$, including the path up to first visit of level h_{l-1} on $0 \leftrightarrow y$. Observe that $\{L_k^x \overset{k}{\sim} n_k\}$ are measurable on $\mathcal{G}_{l+1}^y$ for all $k = 2, \ldots, n$. With $J_i := \{i+1, \ldots, n\}$, we have that

$$\{x, y \text{ are } n\text{-perfect}\} \subset \{L_k^x \overset{k}{\sim} n_k, \ k \in J_1\} \bigcap \{L_k^y \overset{k}{\sim} n_k, \ k \in J_l\} \qquad (4.31)$$

(see also Fig. 4.11). Let $\Gamma(J_i) := \{m_{i+1}, \ldots, m_n : |m_k - n_k| \le k, k \in J_i\}$. Since $q_n = \mathbb{P}(L_k^x \overset{k}{\sim} n_k, \ k \in J_1)$ and given $L_{l+1}^y = m_{l+1}$ the random variables L_k^y for $k \in J_{l+1}$ are independent of $\mathcal{G}_{l+1}^y$, we have that

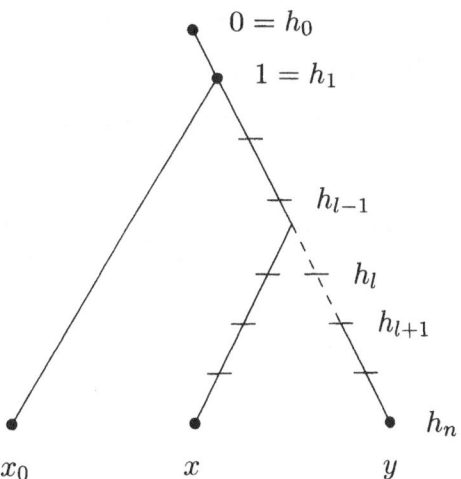

Fig. 4.11. Information about the path on the dashed part of T is ignored in (4.31)

$$\mathbb{P}(x \text{ and } y \text{ are n-perfect })$$

$$\leq \sum_{\Gamma(J_l)} \mathbb{E}\left[\mathbb{P}(L_k^y = m_k, \ k \in J_{l+1} \mid L_{l+1}^y = m_{l+1}, \mathcal{G}_{l+1}^y) \ ; L_k^x \overset{k}{\sim} n_k, \ k \in J_1\right]$$

$$= q_n \left[\sum_{\Gamma(J_l)} \prod_{k=l+1}^{n-1} \mathbb{P}(L_{k+1} = m_{k+1} \mid L_k = m_k)\right] \qquad (4.32)$$

(where $m_1 = 1$). Note that by Lemma 4.6,

$$q_n = \sum_{\Gamma(J_1)} \prod_{k=1}^{n-1} \mathbb{P}(L_{k+1} = m_{k+1} \mid L_k = m_k)$$

$$\geq q_{l+1} \inf_{|m_{l+1}-n_{l+1}|\leq l+1} \sum_{\Gamma(J_{l+1})} \prod_{k=l+1}^{n-1} \mathbb{P}(L_{k+1} = m_{k+1} \mid L_k = m_k)$$

$$\geq q_{l+1}C^{-2} \sup_{|m_{l+1}-n_{l+1}|\leq l+1} \sum_{\Gamma(J_{l+1})} \prod_{k=l+1}^{n-1} \mathbb{P}(L_{k+1} = m_{k+1} \mid L_k = m_k)$$

$$\geq q_{l+1}C^{-2}(2l+3)^{-1} \sum_{\Gamma(J_l)} \prod_{k=l+1}^{n-1} \mathbb{P}(L_{k+1} = m_{k+1} \mid L_k = m_k) . \qquad (4.33)$$

Combining (4.32) and (4.33), we see that

$$\mathbb{P}(x \text{ and } y \text{ are n-perfect}) \leq q_n^2 \frac{C^2(2l+3)}{q_{l+1}} ,$$

and (4.30) follows from the bound $q_l \geq (l!)^{-\varsigma-\delta_l}$.   □

Having Lemma 4.8, we next bound the second moment of the size of the set $\mathcal{S}_n^v(\alpha)$.

Lemma 4.9. *For any $0 < \alpha < 1$, there exists $K(\alpha) < \infty$ such that for all n,*

$$\mathbb{E}(|\mathcal{S}_n^v(\alpha)|^2) \leq K(\alpha) \left(\mathbb{E}|\mathcal{S}_n^v(\alpha)|\right)^2 .$$

Proof. We may and shall assume without loss of generality that $|v| = 1$ and the original tree is of depth h_n. Recall that $\mathbb{E}(|\mathcal{S}_n^v(\alpha)|) = q_n b^{h_n-1}$ and by direct computation,

$$\mathbb{E}(|\mathcal{S}_n^v(\alpha)|^2) = \sum_{x,y \in \partial T^v} \mathbb{P}(x \text{ and } y \text{ are n-perfect })$$

$$= b^{h_n-1} \sum_{y \in \partial T^v} \mathbb{P}(x \text{ and } y \text{ are n-perfect })$$

$$\leq b^{h_n-1} \sum_{l=2}^{n} q_{n,l} b^{h_n-h_{l-1}} = b^{2(h_n-1)} \sum_{l=2}^{n} q_{n,l} b^{1-h_{l-1}}$$

$$\leq \left(\mathbb{E}|\mathcal{S}_n^v(\alpha)|\right)^2 \sum_{l=2}^{n} \frac{q_{n,l}}{q_n^2} b^{1-h_{l-1}} , \qquad (4.34)$$

where we counted the leaves within T^v according to the value of $2 \leq l \leq n$ such that $h_{l-1} \leq |x \wedge y| \leq h_l$. By lemma 4.8 we have the estimates

$$\frac{q_{n,l}}{q_n^2} \leq (l!)^{\zeta + \gamma_l} \leq e^{(\zeta + \gamma_l)l \log l} = b^{(\alpha + \gamma_l)h_l} \tag{4.35}$$

(recall that $\zeta = c\alpha \log b$ and $h_l = [cl \log l]$ absorbing the factor $1/(c \log b)$ into the sequence γ_n). Combining (4.35) with (4.34) we obtain that

$$\mathbb{E}(|\mathcal{S}_n^v(\alpha)|^2) \leq (\mathbb{E}|\mathcal{S}_n^v(\alpha)|)^2 \sum_{l=2}^{\infty} b^{(\alpha + \gamma_l - 1)h_{l-1}} = (\mathbb{E}|\mathcal{S}_n^v(\alpha)|)^2 K(\alpha) ,$$

since $\alpha + \gamma_l - 1 < 0$ is bounded away from zero for all l large enough and h_l grows faster than linear. This completes the proof of the lemma. $\square$

It follows from Lemma 4.9 that for any $\lambda > 0$

$$\mathbb{P}\big(|\mathcal{S}_n^v(\alpha)| \geq (1 - \lambda)\mathbb{E}(|\mathcal{S}_n^v(\alpha)|)\big) \geq \frac{\lambda^2 (\mathbb{E}|\mathcal{S}_n^v(\alpha)|)^2}{\mathbb{E}(|\mathcal{S}_n^v(\alpha)|^2)} \geq \frac{\lambda^2}{K(\alpha)} > 0 , \tag{4.36}$$

for example, see [Kah85, page 8]. Unfortunately, this falls short of the consequence of property (4.18), namely a size bound that is valid with probability that approaches one as $N \to \infty$. To lift (4.36) to such a bound we construct next a somewhat larger set of leaves. To this end, recall that

$$\mathcal{S}_n^v(\alpha) = \{x \in \partial T^v : \text{x is n-perfect between } \tau_v \text{ and } \sigma_{v^f}\} ,$$

where T^v is the subtree of v and all its descendents within T whose depth is $h_n - 1$, v^f denotes the parent of v in T, τ_v denotes the first time the SRW on T visits v and σ_{v^f} is the first hitting time of v^f after τ_v. Fixing N, let $n = n(N) = \sup\{k : h_{k+1} \leq N\}$ and define

$$\mathcal{S}_N(\alpha) = \bigcup \{\mathcal{S}_n^v(\alpha) : |v| = N - (h_n - 1), \tau_v < \tau_0\} . \tag{4.37}$$

This is the set of leaves we alluded to before.

Our next lemma shows that the subset $\mathcal{S}_N(\alpha)$ of ∂T has the property (4.16).

Lemma 4.10. *With $\mathcal{S}_N(\alpha)$ defined as (4.37), for any $\delta > 0$,*

$$\mathbb{P}(\mathcal{S}_N(\alpha) \not\subseteq \mathcal{F}_N(\alpha(1 - 3\delta))) \to 0 \text{ as } N \to \infty .$$

Proof. For each $x \in \partial T$ let L_n^x denote the the number of SRW excursions to x from its ancestor of distance $h_n - h_{n-1}$ prior to the stopping time τ_0. Note that if $x \in \mathcal{S}_N(\alpha)$ then necessarily $L_n^x \geq n_n - n$. Therefore, using Lemma 4.7, and that $b^N = \exp(o(n^2 \log n))$ combined with $n_n - n = O(n^2 \log n)$, we have

$$\mathbb{P}\big(\min\{\widetilde{T}_x \,:\, x \in \mathcal{S}_N(\alpha)\} \le (1-2\delta)n_n(h_n - h_{n-1})\big)$$

$$\le b^N \mathbb{P}\Big(\widetilde{T}_x \le (1-2\delta)n_n(h_n - h_{n-1}) | L_n^x \ge n_n - n\Big)$$

$$\le b^N e^{-\theta(\delta)(n_n - n)} \to 0 \,.$$

Now, observe that for n large,

$$(1-2\delta)n_n(h_n - h_{n-1}) \ge \alpha(1-2\delta)c^2 n^2 (\log n)^2 \log b$$
$$\ge \alpha(1-3\delta)h_{n+2}^2 \log b$$
$$\ge \alpha(1-3\delta)N^2 \log b \,.$$

Thus, by the definition of $\mathcal{F}_N(\alpha(1-3\delta))$ we see that

$$\mathbb{P}\big(\exists x \in \mathcal{S}_N(\alpha), x \notin \mathcal{F}_N(\alpha(1-3\delta))\big) \to 0 \,,$$

as $N \to \infty$, which is the statement of the lemma. $\square$

Our final lemma shows the "moral equivalent" of (4.18), namely, that with high probability $|\mathcal{S}_N(\alpha)|$ is sufficiently large. In view of Lemma 4.10, this completes the proof of Theorem 4.1.

Lemma 4.11. *For any $0 < \alpha < 1$ and N let $\mathcal{S}_N(\alpha)$ be the set of leaves of T defined in (4.37). Then, for any $\delta > 0$,*

$$\mathbb{P}\big(|\mathcal{S}_N(\alpha)| \le b^{N(1-\alpha-\delta)}\big) \to 0 \text{ as } N \to \infty \,.$$

Proof. Set $n = \sup\{k : h_{k+1} \le N\}$ and $M = M(N) = N - (h_n - 1)$, noting that $M = M(N) \to \infty$ as $N \to \infty$. By (4.36), there exists $\rho = \rho(\alpha) < 1$ such that, uniformly in n and for any $v \in T$ with $|v| = M$,

$$\mathbb{P}\big(|\mathcal{S}_n^v(\alpha)| \le b^{h_n(1-\alpha-\delta)}\big) \le \rho \,. \tag{4.38}$$

Let $\mathcal{G}_M$ be the σ-field generated by the SRW path till $\tau_M = \inf\{t \ge 0 : |S_t| = M\}$ and thereafter, whenever for some v at level M the original SRW gets inside T^v we erase the part of the path inside T^v (including possible returns to v) and continue to observe it upon exiting T^v to the parent v^f of v, stopping at the first visit to the root (c.f. Fig. 4.12). Note that the events $\{\tau_v < \tau_0\}$ for $|v| \le M$ are measurable on $\mathcal{G}_M$, and hence so is

$$R_M = \sum_{|v|=M} \mathbf{1}_{\{\tau_v < \tau_0\}} \,.$$

Further, conditional on $\mathcal{G}_M$ the random variables $\{|\mathcal{S}_n^v| : |v| = M, \tau_v < \tau_0\}$ are i.i.d. each having the law of the number of n-perfect leaves in a tree of depth h_n for a SRW that starts at height 1 and terminates at the first visit to the root (see also Fig. 4.13). Consequently, conditioning on $\mathcal{G}_M$ and applying (4.38) we see that for any l,

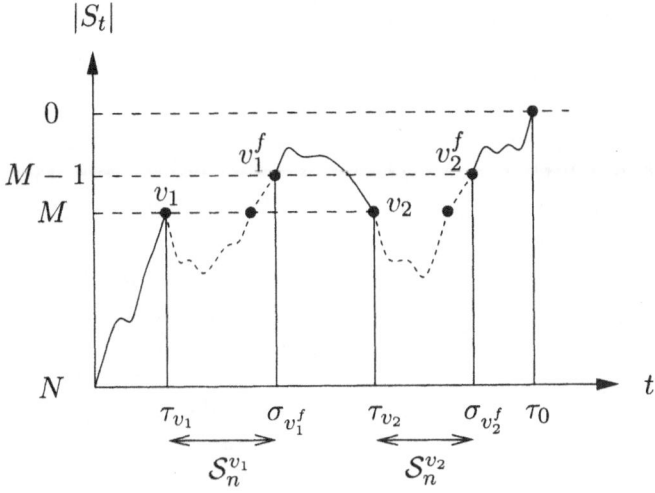

Fig. 4.12. The SRW path within $\mathcal{G}_M$ is marked by a solid line, highlighting the parts of the path that determine $\mathcal{S}_n^{v_i}$. Here $R_M = 2$

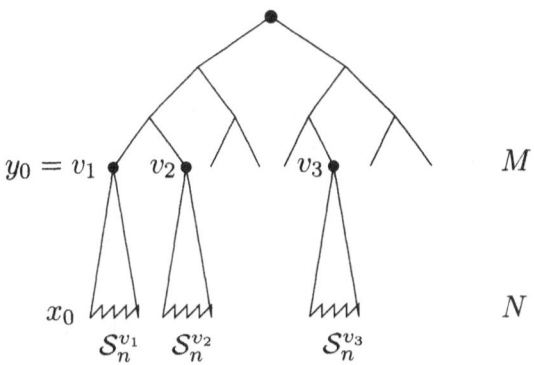

Fig. 4.13. The relation between $|\mathcal{S}_n^{v_i}|$ (here $R_M = 3$)

$$\mathbb{P}\left(|\mathcal{S}_N(\alpha)| \le b^{N(1-\alpha-2\delta)}\right) \le \mathbb{E}\left(\rho^{R_M}\right) \le \rho^{b^l} + \mathbb{P}(R_M < b^l) \,.$$

To complete the proof of the lemma, it thus suffices to show that $\mathbb{P}(R_M < b^l) \to 0$ as $M \to \infty$ for any fixed l. To this end, observe that R_M is exactly the number of leaves of a b-ary tree T_M of depth M visited by a SRW prior to its first visit of the root. Considering $M > l$, let y_0 denote the left-most leaf of T_M which is where the SRW starts and z denote the vertex at level $M - l$ on the ray $0 \leftrightarrow y_0$. The SRW must visit z on its way from y_0 to the root of T_M. Upon doing so, the probability that it thereafter visits the root prior to any specific leaf $y \in T^z$ is exactly l/M. There are b^l leaves in T^z, so by the union bound,

$$\mathbb{P}(R_M < b^l) \leq \mathbb{P}\left(\tau_0 < \max_{y \in \partial T^z} \tau_y\right) \leq b^l l/M \to 0, \quad \text{as } M \to \infty,$$

completing the proof of the lemma. □

Remark 4.12. *We comment on the relation between the method we use here and that of [BDG01], as it would apply in the current setup. Instead of introducing the notion of n-perfect leaves, [BDG01] call the suitably chosen levels h_k successful if enough of the vertices $v \in T$ with $|v| = h_k$ have at least $n_k - k$ excursions leading to them from the previous level h_{k-1}. Starting at level $M = \delta N$ (corresponding to our h_1) provides sufficient independence to start this process with sufficiently many vertices v at a level h_2 down to tree. The notion of a successful level is then propagated by considering the number of excursions to vertices of the next level h_{k+1} during the first $n_k - k$ excursions between their ancestors at levels h_{k-1} and h_k. In this manner, taking the union over the probabilities of failure at the different levels, [BDG01] are able to show that with high probability the last level, consisting of the leaves of the tree, is successful. In contrast, we are dealing here with the profile of excursion along a given ray. This has the advantage of providing a more complete description of the set of leaves of interest, at the cost of slightly more involved computations.*

4.4 Cover Time and Late Points for SRW on *b*-ary Tree

Consider the SRW S_t on the *b*-ary rooted regular tree T of depth N, starting at x_0, the left most leaf of T. How many steps does it take till S_t has visited each vertex v of T at least once? That is, what is the cover time C_N of T by the SRW?

Clearly, $C_N \to \infty$ as $N \to \infty$. Further, it follows from the general considerations of [Ald91b] that $C_N/\mathbb{E}(C_N) \to 1$ in probability, as $N \to \infty$. Considering $\mathbb{E}(C_N)$, it is not hard to show that it has the same growth asymptotic as for the case where the SRW starts at the root, which is also the same growth asymptotic as the expectation of C_N^+, the time it takes the latter SRW to cover T and return to the root thereafter. The number of steps the SRW takes between consecutive visits to the root are i.i.d. random variables, each of mean $a_N = 2(b^N - 1)/(b - 1)$. Hence, by Wald's lemma, $\mathbb{E}(C_N^+) = a_N \mathbb{E}(\widetilde{C}_N^+)$, where $\widetilde{C}_N$ and $\widetilde{C}_N^+$ denote the number of returns to the root till times C_N and C_N^+, respectively, by the SRW which starts at the root. Obviously, also $\mathbb{E}(\widetilde{C}_N^+) = \mathbb{E}(\widetilde{C}_N)(1 + o(1))$ (c.f. [Per03]).

As mentioned before, the asymptotic growth $\mathbb{E}(\widetilde{C}_N^+) = N^2 b \log b(1 + o(1))$ is provided in [Ald91a] and also in [Per03]. Both do so by relying on an embedded branching process argument that exploits the tree structure, hence are of little help in proving the corresponding Brownian result (1.7), or its analog (1.5) for the simple random walk on the two dimensional lattice torus.

We shall next outline how to adapt the method we used for proving Theorem 4.1 so as to provide the growth asymptotic of $\widetilde{C}_N$. As shown in Chap. 5, this approach is robust enough to provide the analogs of Theorem 4.1 for the Brownian motion (and for the two dimensional random walk). Indeed, though we do not provide details here, the same ideas as in Chap. 5 are applied in [DPRZ04] en-route to the proof of (1.5) and (1.7).

Turning to the cover time of the b-ary tree T, let $\widetilde{T}_x$ denote the time of the first visit to $x \in T$ measured in terms of number of returns to the root by the SRW that starts at the root, before its first visit of x, and use

$$\mathcal{L}_N(\alpha) = \{x \in \partial T : \widetilde{T}_x \geq \alpha N^2 b \log b\} \, ,$$

to denote the set of α-late leaves of T, for $\alpha > 0$. Observe that

$$\widetilde{C}_N = \max_{x \in \partial T}\{\widetilde{T}_x\} \, ,$$

hence the analog of Theorem 4.1 is the statement that for any $0 < \alpha < 1$,

$$\lim_{N \to \infty} \frac{1}{N} \log |\mathcal{L}_N(\alpha)| = (1 - \alpha) \log b \quad \text{in probability} \, ,$$

whereas for any $\alpha > 1$,

$$\lim_{N \to \infty} \mathbb{P}(\mathcal{L}_N(\alpha) = \varnothing) = 1 \, ,$$

which implies also that $N^{-2}\widetilde{C}_N \to b \log b$. It is easy to see that at each visit to the root the probability to hit a specific leaf x before returning to the root is $1/(bN)$, hence the identity $\mathbb{P}(\widetilde{T}_x \geq t) = (1 - \frac{1}{bN})^t$, replacing (4.4). This clearly yields the estimate

$$\mathbb{E}(|\mathcal{L}_N(\alpha)|) = b^{N(1-\alpha)(1+o(1))} \, ,$$

which in turn provides the upper bounds on the probability that $|\mathcal{L}_N(\alpha)|$ is too large (see (4.3) and (4.5) for such a derivation). It thus remains to provide a lower bound on the size of $\mathcal{L}_N(\alpha - \delta)$ by constructing the appropriate modified set of leaves $\mathcal{S}_N(\alpha)$ for which $\mathcal{S}_N(\alpha) \subset \mathcal{L}_N(\alpha - \delta)$ with high probability, $\mathbb{E}(|\mathcal{S}_N(\alpha)|) = b^{N(1-\alpha)(1+o(1))}$, while the value of $|\mathcal{S}_N(\alpha)|$ does not fall much below its mean. Setting $h_1 = 1$, $h_k = [ck \log k]$ and $n_k = \zeta k^2 \log k$, for $2 \leq k \leq n$, at the core of this construction are the sets $\mathcal{S}_n^v(\alpha)$ of n-*perfectly-late* leaves of the sub-trees T^v of depth h_n, for the vertices v at level $M = M(N) = N - (h_n - 1) \to \infty$ of T. A leaf x of a sub-tree T^v is n-perfectly-late if the birth-death process $\{X_l\}$ obtained by observing only the excursions of the SRW between the discrete levels h_k along the ray $v \leftrightarrow x$ is such that $L_2^x = 0$ and $|L_k^x - n_k| \leq k$ for $k = 3, \ldots, n - 1$, where

$$L_k^x = \sum_{l=0}^{\mathcal{R}_n} \mathbf{1}_{\{X_l=k, X_{l+1}=k-1\}} \, ,$$

and $\mathcal{R}_n$ is the time of the n_n return of $\{X_l\}$ to its starting level $X_0 = n$ (i.e. for the original SRW to complete n_n excursions between levels h_n and h_{n-1} on the ray to x).

Comparing Fig. 4.7 with Fig. 4.14 we see that the n-perfect leaves are in duality with the n-perfectly-late leaves. Whereas in the former case $\{X_l\}$ starts at $X_0 = 1$ (corresponding to vertex $v \in T$) and seeks about n_n excursions to level n (i.e. $x \in \partial T$), prior to absorption at level 0 (i.e. visiting v^f), here $\{X_l\}$ starts at $X_0 = n$ (which is now the designation of v) and seeks to complete n_n returns to this level prior to first visit of X_l to level 1 (which is now the designation of the leaf x).

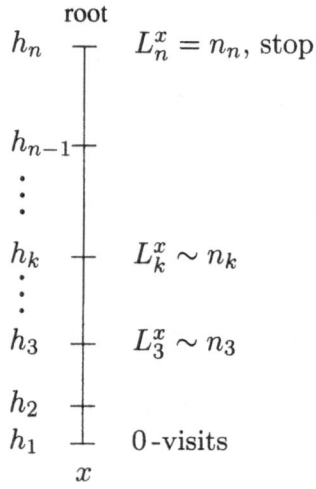

Fig. 4.14. Excursion counts L_k^x for the cover time of SRW on T

We skip the detailed computations for this case, noting only that for any choice of $c, \zeta > 0$ we have as $n \to \infty$ that $\bar{q}_n = \mathbb{P}(x$ is n-perfectly-late $) = (n!)^{-\zeta(1+o(1))}$ by an argument very similar to that of Proposition 4.4. Hence, the same choice of $\zeta = c\alpha \log b$ as for n-perfect leaves, yields the correct growth of $\mathbb{E}(|\mathcal{S}_n^v(\alpha)|)$, namely $\bar{q}_n = b^{-h_n \alpha(1+o(1))}$. Adapting the proof of Lemma 4.8, we find that for some $C < \infty$ and any $l = 2, \dots, n$,

$$\bar{q}_{n,l} \triangleq \sup_{h_{l-1} < |x \wedge y| \leq h_l} \mathbb{P}(x \text{ and } y \text{ are n-perfectly-late }) \leq C l \bar{q}_n \bar{q}_{l-2} \,.$$

Though in this case $n \mapsto \mathbb{E}(|\mathcal{S}_n^v(\alpha)|^2)/\mathbb{E}(|\mathcal{S}_n^v(\alpha)|)^2$ is no longer bounded, its polynomial growth is dealt with by choosing $M(N) = \rho(1 + o(1)) \log n$ for sufficiently large $\rho < \infty$, since $|\mathcal{S}_N(\alpha)|$ which is now the sum of the b^M i.i.d. random variables $|\mathcal{S}_n^v(\alpha)|$ then satisfies (4.18). To complete the proof it remains to derive the analog of Lemma 4.7, namely to consider the number of returns

$R_{v,u}$ to the root of T during the first $n_n - n$ excursions from a specific vertex v at level $M(N)$ to a specific vertex $u \in T^v$ which is $h_n - h_{n-1} = c(1+o(1)) \log n$ levels further from the root of T. Since $R_{v,u}$ is the sum of $n_n - n$ i.i.d. random variables, each of which has exponential tails, the probability that it is less than $(1 - \delta)$ of its mean decays fast enough that even the union of such events over all choices of u and v is negligible. As for $\mathbb{E}(R_{v,u})$, this is a product of the probability $c/(c + \rho)(1 + o(1))$ that the SRW visits the root before u, when starting at v, and the expected number of returns to the root thereafter, before first visiting u. The latter is a Geometric random variable of success probability $(1+o(1))/(b(\rho+c) \log n)$, hence $\mathbb{E}(R_{v,u}) = bc(1+o(1)) \log n$. Note that $(n_n - n)bc \log n = \alpha b N^2 (1 + o(1)) \log b$ for our choice of $\zeta = c\alpha \log b$, resulting with the desired connection between $\mathcal{S}_N(\alpha)$ and $\mathcal{L}_N(\alpha(1 - 3\delta))$.

5 From Trees to Walks Via Brownian Motion

We start by showing in Sect. 5.1 that results about favorite points for planar (simple) random walks are consequences of the corresponding results for the planar Brownian motion. Essential to the success of this approach is the sharp strong approximation theorem of Komlós-Major-Tusnády [KMT75] (or its multidimensional version due to Einmahl [Ein89]; for recent developments in sharp multidimensional strong approximation c.f. [Zai03] and the references therein). We provide here the details of one derivation, that of the Erdős and Taylor conjecture (1.1) out of the corresponding conjecture of Perkins and Taylor. The same strategy has been employed in [DPRZ02] when dealing with favorite points for the intersection of two planar simple random walks and in [DPRZ04] when deriving information (1.5) about the cover time of the two dimensional lattice torus by simple random walk out of the corresponding result (1.7) about the Brownian ε-cover time of the unit torus $\mathbb{T}^2$.

We note in passing that one may alternatively work directly with the random walk, as done for example in [DPRZ05] and [DPR05]. One needs then to replace the PDEs that provide the basic hitting time, hitting probabilities and occupation measure estimates for the Brownian motion (c.f. Sects. 2.3, 3.3 and 5.2), with the corresponding approximations for the random walk, many of which are available in [Law91].

In Sect. 5.2 we outline the derivation of (1.3), that is, we determine the Hausdorff dimension of the set of thick points for the planar Brownian motion, which is more than enough for what we do in Sect. 5.1. This derivation reduces to the core problem of finding sufficiently many n-perfect points. In Sect. 5.3 we explain the relation of the latter to the corresponding result about the favorite leaves of b-ary regular trees, the proof of which we detailed already in Chap. 4.

5.1 From Brownian Motion to Simple Random Walk

Our starting point is (1.3), namely that for any $a \leq 2$,

$$\dim_{\mathrm{H}} \left\{ x \in D(0,1) : \lim_{\varepsilon \to 0} \frac{\mu_{\bar\theta}^w(D(x,\varepsilon))}{\varepsilon^2 |\log \varepsilon|^2} = a \right\} = 2 - a \qquad a.s. \qquad (5.1)$$

where $\dim_{\mathrm{H}}(A)$ denotes the Hausdorff dimension of the set A, $\bar\theta = \inf\{t : |w_t| = 1\}$ is the hitting time of the unit disc by a planar Brownian motion $\{w_t\}_{t \geq 0}$ started at the origin, and

$$\mu_t^w(D(x,r)) = \int_0^t \mathbf{1}_{D(x,r)}(w_s)\, ds \, ,$$

is the Brownian occupation measure of the open disc $D(x,r)$ in $\mathbb{R}^2$ of radius r centered at x. As we already mentioned, the derivation of (5.1) shall be outlined in Sects. 5.2 and 5.3. So, assuming (5.1) has been established, our goal here is to derive out of it the Erdős and Taylor conjecture (1.1), namely

Theorem 5.1. *Let $T_n(x)$ denote the number of visits to x by the path of the simple random walk in $\mathbb{Z}^2$ up to time n. We have for $T_n^* := \max_{x \in \mathbb{Z}^2} T_n(x)$ – the number of visits to the most frequently visited site, that*

$$\lim_{n \to \infty} \frac{T_n^*}{(\log n)^2} = \frac{1}{\pi} \, , \qquad a.s. \qquad (5.2)$$

Remark 5.2. *This is just one part of Theorem 1.1 of [DPRZ01], which establishes in a similar manner also the relation (1.2). As shown in Theorem 5.1 of [DPRZ01], these results hold in greater generality, whenever the increments of the two-dimensional random walk we consider are centered, with all moments finite, and are not supported on a proper subgroup of $\mathbb{Z}^2$. In doing so, one uses [Ein89, Theorem 1], which is the multidimensional extension of the results of [KMT75].*

Proof. We first prove the harder part of (5.2), which is the lower bound

$$\liminf_{n \to \infty} \frac{T_n^*}{(\log n)^2} \geq \frac{1}{\pi} \; a.s. \qquad (5.3)$$

It is in this part that we make use of (5.1) and the strong approximation results of [Ein89] or [KMT75]. To this end, let $h(\varepsilon) = \varepsilon^2 |\log \varepsilon|^2$ and

$$R_{t,\varepsilon} = \sup_{|z| < 1} \left\{ \frac{\mu_t^w(D(z,\varepsilon))}{\varepsilon^2 |\log \varepsilon|^2} \right\} .$$

Fixing $\delta > 0$, the set considered in (5.1) for $a = 2 - \delta/2$ is almost surely non-empty (which is all we need here). Hence, almost surely,

$$\liminf_{\varepsilon \to 0} R_{\bar\theta,\varepsilon} \geq \sup_{|z| < 1} \liminf_{\varepsilon \to 0} \frac{\mu_{\bar\theta}^w(D(z,\varepsilon))}{\varepsilon^2 |\log \varepsilon|^2} \geq 2 - \delta/2 \, ,$$

and consequently, also

$$\lim_{\varepsilon \to 0} \mathbb{P}(R_{\bar{\theta},\varepsilon} \geq 2 - \delta) = 1 \ .$$

Fixing $0 < \eta < 1/2$ let $\varepsilon_n = n^{\eta - 1/2}$. Since $\mathbb{P}(\bar{\theta} \leq 1) > 0$ and the mapping $t \mapsto R_{t,\varepsilon}$ is non-decreasing, it follows that for some $\bar{p}_0 > 0$, and all n large enough,

$$\mathbb{P}(R_{1,\varepsilon_n} \geq 2(1 - \delta)) \geq 3\bar{p}_0 \ . \tag{5.4}$$

With $\delta\varepsilon_n \gg \sqrt{\log n / n}$, we have by Lévy's modulus of continuity theorem (see (3.20)), that $\mathbb{P}(\mathcal{A}_n) \to 1$ as $n \to \infty$, where

$$\mathcal{A}_n = \left\{ \sup_{0 \leq t \leq 1} |w_{[nt]/n} - w_t| \leq \delta\varepsilon_n \right\} \ .$$

Let

$$\overline{R}_n = \sup_{|z| < 1} \left\{ \sum_{j=1}^{n} \mathbf{1}_{w_{j/n} \in D(z,(1+\delta)\varepsilon_n)} \right\} \ .$$

When the event $\mathcal{A}_n$ occurs, if $w_t \in D(z, \varepsilon_n)$ for some $t \in [\frac{j-1}{n}, \frac{j}{n}]$ then $w_{j/n} \in D(z, (1+\delta)\varepsilon_n)$, implying that $n^{-1}\overline{R}_n \geq h(\varepsilon_n)R_{1,\varepsilon_n}$. Hence, by (5.4) it follows that for all n large enough,

$$\mathbb{P}(\overline{R}_n \geq 2(1 - \delta)n\varepsilon_n^2 |\log \varepsilon_n|^2) \geq 2\bar{p}_0 \ . \tag{5.5}$$

By Komlós-Major-Tusnády [KMT75] strong approximation theorem, there exists a universal constant $C < \infty$ such that we can, for each n, construct the simple random walk $\{S_k\}_{k=1}^{n}$ on $\mathbb{Z}^2$ and the planar Brownian motion $\{w_t\}_{0 \leq t \leq 1}$ on the same probability space, so that

$$\lim_{n \to \infty} \mathbb{P}\left(\max_{k=1,\dots,n} |w_{k/n} - \frac{\sqrt{2}}{\sqrt{n}}S_k| \leq C\frac{\log n}{\sqrt{n}} \right) = 1 \ . \tag{5.6}$$

The one dimensional construction of [KMT75] suffices for (5.6), since rotating the axes by $\pi/4$, one may view $\{S_k\}$ as two independent one-dimensional simple random walks of step size $1/\sqrt{2}$ (see Fig. 5.1). We note in passing that the error rate in (5.6) when using the classical Skorohod embedding construction is at best $n^{-1/4} \gg \varepsilon_n$ which would not have been sufficient for us. Taking n large enough so that $\delta\varepsilon_n \geq C \log n/\sqrt{n}$, the event considered in (5.6) implies that

$$\widetilde{R}_n = \sup_{y \in \mathbb{R}^2} \left\{ \sum_{j=1}^{n} \mathbf{1}_{S_j \in D(y,(1+2\delta)\varepsilon_n \sqrt{n/2})} \right\} \geq \overline{R}_n \ .$$

Therefore, combining (5.5) and (5.6), it follows that for all n large enough

$$\mathbb{P}\left(\widetilde{R}_n \geq 2(1 - \delta)n\varepsilon_n^2 |\log \varepsilon_n|^2 \right) \geq \bar{p}_0 \ . \tag{5.7}$$

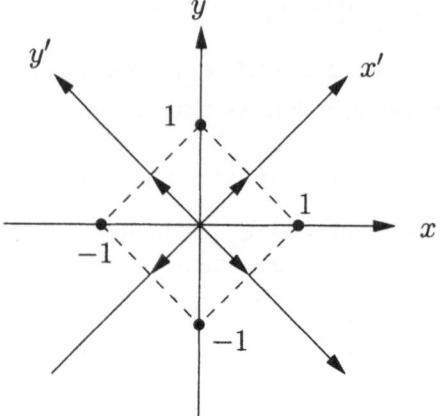

Fig. 5.1. Increments of S_k are the sum of independent steps along the x' and y' coordinates

By volume considerations, the number of integer lattice points in any disc of radius r in $\mathbb{R}^2$ is at most $\pi(r + \sqrt{2})^2$ (see Fig. 5.2), which for $r_n = (1 + 2\delta)\varepsilon_n\sqrt{n/2}$ is at most $\kappa_n = \frac{\pi}{2}(1+2\delta)^3 n^{2\eta}$, provided n is sufficiently large. The number of visits by the walk to each disc $D(y, r_n)$ is thus at most κ_n times the maximum of the number of visits to the integer lattice points within this disc (this argument is often called the pigeonhole principle). Consequently, $T_n^* \geq \widetilde{R}_n/\kappa_n$ and since $|\log \varepsilon_n| = (\frac{1}{2} - \eta)\log n$, we infer, using (5.7) that for all n large enough,

$$\mathbb{P}\left(T_n^* \leq \frac{(1 - \delta)(1 - 2\eta)^2}{(1 + 2\delta)^3 \pi}(\log n)^2\right) \leq 1 - \bar{p}_0 \,. \tag{5.8}$$

For notational simplicity let us denote

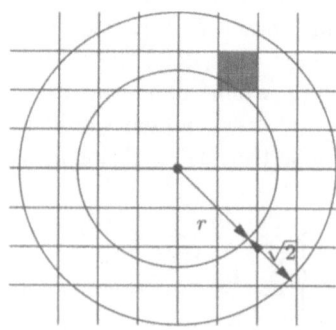

Fig. 5.2. Total area of boxes intersecting the inner disc can not exceed the area of the outer disc

$$c_{\delta,\eta} = \frac{(1-\delta)}{(1+2\delta)^3}(1-2\eta)^2 .$$

A path of length n contains $k = [n^\delta]$ disjoint segments of length $m = [n^{1-\delta}]$ each. For $i = 1, \ldots, k$ let $T_m^{(i)}$ denote the maximal number of visits to any lattice site during the i-th such segment. Note that $T_n^* \geq \max_{i=1}^k T_m^{(i)}$. Since the increments of the random walk are independent, it follows that the random variables $T_m^{(i)}$, $i = 1, \ldots, k$ are i.i.d. copies of T_m^*. Therefore, we deduce from (5.8) that for any n large enough

$$\mathbb{P}\left(T_n^* \leq \frac{c_{\delta,\eta}}{\pi}(\log[n^{1-\delta}])^2\right) \leq \left[\mathbb{P}\left(T_m^* \leq \frac{c_{\delta,\eta}}{\pi}(\log m)^2\right)\right]^{[n^\delta]} \leq (1 - \bar{p}_0)^{[n^\delta]} .$$

Since the right hand side is summable, by the Borel-Cantelli lemma we see that almost surely

$$\liminf_{n\to\infty} \frac{T_n^*}{(\log n)^2} \geq \frac{1}{\pi}(1-\delta)^3 c_{\delta,\eta} .$$

Finally, take the limit as $\delta, \eta \downarrow 0$ to complete the proof of the lower bound (5.3).

To establish the corresponding upper bound we proceed by direct computation. Recall that by the local CLT we have $\mathbb{P}(S_k = 0) = \frac{1}{\pi k}(1 + o(1))$ as $k \to \infty$ (for example, this is done in more generality in [Spi64, P9 of Chap. 7]). Therefore, as $n \to \infty$ we have that

$$\sum_{j=0}^n \mathbb{P}(S_j = 0) = \frac{\log n}{\pi}(1 + o(1)) .$$

Observe that $\{S_k = 0\}$ are renewal events. More generally, let $u_k = \mathbb{P}(\text{renewal at } k)$ and $f_k = \mathbb{P}(\text{no renewal at } 1, \ldots, k)$, with $f_0 = 1$. Let θ_n denote the time of the last renewal in $\{1, 2 \ldots, n\}$, setting $\theta_n = 0$ if no renewal occurred in this set. Clearly,

$$\mathbb{P}(\theta_n = k) = u_k f_{n-k} \quad k \geq 1, \qquad \mathbb{P}(\theta_n = 0) = f_n ,$$

hence with k enumerating all possible values of θ_n we have that

$$f_n + \sum_{k=1}^n u_k f_{n-k} = 1 . \tag{5.9}$$

We use Theorem 8.7.3 of [BGT87], which states that if $L_n = \sum_{k=1}^n u_k$ is a slowly varying function and (5.9) holds, then $f_n L_n \to 1$ as $n \to \infty$. In our case $L_n = \frac{\log n}{\pi}(1 + o(1))$, so for all n sufficiently large,

$$f_n = \mathbb{P}(S_k \neq 0 \text{ for all } k = 1, \ldots, n) \geq \frac{1-\delta}{\log n}\pi . \tag{5.10}$$

The event $\{T_n(0) \geq \ell\}$ implies that upon the i-th return to the origin of $\{S_k\}$, the next return to the origin followed within at most n steps, for $i = 0, 1, \ldots, \ell - 1$. Thus, applying the strong Markov property at the successive returns of $\{S_k\}$ to 0 we get from (5.10) that for any $\alpha > 0$,

$$\mathbb{P}(T_n(0) \geq \frac{\alpha}{\pi}(\log n)^2) \leq (1 - f_n)^{(\alpha/\pi)(\log n)^2}$$

$$\leq e^{-(1-\delta)\alpha \log n} = n^{-(1-\delta)\alpha} . \tag{5.11}$$

Let $\mathcal{R}_k = \{S_1, \ldots, S_{k-1}\}$ be the set of sites visited prior to time k and define

$$M_n(\alpha) = |\{x \in \mathbb{Z}^2 : T_n(x) \geq \frac{\alpha}{\pi}(\log n)^2\}|$$

$$= |\{k \leq n : S_k \notin \mathcal{R}_k, T_n(S_k) \geq \frac{\alpha}{\pi}(\log n)^2\}| .$$

By the definition of T_n^* and $M_n(\alpha)$, the Markov property of the SRW at time k and (5.11),

$$\mathbb{P}(T_n^* \geq \frac{\alpha}{\pi}(\log n)^2) \leq \mathbb{E}(M_n(\alpha)) = \sum_{k=1}^{n} \mathbb{P}(S_k \notin \mathcal{R}_k, T_n(S_k) \geq \frac{\alpha}{\pi}(\log n)^2)$$

$$\leq n\mathbb{P}(T_n(0) \geq \frac{\alpha}{\pi}(\log n)^2) \leq n^{1-(1-\delta)\alpha} . \tag{5.12}$$

Fixing $\alpha > 1$, by taking $\delta > 0$ small enough we ensure that the right hand side of (5.12) is summable on the subsequence $n_m = 2^m$. Hence, by the Borel-Cantelli lemma, almost surely

$$\limsup_{m \to \infty} \frac{T_{n_m}^*}{(\log n_m)^2} \leq \frac{\alpha}{\pi} . \tag{5.13}$$

Since $n \mapsto T_n^*$ and $n \mapsto (\log n)^2$ are both non-decreasing, for any $n \in (n_m, n_m + 1)$,

$$\frac{T_{n_m}^*}{(\log n_{m+1})^2} \leq \frac{T_n^*}{(\log n)^2} \leq \frac{T_{n_{m+1}}^*}{(\log n_m)^2} ,$$

and with $\frac{\log n_{m+1}}{\log n_m} \to 1$ as $m \to \infty$, the event (5.13) implies that

$$\limsup_{n \to \infty} \frac{T_n^*}{(\log n)^2} \leq \frac{\alpha}{\pi}$$

as well. Taking $\alpha \downarrow 1$ completes the proof of the stated upper bound and with it, that of the theorem.   □

5.2 Thick Points of Two Dimensional Brownian Motion

Let $\{w_t\}_{t \geq 0}$ denote the Brownian motion in $\mathbb{R}^2$ starting at the origin, and $\bar{\theta}_c = \bar{\theta}_c(w) = \inf\{t : |w_t| = c\}$, denote the exit time of the path from $D(0, c)$ for each $c > 0$ (using the convention that $\bar{\theta} = \bar{\theta}_1$). In this section we outline the proof of (5.1) (see [DPRZ01] for the complete proof). Indeed, it is the direct consequence of the following two propositions:

Proposition 5.3. *For any* $0 < a < 2$, *almost surely* $\dim_{\mathrm{H}}(\mathsf{Thick}_{\geq a}) \leq 2 - a$, *where*

$$\mathsf{Thick}_{\geq a} = \left\{ x \in D(0,1) : \limsup_{\varepsilon \to 0} \frac{\mu_{\bar{\theta}}^{w}(D(x,\varepsilon))}{\varepsilon^2 |\log \varepsilon|^2} \geq a \right\}.$$

Proposition 5.4. *Let* $\mathcal{B}_c := \{w : \dim_{\mathrm{H}}(\Gamma_c(w)) \geq \alpha\}$, *where*

$$\Gamma_c = \Gamma_c(w) := \left\{ x \in D(0,c) : \lim_{\varepsilon \to 0} \frac{\mu_{\bar{\theta}_c}^{w}(D(x,\varepsilon))}{\varepsilon^2 |\log \varepsilon|^2} = a \right\}.$$

Then, $\mathbb{P}(\mathcal{B}_1) = 1$ *for any* $0 < \alpha < 2 - a$.

We start by computing the expected value of $\mu_{\bar{\theta}_c}^{w}(D(0,\rho))$ and bounding its moment generating function, both of which we need for proving these propositions.

Lemma 5.5. *Fix* $c \geq \rho > 0$. *Let*

$$\bar{\tau} = \mu_{\bar{\theta}_c}^{w}(D(0,\rho)) = \int_0^{\bar{\theta}_c} \mathbf{1}_{D(0,\rho)}(w_s)\,ds\,,$$

and $\mathbb{E}^z(\cdot)$ *denote the expectation when the two dimensional Brownian motion* $\{w_t\}_{t \geq 0}$ *starts at* $w_0 = z \in D(0,c)$. *Then,*

$$\mathbb{E}^z(\bar{\tau}) = \rho^2 \log(c/|z|)\,, \qquad for \quad \rho \leq |z| \leq c\,. \tag{5.14}$$

Further, with $\zeta = \rho^2(\log(c/\rho) + \frac{1}{2})$,

$$\mathbb{E}^z(\bar{\tau}^k) \leq k! \zeta^k\,, \tag{5.15}$$

for all $k \geq 1$ *and* $z \in D(0,c)$, *implying that for any* $0 \leq \lambda < \zeta^{-1}$ *and all* $z \in D(0,c)$,

$$\mathbb{E}^z(e^{\lambda \bar{\tau}}) \leq \frac{1}{1 - \lambda \zeta}\,. \tag{5.16}$$

Proof of Lemma 5.5. By Lemma 2.10, the function $u(z) = \mathbb{E}^z(\bar{\tau})$ is the bounded solution of the PDE $\frac{1}{2}\Delta u = -\mathbf{1}_{|z| \leq \rho}$ in $D(0,c)$ subject to the boundary condition $u = 0$ on $\partial D(0,c)$ and such that u and its first derivatives are continuous at $\partial D(0,\rho)$. Due to radial symmetry, $u(z) = \widehat{u}(|z|)$ with $\widehat{u}(r)$ the bounded solution of the corresponding ODE

$$\begin{cases} \frac{1}{2}\widehat{u}'' + \frac{1}{2r}\widehat{u}' = -\mathbf{1}_{r \leq \rho} & \text{for } 0 \leq r \leq c, \\ \widehat{u}(c) = 0\,. \end{cases} \tag{5.17}$$

Solving (5.17) subject to continuity of $\widehat{u}$ and $\widehat{u}'$ at $r = \rho$, one finds that

$$\widehat{u}(r) = \begin{cases} -\frac{r^2}{2} + \frac{\rho^2}{2} + \rho^2 \log(c/\rho)\,, & r \leq \rho \\ \rho^2 \log(c/r)\,, & \rho \leq r \leq c, \end{cases}$$

proving (5.14). Noting that $\hat{u}(r) \leq \hat{u}(0) = \zeta$ we get (5.15) by applying Kac's moment formula (see Proposition 6.6 in the context of two dimensional Brownian motion, with $v(x) = \mathbf{1}_{|x| \leq \rho}$ and killing time $\bar{\theta}_c$, for which the positive operator G_v there is bounded in $L^\infty(D(0,c))$ by ζ). We then get the bound (5.16) by the bound we use when proving the Feynman-Kac formula of Theorem 6.7. $\qquad\square$

Equipped with Lemma 5.5, we turn to the easier proof of Proposition 5.3, which is similar to the proof of the upper bound on $\dim_{\mathrm{H}}(\mathsf{Thick}_{\geq a})$ in the context of Theorem 3.20 (where we dealt with $d > 2$).

Proof of Proposition 5.3. Let $h(\varepsilon) = \varepsilon^2 |\log \varepsilon|^2$. Fixing $\delta > 0$ small enough ($\delta < 1/22$ will do) choose a monotone decreasing sequence $\tilde{\varepsilon}_n \downarrow 0$ as $n \to \infty$ in such a way that $\tilde{\varepsilon}_1 < e^{-1}$ and

$$h(\tilde{\varepsilon}_{n+1}) = (1 - \delta)h(\tilde{\varepsilon}_n),$$

implying that $\tilde{\varepsilon}_n$ is monotone decreasing in n. By monotonicity of $\varepsilon \mapsto \mu_{\bar{\theta}}^w(D(x, \varepsilon))$ and our choice of $\tilde{\varepsilon}_n$, it is easy to see that

$$\mathsf{Thick}_{\geq a} \subseteq D_a = \left\{ x \in D(0,1) : \limsup_{n \to \infty} \frac{\mu_{\bar{\theta}}^w(D(x, \tilde{\varepsilon}_n))}{h(\tilde{\varepsilon}_n)} \geq (1 - \delta)a \right\}.$$

Moving next to discretize the space, let $\mathcal{K}_n$ denote a maximal collection of points $x_{j,n} \in D(0,1)$ such that $\inf_{\ell \neq j} |x_{\ell,n} - x_{j,n}| \geq \delta\tilde{\varepsilon}_n$. Fixing $0 < a < 2$, let $\mathcal{A}_n$ be the random subset of those $x_{j,n} \in \mathcal{K}_n$ such that

$$\mu_{\bar{\theta}}^w(D(x_{j,n}, (1+\delta)\tilde{\varepsilon}_n)) \geq (1 - 2\delta)ah(\tilde{\varepsilon}_n).$$

Since $D(0,1) \subseteq D(x, 2)$ for all $x \in D(0,1)$ and the Brownian law is invariant under translation, for any $\rho, b, \lambda \geq 0$ we have that

$$\mathbb{P}(\mu_{\bar{\theta}}^w(D(x, \rho)) \geq b) \leq \mathbb{P}^{-x}(\mu_{\bar{\theta}_2}^w(D(0, \rho)) \geq b)$$
$$\leq e^{-\lambda b}\mathbb{E}^{-x}\left(e^{\lambda\mu_{\bar{\theta}_2}^w(D(0,\rho))}\right). \qquad (5.18)$$

Taking $b = (1 - 2\delta)ah(\tilde{\varepsilon}_n)$ and applying (5.16) for $c = 2$, $\rho = (1+\delta)\tilde{\varepsilon}_n$ and $\lambda = (1+\delta)^{-1}\zeta^{-1}$ one easily checks that by (5.18), for any n large enough

$$\sup_{x \in D(0,1)} \mathbb{P}(\mu_{\bar{\theta}}^w(D(x, (1+\delta)\tilde{\varepsilon}_n)) \geq (1 - 2\delta)ah(\tilde{\varepsilon}_n)) \leq \tilde{\varepsilon}_n^{(1-10\delta)a}.$$

By volume considerations, $\#\mathcal{K}_n \leq \tilde{\varepsilon}_n^{-2-\delta a}$ for all n sufficiently large, and hence

$$\mathbb{E}(\#\mathcal{A}_n) \leq \tilde{\varepsilon}_n^{(1-11\delta)a-2}. \qquad (5.19)$$

For any $x \in D(0,1)$ there exists $x_{j,n} \in \mathcal{K}_n$ such that $|x - x_{j,n}| < \delta\tilde{\varepsilon}_n$, hence $D(x, \tilde{\varepsilon}_n) \subseteq D(x_{j,n}, (1+\delta)\tilde{\varepsilon}_n)$. Consequently,

$$\mathcal{C}_m = \bigcup_{n \geq m} \bigcup_{x_{j,n} \in \mathcal{A}_n} D(x_{j,n}, \delta \tilde{\varepsilon}_n)$$

forms a cover of D_a by sets of maximal diameter $2\delta\tilde{\varepsilon}_m$. Considering the decreasing in m covers $\mathcal{C}_m$ we see that for any $\gamma > 0$ and all m,

$$\mathcal{H}^\gamma(D_a) \leq Z_{m,\gamma} := \sum_{n=m}^{\infty} \sum_{x_{j,n} \in \mathcal{A}_n} |D(x_{j,n}, \delta \tilde{\varepsilon}_n)|^\gamma \qquad (5.20)$$

For $\gamma = 2 - (1 - 12\delta)a > 0$ and all large m it follows from (5.19) that

$$\mathbb{E}(Z_{m,\gamma}) \leq (2\delta)^\gamma \sum_{n=m}^{\infty} \tilde{\varepsilon}_n^{\,\delta a} < \infty$$

(by our choice of $\tilde{\varepsilon}_n$). Consequently, $Z_{m,\gamma}$ is finite almost surely which in turn implies by (5.20) that $\dim_{\mathrm{H}}(\mathrm{Thick}_{\geq a}) \leq \dim_{\mathrm{H}}(D_a) \leq \gamma$ almost surely. Taking $\delta \downarrow 0$ completes the proof of the proposition. $\qquad \square$

We conclude this section by reducing the harder proof of Proposition 5.4 to the question of controlling the first two moments of n-perfect points, which we further address in Sect. 5.3. To this end, our first step is a zero-one law, which serves here the same purpose as Lemma 4.11 did in the course of proving Theorem 4.1 about the α-favorite leaves of b-ary trees.

Lemma 5.6. *If* $\mathbb{P}(\mathcal{B}_1) > 0$, *then necessarily* $\mathbb{P}(\mathcal{B}_1) = 1$.

Proof. It is not hard to check that $\Gamma_c(w) = c\Gamma_1(w^c)$ for $w_t^c := c^{-1}w_{c^2 t}$. By Brownian scaling $w_\cdot^c$ has the same law as $w_\cdot$, hence $\mathbb{P}(\mathcal{B}_c) = p$ is independent of c.

Define $\mathcal{B} := \limsup_{n \to \infty} \mathcal{B}_{n^{-1}}$, so that $\mathbb{P}(\mathcal{B}) \geq p$. Since $\mathcal{B}_c$ is in the σ-field generated by $\{w_t, t \in [0, \bar{\theta}_c]\}$ and $\bar{\theta}_{n^{-1}} \downarrow 0$ as $n \to \infty$, we have by Blumenthal's zero-one law (c.f. [Bas99, Corollary I.3.6] or more generally [RY99, Theorem III.2.15]) that $\mathbb{P}(\mathcal{B}) \in \{0, 1\}$. Thus, $p > 0$ yields $\mathbb{P}(\mathcal{B}) = 1$.

Fixing $0 < b < c$ we have that $\Gamma_b(w) \setminus \{w_t : \bar{\theta}_b \leq t \leq \bar{\theta}_c\} \subset \Gamma_c(w)$. With $[w]$ denoting the range of the planar Brownian motion w started at any fixed point, a fractal geometry argument shows that for any fixed analytic set A, almost surely $\dim_{\mathrm{H}}(A \setminus [w]) = \dim_{\mathrm{H}}(A)$ (c.f. [DPRZ01, (3.2)]). Applying the strong Markov property at $\bar{\theta}_b$ and checking that the set $\Gamma_b(w)$ is almost surely an analytic set we thus have that the events $\mathcal{B}_c$ are essentially increasing in c in the sense that for any $0 < b < c$

$$\mathbb{P}(\mathcal{B}_b \setminus \mathcal{B}_c) = 0 .$$

Consequently, $\mathbb{P}(\mathcal{B} \setminus \mathcal{B}_1) \leq \mathbb{P}(\bigcup_n \{\mathcal{B}_{n^{-1}} \setminus \mathcal{B}_1\}) = 0$, so if $p = \mathbb{P}(\mathcal{B}_1) > 0$ then also $\mathbb{P}(\mathcal{B}_1) = 1$. $\qquad \square$

Fixing $0 < \alpha < 2 - a$, our next step is to construct a *random* set $F \subseteq \Gamma_1$ and a *random* positive Borel measure $\nu = \nu_w$ supported on F such that $\mathbb{P}(\mathcal{E}_\alpha(\nu) < \infty) > 0$ where

$$\mathcal{E}_\alpha(\mu) = \int \int |x - y|^{-\alpha} \, d\mu(x) \, d\mu(y)$$

denotes the α-energy of the measure μ. Then, by the potential theoretic lower bound of Proposition 3.6

$$\mathbb{P}(\mathcal{B}_1) = \mathbb{P}\left(\dim_{\mathrm{H}}(\Gamma_1) \geq \alpha\right) \geq \mathbb{P}\left(\dim_{\mathrm{H}}(F) \geq \alpha\right) \geq \mathbb{P}(\mathcal{E}_\alpha(\nu) < \infty) > 0 ,$$

which in view of Lemma 5.6 completes the proof of Proposition 5.4.

Proceeding to construct F, we fix $\epsilon_1 > 0$ small enough ($\epsilon_1 = 1/8$ will do), and the square $S = [\epsilon_1, 2\epsilon_1]^2 \subset D(0, 1)$. Our choice of ϵ_1 is such that for all $x \in S$ and $y \in S \cup \{0\}$ both $0 \notin D(x, \epsilon_1)$ and $0 \in D(x, 1/2) \subset D(y, 1) \subset D(x, 2)$, see Fig. 5.3. We take $\varepsilon_k = \epsilon_1(k!)^{-3}$, so $\log(1/\varepsilon_k) = 3k \log k(1 + o(1))$ corresponds to the levels h_k we considered in the course of proving Theorem 4.1. For $x \in S$, $k \geq 2$ and $\rho > \epsilon_1$ let $N_k^x(\rho)$ denote the number of excursions from $\partial D(x, \varepsilon_{k-1})$ to $\partial D(x, \varepsilon_k)$ prior to hitting $\partial D(x, \rho)$, for example, see Fig. 5.4. Setting $n_k = 3ak^2 \log k$ as in Sect. 4.2, we call a point $x \in S$ *n-perfect* if

$$n_k - k \leq N_k^x(1/2) \leq N_k^x(2) \leq n_k + k , \qquad \text{for all} \quad k = 2, \ldots, n . \quad (5.21)$$

The relation with Sect. 4.2 is that conditional on hitting $\partial D(x, \epsilon_1)$ before $\partial D(x, \rho)$ the random variables $N_k^x(\rho)$, $k = 2, \ldots, n$ have the same joint law as that of the non-homogeneous Markov chain $\{L_k\}$ of (4.20) for

$$p_k = \frac{\log(\varepsilon_{k-1}/\varepsilon_k)}{\log(\varepsilon_{k-1}/\varepsilon_{k+1})} = \frac{1}{2} + O\left(\frac{1}{k \log k}\right) \quad (5.22)$$

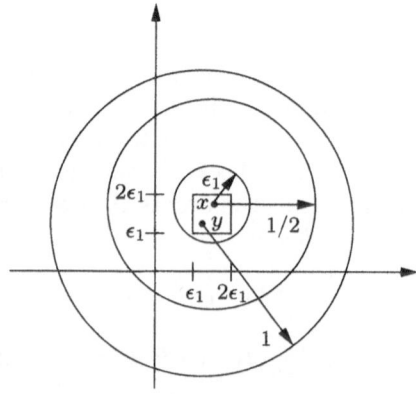

Fig. 5.3. Discs centered at $x, y \in S$

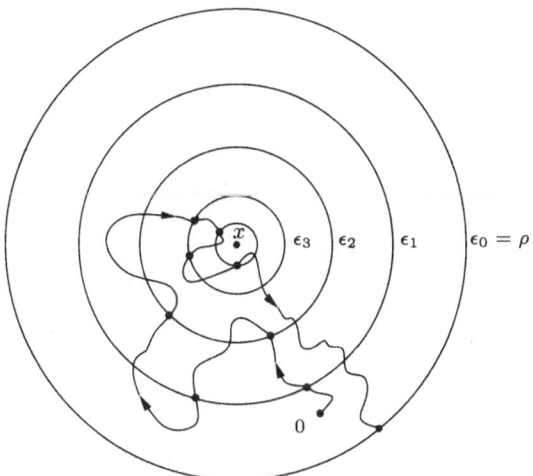

Fig. 5.4. Path realization with $N_2^x = 2$, $N_3^x = 1$ and $N_4^x = 2$

with $\varepsilon_0 := \rho$ (compare with (4.19)). Successively partitioning S into $M_n = \epsilon_1^2/(2\varepsilon_n)^2$ non overlapping squares $S(n,i)$ of edge length $2\varepsilon_n$ each, let $x_{n,i}$ denote the center of $S(n,i)$ and consider the random sets

$$A_n = \bigcup \{S(n,i) : x_{n,i} \text{ is n-perfect}\},$$

which are the analogous of the set $\mathcal{S}_n^v(\alpha)$ of Sect. 4.3.

The next lemma, whose proof we defer, relates the n-perfect property of a point x with the Brownian occupation measures $\mu_\theta^w(D(x,\varepsilon))$ for $\varepsilon \geq \varepsilon_n$. In this respect, it is analogous to Lemma 4.10 of Sect. 4.3.

Lemma 5.7. *There exists an $\eta(\varepsilon) = \eta(\varepsilon, w) \to 0$ almost surely such that for all m and all $x \in S$, if x is m-perfect then*

$$a - \eta(\varepsilon) \leq \frac{\mu_\theta^w(D(x,\varepsilon))}{h(\varepsilon)} \leq a + \eta(\varepsilon), \qquad \text{for all } \varepsilon \geq \varepsilon_m. \qquad (5.23)$$

Note that the n-perfect property does not control the occupation measure at scales smaller than ε_n. Nevertheless, in view of Lemma 5.7, the closed random set

$$F = F(\omega) = \bigcap_m F_m := \bigcap_m \overline{\bigcup_{n \geq m} A_n},$$

is a subset of Γ_1 and is the candidate for the support of the measure ν we wish to construct. Indeed, by definition, an n-perfect point is also m-perfect for all $m \leq n$. Hence, any $x \in F_m$ is the limit of a sequence of m-perfect points $y_k \in S$. Since

$$\mu_\theta^w(D(x, \varepsilon - |x - y_k|)) \leq \mu_\theta^w(D(y_k, \varepsilon)) \leq \mu_\theta^w(D(x, \varepsilon + |x - y_k|)),$$

 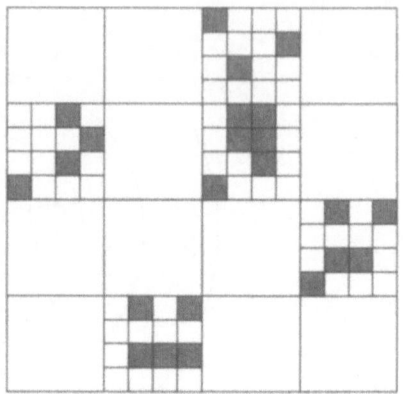

Fig. 5.5. Shading of A_n (*left*) and partial shading of F_m, $m \geq n$ (*right*)

applying (5.23) for y_k such that $|x - y_k| \leq \delta\varepsilon$ we see that for any $\delta > 0$, $\varepsilon' = \varepsilon/(1-\delta)$ and $\varepsilon'' = \varepsilon/(1+\delta)$

$$(a - \eta(\varepsilon''))h(\varepsilon'') \leq \mu_\theta^w(D(x,\varepsilon)) \leq (a + \eta(\varepsilon'))h(\varepsilon') \qquad (5.24)$$

provided $\varepsilon \geq \varepsilon_m$. If $x \in F$ then the inequalities (5.24) apply for all $\varepsilon > 0$, so considering first $\varepsilon \to 0$ then $\delta \downarrow 0$ we see that $x \in \Gamma_1$ as well (recall that both $h(\varepsilon'') \geq (1+\delta)^{-2}h(\varepsilon)$ and $h(\varepsilon') \leq (1-\delta)^{-2}h(\varepsilon)$ for $\varepsilon \leq 1-\delta$).

Having constructed F we turn to construct the random measure ν. In Sect. 5.3 we relate the correlation structure of $\{Y(n,i) = \mathbf{1}_{\{x_{n,i} \text{ is n-perfect}\}}\}$ to that of Sect. 4.3 and conclude that

Lemma 5.8. *Let* $\nu_n = \nu_{n,w}$ *for* $n \geq 2$ *be the random measure supported on* A_n *whose density with respect to Lebesgue measure is*

$$\frac{d\nu_n}{dx} = \epsilon_1^{-2} \sum_{i=1}^{M_n} Q_{n,i}^{-1} Y(n,i) \mathbf{1}_{S(n,i)}(x) ,$$

where $Q_{n,i} = \mathbb{P}(Y(n,i) = 1)$. *Then,*

$$\sup_n \mathbb{E}\big(\nu_n(S)^2\big) < \infty , \qquad (5.25)$$

and

$$\sup_n \mathbb{E}\big(\mathcal{E}_\alpha(\nu_n)\big) < \infty . \qquad (5.26)$$

In particular, (5.26) suggests that ν_n could have been a candidate for ν, except that A_n is not a subset of F. However, since $\nu_n(G_m) = 0$ for $G_m = F_m^c$ (the complement of the closed set F_m) and all $n \geq m$, it follows that if ν is a limit point of ν_n with respect to the topology of weak convergence of positive Borel measures on S, then $\nu(G_m) = 0$ for all m. Therefore, any such ν is

supported on the subset F of Γ_1. For any $b < \infty$ the set $\mathcal{M}_b$ of non-negative
Borel measures μ on S with $\mu(S) \in [b^{-1}, b]$ is compact when equipped with
this topology. Further, the α-energy $\mathcal{E}_\alpha(\mu)$ is a linear functional on the product
measure $\mu \times \mu(x, y)$, whose kernel $|x - y|^{-\alpha}$ is the supremum of the bounded
continuous kernels $|x - y|^{-\alpha} \wedge M$. With $\mu \mapsto \mu \times \mu$ continuous on $\mathcal{M}_b$, it
thus follows that $\mu \mapsto \mathcal{E}_\alpha(\mu)$ is lower semi-continuous on $\mathcal{M}_b$. Consequently,
$\mathcal{M}_{b,e} = \{\mu \in \mathcal{M}_b : \mathcal{E}_\alpha(\mu) \leq e\}$ are compact subsets of $\mathcal{M}_b$ for any $e < \infty$.
Let $\mathcal{C}_n = \{w : \nu_{n,w} \in \mathcal{M}_{b,e}\}$ and set $\mathcal{C} = \limsup_n \mathcal{C}_n$. For any $w \in \mathcal{C}$ there
is a subsequence $n_k = n_k(w) \to \infty$ such that $\nu_{n_k,w} \in \mathcal{M}_{b,e}$ for all k. By
compactness of $\mathcal{M}_{b,e}$, the sequence $\nu_{n_k,w}$ has at least one limit point ν_w in
$\mathcal{M}_{b,e}$. Such ν_w is thus a finite measure supported on F, having positive mass
and finite α-energy, as needed. By Fatou's lemma,

$$\mathbb{P}(\mathcal{C}) \geq \liminf_{n \to \infty} \mathbb{P}(\mathcal{C}_n),$$

so it only remains to find $b, e < \infty$ such that $\mathbb{P}(\mathcal{C}_n)$ is bounded away from zero.
To this end, note that $\mathbb{E}(Y(n, i)) = Q_{n,i}$, so $\nu_n(S)$ has expected value one,
independent of n, and uniformly bounded second moment (by (5.25)). Hence,
the same inequality we used in (4.36) shows that $\inf_n \mathbb{P}(\nu_n(S) \geq 1/2) = 3v > 0$
(see [Kah85, page 8]). Further, by Markov's inequality, $\mathbb{P}(\nu_n(S) \geq b) \leq 1/b$
for any n and b. So, taking $b < \infty$ large enough guarantees that

$$\mathbb{P}(b \geq \nu_n(S) \geq b^{-1}) \geq 2v > 0, \qquad \forall n. \tag{5.27}$$

By (5.26) we know that $\mathbb{E}(\mathcal{E}_\alpha(\nu_n))$ is uniformly bounded, so applying Markov's
inequality we can find $e < \infty$ such that

$$\mathbb{P}(\mathcal{E}_\alpha(\nu_n) \leq e) \geq 1 - v > 0, \qquad \forall n,$$

which when combined with (5.27) shows that for all n,

$$\mathbb{P}(\mathcal{C}_n) = \mathbb{P}(b \geq \nu_n(S) \geq b^{-1}, \mathcal{E}_\alpha(\nu_n) \leq e) \geq v > 0,$$

as needed to complete the proof of Proposition 5.4.

5.3 From Trees to Brownian Motion: n-Perfect Points

We chose the square S close enough to 0 so if $x \in S$ is n-perfect, then the
number of excursions from $\partial D(x, \varepsilon_{k-1})$ to $\partial D(x, \varepsilon_k)$ *prior to $\bar\theta$* is also between
$n_k - k$ and $n_k + k$. It is this property that leads to Lemma 5.7, but the use
of a stopping time related to the x-concentric discs in the definition of $N_k^x(\rho)$
simplifies the task of estimating first and second moments of $Y(n, i)$ which we
need for Lemma 5.8.

Lemma 5.7 serves here the same purpose as Lemma 4.10 did for favorite
leaves of SRW on b-ary trees, that is, relating excursion counts to occupation
times. Recall that Lemma 4.10 is based on Lemma 4.7 about the concentration

of sum of i.i.d. Geometric random variables. Here the corresponding object is the sum of occupation measures of $D(x', \varepsilon'_{k,j})$ during the different Brownian excursions between $\partial D(x', \varepsilon'_k)$ and $\partial D(x', \varepsilon''_{k-1})$. Indeed, thanks to (5.16), these i.i.d random variables have the necessary exponential tails to provide an $e^{-O(k^2/\log k)}$ bound on the probability that their sum deviates from its mean, for each fixed k and x'. The one complication we now face is that we need to relate the n-perfect property to occupation measure for a *continuum* of points $x \in S$. Nevertheless, it can be done while considering such deviations only on a fine, but finite, δ_k-net $\mathcal{D}_k$ of points x' such that $\#\mathcal{D}_k = e^{o(k^2/\log k)}$. This, of course, lets us get the desired conclusion via the union bound (over x') and the Borel-Cantelli lemma (over k). To make it all work, we take $\varepsilon''_{k-1} \leq \varepsilon_{k-1} - \delta_k$ and $\varepsilon'_k \geq \varepsilon_k + \delta_k$. As depicted in Fig. 5.6, if x is an n-perfect point whose nearest neighbor in $\mathcal{D}_k$ is x', then the lower bound on the number of excursions between $\partial D(x, \varepsilon_k)$ and $\partial D(x, \varepsilon_{k-1})$ implies the same bound for the x' excursions. With high enough probability the latter translates to a lower bound on the Brownian occupation measure of $D(x', \varepsilon'_{k,j})$. The same lower bound then applies to the Brownian occupation measure of $D(x, \varepsilon_{k,j})$, with $\varepsilon_{k,j} \geq \varepsilon'_{k,j} + \delta_k$. Finally, the $\varepsilon_{k,j}$ provide for $j = 0, 1, \ldots, 3k \log k$ a fine enough discrete interpolation between ε_{k+1} and ε_k to keep $\mu^w_\theta(D(x, \varepsilon))/h(\varepsilon)$ for $\varepsilon \in [\varepsilon_{k,j+1}, \varepsilon_{k,j}]$ within $1 \pm o(1)$ of its values at $\varepsilon = \varepsilon_{k,j}$ and at $\varepsilon = \varepsilon_{k,j+1}$. A similar construction gives the complementary upper bound on the occupation measure of n-perfect points, using this time radii $\bar{\varepsilon}''_{k-1} \geq \varepsilon_{k-1} + \delta_k$, $\bar{\varepsilon}'_k \leq \varepsilon_k - \delta_k$ and $\bar{\varepsilon}'_{k,j} \geq \varepsilon_{k,j} + \delta_k$ (see [DPRZ01, Sect. 6] for details).

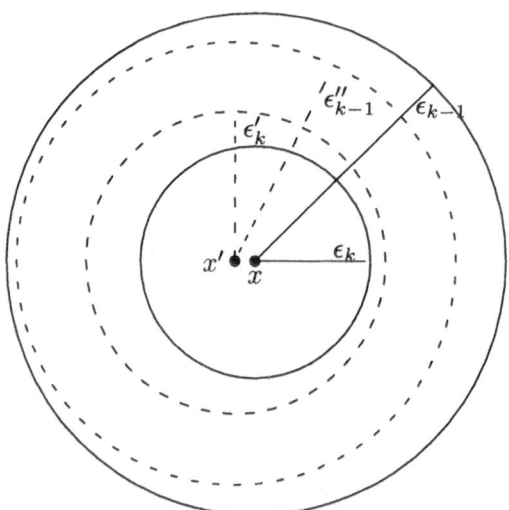

Fig. 5.6. Approximating the occupation measure at x by that of the nearest x' on a δ_k-net

The proof of Lemma 5.8 relies on first and second moment estimates that are very much like those derived in Sects. 4.2–4.3 for n-perfect leaves.

Starting with the first moment, note that the order of visits to concentric discs of radii ε_k by the Brownian motion has the law of the birth-death process $\{X_l\}$ of Sect. 4.2, taking here the transition probabilities p_k of (5.22). Let

$$\mathcal{C}_n = \{n_k - k \le N_k^x(1/2) \le n_k + k \qquad \text{for all } 2 \le k \le n\} ,$$

and $\tau_{x,r} = \inf\{t \ge 0 : w_t \in \partial D(x,r)\}$. Since $0 \notin D(x,\epsilon_1)$ for all $x \in S$, we have that

$$\mathbb{P}(\mathcal{C}_n \,|\, \tau_{x,\epsilon_1} < \tau_{x,1/2}) = q_n$$

of Proposition 4.4, now for $\zeta = 3a$. For x to be n-perfect, necessarily $\tau_{x,\epsilon_1} <$

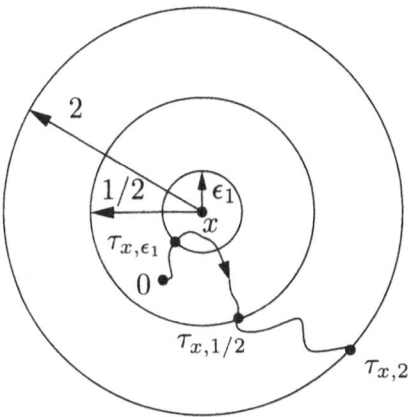

Fig. 5.7. A path for which $\mathcal{C}_n$ implies that x is n-perfect

$\tau_{x,1/2}$. So, $q_n \ge Q_{n,i}$ for all i. Further, if after time $\tau_{x,1/2}$ the Brownian motion hits $\partial D(x,2)$ before it hits $\partial D(x,\epsilon_1)$, then $N_k^x(1/2) = N_k^x(2)$ and the event $\mathcal{C}_n$ results with x being n-perfect (c.f. Fig. 5.7). As $\sup\{|x| : x \in S\} < 1/2$, we deduce that $Q_{n,i} \ge c_0 q_n$ for some $c_0 > 0$ and all n, i. So, the first moment estimates for n-perfect points are, for all practical purpose, the same as those for the n-perfect leaves of Sect. 4.2.

Moving to the second moment estimates, we reproduce the conclusion of Lemma 4.8, which translates here to

$$q_{n,l} \stackrel{\triangle}{=} \sup_{\substack{2\epsilon_l < |x-y| \le 2\epsilon_{l-1} \\ x \in S, y \in S}} \mathbb{P}\Big(x \text{ and } y \text{ are n-perfect}\Big) \le q_n^2 \, (l!)^{\zeta+\gamma_l} , \qquad (5.28)$$

for some $\gamma_l \to 0$. The proof of (5.28) is a re-run of that of Lemma 4.8, taking here $L_k^x = N_k^x(1/2)$, $L_k^y = N_k^y(1)$, and $\mathcal{G}_l^y$ to be the σ-field generated by the Brownian excursions from $\partial D(y,\varepsilon_{l-1})$ to $\partial D(y,\varepsilon_l)$, including the path up to $\tau_{y,\varepsilon_{l-1}}$. Indeed, our choice of l guarantees that $\partial D(x,\varepsilon_k)$ does not intersect

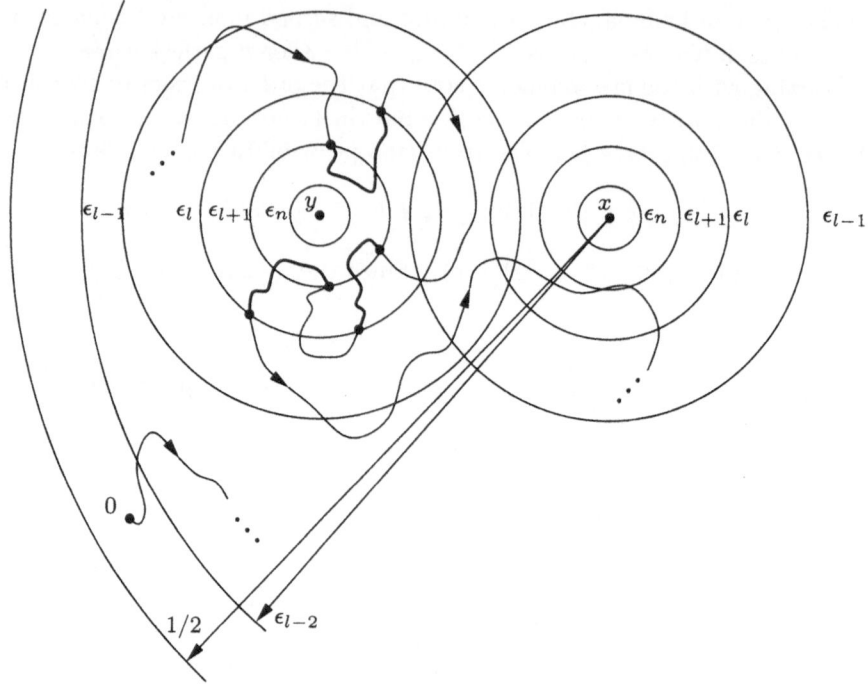

Fig. 5.8. Discs centered at x and y play a role in deriving (5.28). The portion of the Brownian path that is not in $\mathcal{G}^y_{l+1}$ is darker

$D(y, \varepsilon_l)$ when $k \neq l - 1$ (recall that $\varepsilon_{l-2} > 2\varepsilon_{l-1} + \varepsilon_l$, see also Fig. 5.8). Further, $0 \in D(x, 1/2) \subset D(y, 1)$, so moving to radius 1 for the y-centered excursions, we know that $\tau_{x,1/2} < \tau_{y,1}$. As a result, the random variables $N^x_k(1/2)$ are measurable on $\mathcal{G}^y_{l+1}$, with the possible exception of $k = l - 1, l$. Excluding these two values of k does not matter for the conclusion of (5.28), as it is easily absorbed in the $(l!)^{\gamma_l}$ "fudge factor". However, we encounter here an annoying technical problem we did not have in Lemma 4.8. That is, even after fixing $N^y_{l+1}(1) = m_{l+1}$, the random variables $\{N^y_k(1)\}_{k \geq l+2}$ depend on $\mathcal{G}^y_{l+1}$ via the *knowledge of the initial and final points* of each of the m_{l+1} Brownian excursions from $\partial D(y, \varepsilon_{l+1})$ to $\partial D(y, \varepsilon_l)$. Luckily, we are able to control this effect by showing that for some $\kappa < \infty$, any $2 \leq l \leq n - 2$, and all $y \in S$,

$$\mathbb{P}(N^y_k(1) = m_k; k \in J_{l+1} \mid N^y_{l+1}(1) = m_{l+1}, \mathcal{G}^y_{l+1})$$

$$\leq \left(1 + \kappa \frac{\varepsilon_{l+1}}{\varepsilon_l}\right)^{m_{l+1}} \prod_{k=l+1}^{n-1} \mathbb{P}(L_{k+1} = m_{k+1} \mid L_k = m_k), \qquad (5.29)$$

and since $m_{l+1} \ll (l+1)^3 = \varepsilon_l/\varepsilon_{l+1}$, this is also easily absorbed in the $(l!)^{\gamma_l}$ "fudge factor" (whereas $h_k = [ck \log k]$ works in Sect. 4.2 for any fixed $c > 0$, we set $|\log \varepsilon_k| = 3k \log k (1 + o(1))$ in Sect. 5.2 in anticipation of (5.29)). To see why (5.29) holds, note that given their initial and final points, the m_{l+1} Brownian excursions from $\partial D(y, \varepsilon_{l+1})$ to $\partial D(y, \varepsilon_l)$ are mutually independent and the events in (5.29) are in the σ-field of the excursions of the path from $\partial D(y, \varepsilon_{l+2})$ to $\partial D(y, \varepsilon_{l+1})$ during these m_{l+1} excursions from $\partial D(y, \varepsilon_{l+1})$ to $\partial D(y, \varepsilon_l)$. The next lemma states that for $R' \gg R \gg r$, the σ-field of excursions of the Brownian path from $\partial D(y, r)$ to $\partial D(y, R)$ prior to $\tau_{y, R'}$, is almost independent of the initial point $z \in \partial D(y, R)$ and the final point $v \in \partial D(y, R')$ of the path. In particular, we obtain (5.29) by applying this lemma for $R' = \varepsilon_l$, $R = \varepsilon_{l+1}$ and $r = \varepsilon_{l+2}$ (c.f. Fig. 5.9).

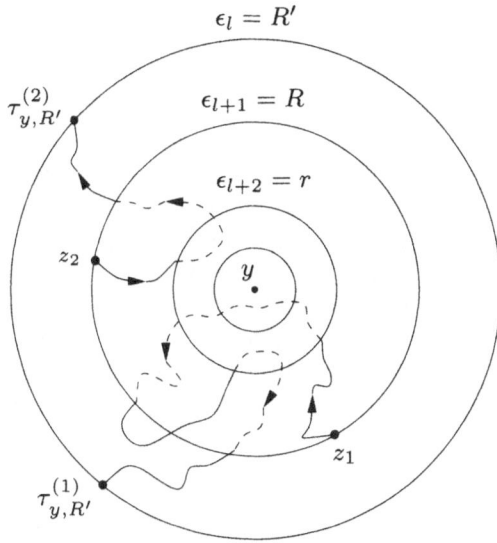

Fig. 5.9. Dashed portion of each Brownian excursion is $\mathcal{H}$-measurable. The end points of each excursion are $\mathcal{G}_{l+1}^y$-measurable

Lemma 5.9. *Let $\mathbb{P}^z(\cdot)$ denote the law of two dimensional Brownian motion $\{w_t\}$ starting at $w_0 = z$. Fixing y and $0 < r < R < R'$, let $\mathcal{H}$ denote the σ-field generated by the excursions of the path from $\partial D(y, r)$ to $\partial D(y, R)$, prior to $\tau_{y, R'}$. Then, uniformly in $z, z' \in \partial D(y, R)$, $v \in \partial D(y, R')$, and $B \in \mathcal{H}$,*

$$\mathbb{P}^z(B \,|\, w_{\tau_{y, R'}} = v) = \left(1 + O\left(\frac{R}{R'} \right) \right) \mathbb{P}^z(B), \tag{5.30}$$

and

$$\mathbb{P}^z(B) = (1 + O(\frac{r}{R} \log \frac{R}{r})) \mathbb{P}^{z'}(B). \tag{5.31}$$

Remark 5.10. *A special case of Lemma 5.9 is stated and proved as Lemma 7.4 of [DPRZ01]. Decomposing the event B according to the number of excursions from $\partial D(y, r)$ to $\partial D(y, R)$ (prior to $\tau_{y, R'}$), the proof of (5.30) relies on* *the Poisson kernel $(R'^2 - |x|^2)/|v - x|^2$ of the density of $w_{\tau_{y,R'}}$ at v starting at $w_0 = x \in D(y, R')$ (c.f. [Bas99, Theorem II.1.17]). We then get (5.31) by considering the first exit from the annulus $D(y, R') \setminus D(y, r)$ and using the corresponding Poisson kernel. The bounds (5.30) and (5.31) apply also for the simple random walk in $\mathbb{Z}^2$ provided in addition $r \gg 1$ and $\log(R/r)$ is comparable to $\log(R'/R)$. See [DPRZ05, Lemma 2.3] for the proof in the random walk context (relying on the analogous Poisson kernel estimates from [Law91, Chapter 1]).*

Having established the first and second moment estimates, we are ready to prove Lemma 5.8. To this end, recall that

$$\mathbb{E}\left(\nu_n(S)^2\right) = M_n^{-2} \sum_{i=1}^{M_n} Q_{n,i}^{-1} + M_n^{-2} \sum_{i \neq j = 1}^{M_n} Q_{n,i}^{-1} Q_{n,j}^{-1} \mathbb{E}(Y(n,i)Y(n,j)) , \quad (5.32)$$

with $Y(n,i) \in \{0,1\}$ and $Q_{n,i} = \mathbb{P}(Y(n,i) = 1) = \mathbb{P}(x_{n,i}$ is n-perfect). The first term in (5.32) is negligible, since $Q_{n,i} \geq c_0 q_n$ and $M_n q_n \to \infty$ by Proposition 4.4 (recall that $\zeta = 3a < 6$). There are at most $O(M_n/((l-1)!)^6)$ points $x_{n,j}$ in a ball of radius $2\varepsilon_{l-1}$ centered at $x_{n,i}$. So, counting the pairs $i, j \in \{1, \ldots, M_n\}$ according to the value of $l = l_{i,j} \in \{2, \ldots, n\}$ such that $2\varepsilon_l < |x_{n,i} - x_{n,j}| \leq 2\varepsilon_{l-1}$, we see that for some $c_1 < \infty$ and all n, the second term in (5.32) is at most $c_1 \sum_{l=2}^{n} q_n^{-2} q_{n,l}((l-1)!)^{-6}$ (compare with (4.34)). Since $\zeta < 6$, the uniform bound of (5.25) is thus a direct consequence of (5.28).

Fixing $0 < \alpha < 2 - a < 2$, it is not hard to check that for some $c_2 < \infty$ and all i, j

$$\int_{S(n,i)} \int_{S(n,j)} |x - y|^{-\alpha} \, dx \, dy \leq c_2 \varepsilon_n^4 (|x_{n,i} - x_{n,j}| + \varepsilon_n)^{-\alpha} .$$

Consequently, counting the pairs i, j according to $l(i,j)$ and using (5.28), we find that for some $c_3, c_4 < \infty$ and all n,

$$\mathbb{E}\left(\mathcal{E}_\alpha(\nu_n)\right) \leq c_2 \varepsilon_n^{4-\alpha} \sum_{i=1}^{M_n} Q_{n,i}^{-1} + c_2 \varepsilon_n^4 \sum_{i \neq j = 1}^{M_n} \frac{\mathbb{E}(Y(n,i)Y(n,j))}{Q_{n,i} Q_{n,j}} \varepsilon_{l(i,j)}^{-\alpha}$$

$$\leq c_3 \varepsilon_n^{2-\alpha} q_n^{-1} + c_3 \sum_{l=2}^{n} l^6 \frac{q_{n,l}}{q_n^2} \varepsilon_l^{2-\alpha} \leq c_4 \sum_{l=2}^{\infty} l^6 \varepsilon_l^{2-\alpha-a-\gamma_l} < \infty .$$

Thus, $\mathbb{E}\left(\mathcal{E}_\alpha(\nu_n)\right)$ is a bounded sequence, as stated in (5.26).

6 Kac's Moment Formula and Ciesielski-Taylor Identities

In this chapter we explore various aspects of the derivation of (3.18). In Sect. 6.1 we follow the proof provided in Lemma 2.1 of [DPRZ99] in the wider context of transient symmetric stable processes. In Sect. 6.2 we follow [FP99] in putting this result in the perspective of Kac's moment formula for additive functionals of a Markov process, which we have used already when deriving (5.15) and (5.16) in Sect. 5.2. Finally, in Sect. 6.3 we follow Yor's derivation (in [Yor91]) of Ciesielski-Taylor identities out of the Ray-Knight theorem (c.f. [PY03] for a host of relations in the spirit of the Ciesielski-Taylor identities). In the context of (3.18) it is here that we identify θ_d as half the minimal eigenvalue of the Laplacian in the unit ball of $\mathbb{R}^{d-2}$.

6.1 Occupation Measure of Transient Symmetric Stable Processes

Let $p_t(x)$ be the probability density function of the symmetric stable process $\{X_t\}$ of index $\beta < d$, in $\mathbb{R}^d$, started at $X_0 = 0$. Define the 0-potential density for $\{X_t\}$ as $u^0(x) = \int_0^\infty p_t(x)\,dt$. The Brownian motion in $\mathbb{R}^d$ ($d \geq 3$) is the special case of $\beta = 2$. By stable scaling (X_t has the same law as $b^{-1}X_{b^\beta t}$) and radial symmetry, we have that

$$u^0(x) = c_{\beta,d}|x|^{\beta-d} \, ,$$

for some constants $c_{\beta,d} < \infty$. In the special case $\beta = 2$, we have that

$$p_t(x) = \frac{e^{-\frac{|x|^2}{2t}}}{(2\pi t)^{d/2}} \, ,$$

so change of variables to $s = |x|^2/(2t)$, $x \neq 0$, shows that

$$c_{2,d} = \frac{1}{2\pi^{d/2}} \int_0^\infty s^{d/2-2}e^{-s}ds = \frac{\Gamma(d/2-1)}{2\pi^{d/2}} \, .$$

Define formally the operator $K = K_{\beta,d}$ by

$$(Kf)(x) = \int_{B(0,1)} u^0(x-y)f(y)\,dy \, .$$

In the following, $\langle \cdot | \cdot \rangle$ denotes the inner product on $L^2(B(0,1))$ (with d-dimensional Lebesgue measure), $\|\cdot\|_2$ is the associated norm and $\mathbf{1}$ denotes the function $1_{\{|x|\leq 1\}}$. We implicitly extend each function $f \in L^2(B(0,1))$ to $L^2(\mathbb{R}^d)$ by setting it to zero outside $B(0,1)$.

Note that u^0 is in general only in $L^1(B(0,1))$. However, we start by checking that

Lemma 6.1. *For any $\beta < d$, the operator $K = K_{\beta,d}$ is bounded from $L^2(B(0,1))$ to itself. Moreover, $K^m u^0$ is bounded (hence in $L^2(B(0,1))$) for all m large enough and if f is bounded then Kf is continuous in $B(0,1)$ as are all the eigenvectors of K in $L^2(B(0,1))$ which correspond to non-zero eigenvalues.*

Proof. Let $\Lambda = \Lambda_{\beta,d}$ denote the operator norm

$$\Lambda := \sup_{f \in L^2(B(0,1))} \frac{\|Kf\|_2}{\|f\|_2} ,$$

of K from $L^2(B(0,1))$ to itself. Because K is symmetric, positive, and using Fubini's theorem, this reduces to

$$\Lambda = \sup \left\{ \int_{B(0,1)^2} u^0(x-y)f(x)f(y)dxdy : f > 0,\ \|f\|_2 = 1 \right\} .$$

Fixing $f \in L^2(B(0,1))$ with $f > 0$ and $\|f\|_2 = 1$, by Fubini and the positivity of f and u^0 we have that

$$\int_{B(0,1)^2} u^0(x-y)f(x)f(y)dxdy \leq \int_{B(0,2)} u^0(z)dz \int_{\mathbb{R}^d} f(x)f(x-z)dx$$

$$\leq \int_{B(0,2)} u^0(z)dz ,$$

using Cauchy-Schwartz to get the second inequality. Hence, $\Lambda \leq \int_{B(0,2)} u^0(z)dz$. Now, in polar coordinates,

$$\int_{B(0,2)} u^0(z)dz = C \int_0^2 r^{\beta-1}dr < \infty ,$$

since $\beta > 0$. Consequently, $\Lambda < \infty$ as claimed.

It is easy to see by scaling that if $0 < \beta < h < d$ then for some $\kappa = \kappa(h,\beta,d) < \infty$,

$$\int_{B(0,1)} |x-y|^{\beta-d}|y-z|^{-h}\,dy \leq \int_{\mathbb{R}^d} |x-y|^{\beta-d}|y-z|^{-h}\,dy$$

$$= \kappa|x-z|^{\beta-h} . \tag{6.1}$$

Consequently, with $u^z(y) = u^0(z-y)$, starting at $h \geq d - \beta$, each application of K lowers the divergence by β until $K^{m-2}u^z(x) \leq c|x-z|^{-h}$ for some $c < \infty$, $h < \beta$ and all z. Taking $q > 1$ such that $\beta > d/q > h$ we have that $K^{m-2}u^z$ is uniformly (in z) bounded in $L^q(B(0,1))$ and $u^0 \in L^p(B(0,2))$ for $p = q/(q-1)$, hence by Hölder's inequality, $K^{m-1}u^z$ is bounded, uniformly in z.

Suppose $K\phi = \lambda\phi$ for some $\phi \in L^2(B(0,1))$ and $\lambda \neq 0$. Then, $\lambda^m\phi(z) = K^m\phi(z) = \langle K^{m-1}u^z|\phi\rangle$ is also bounded, hence so is ϕ.

If f is bounded on $B(0,1)$, then the family of functions $g_x(y) = c_{\beta,d}|x-y|^{\beta-d}f(y)$ is uniformly bounded in $L^p(B(0,1))$ for p as above, hence uniformly integrable, and $g_x(y) \to g_{x'}(y)$ a.e. in $B(0,1)$ as $x \to x'$. Consequently, $Kf(x) = \int_{B(0,1)} g_x(y)dy \to Kf(x')$ as well. In particular, the eigenvectors ϕ corresponding to non-zero eigenvalues of K are bounded, so for such ϕ we have that $K\phi = \lambda\phi$ is continuous, hence so is ϕ. $\qquad\square$

The following is Lemma 2.1 of [DPRZ99], where (3.18) is (6.3) for $\beta = 2$ (with $\theta_d = 1/\Lambda_{2,d}$).

Theorem 6.2. *Let $\{X_t\}$ be a symmetric stable process of index $\beta < d$ in $\mathbb{R}^d$. Then, for any $u > 0$,*

$$\mathbb{P}\left(\mu_\infty^X\left(B(0,1)\right) > u\right) = \sum_{j=1}^\infty \psi_j e^{-u/\lambda_j} \,, \tag{6.2}$$

where $\lambda_1 > \lambda_2 \geq \cdots \geq \lambda_j \geq \cdots > 0$ are the eigenvalues of the operator $K_{\beta,d}$ with the corresponding continuous, orthonormal eigenvectors $\phi_j(y)$, $\psi_j := \phi_j(0)\langle 1|\phi_j\rangle$ and the infinite sum on the right converges uniformly in u away from 0. In particular,

$$\lim_{u\to\infty} e^{u/\Lambda_{\beta,d}}\, \mathbb{P}(\mu_\infty^X(B(0,1)) > u) = \psi_1 > 0\,. \tag{6.3}$$

Proof of Theorem 6.2. For ease of exposition let $\mathcal{I} = \mu_\infty^X(B(0,1))$. We first compute the moments of $\mathcal{I}$.

Lemma 6.3. *With $v_m = K^{m-2}u^0 \in L^2(B(0,1))$, we have for any $n \geq m$,*

$$\mathbb{E}\left(\mathcal{I}^n\right) = n!\langle 1|K^{n-m+1}v_m\rangle\,.$$

Proof. Upon ordering, since $\{X_t\}$ has stationary independent increments, we have by Fubini that

$$\mathbb{E}\left(\mathcal{I}^n\right) = \mathbb{E}\left(\left\{\int_0^\infty \mathbf{1}_{B(0,1)}(X_s)\,ds\right\}^n\right)$$

$$= n!\int_{0\leq t_1\leq\cdots\leq t_n} \mathbb{P}\left(\bigcap_{i=1}^n\{X_{t_i}\in B(0,1)\}\right)dt_1\cdots dt_n$$

$$= n!\int_{0\leq t_1\leq\cdots\leq t_n}\int_{B(0,1)^n}\prod_{j=1}^n p_{t_j-t_{j-1}}(x_j - x_{j-1})\,dx_1\cdots dx_n\, dt_1\cdots dt_n$$

$$= n!\int_{B(0,1)^n}\prod_{j=1}^n u^0(x_j - x_{j-1})\,dx_1\cdots dx_n$$

$$= n!\langle 1|K^{n-1}u^0\rangle = n!\langle 1|K^{n-m+1}v_m\rangle\,,$$

where $x_0 = 0$ and we have used the symmetry of u^0 and that of K.   □

Taking hereafter m large enough that v_m is bounded (see Lemma 6.1), we define

$$g_m(z,u) := \sum_{n=m}^\infty \frac{z^{n-m}}{n!}u^n\,,$$

and express $\mathbb{E}\left(g_m(z,\mathcal{I})\right)$ in two different ways.

Lemma 6.4. *If $v_m \in L^2(B(0,1))$ then for $z \in \mathbb{C}$ with $|z| < 1/\Lambda$,*

$$\mathbb{E}\left(g_m(z,\mathcal{I})\right) = \langle \mathbf{1} | (\mathbf{I} - zK)^{-1} K v_m \rangle \, . \tag{6.4}$$

If in addition $\Re(z) < 0$ then,

$$\mathbb{E}\left(g_m(z,\mathcal{I})\right) = \int_0^\infty e^{zu} f_m(u) du \, , \tag{6.5}$$

where $f_m(u)$ is the $(m-1)$-fold integral from u to ∞ of $\mathbb{P}(\mathcal{I} > \cdot)$.

Proof. Starting with (6.4), we have for $|z| < 1/\Lambda$ by Lemma 6.3, that

$$\frac{|z|^{n-m}}{n!} \mathbb{E}[\mathcal{I}^n] \leq (|z|\Lambda)^{n-m} \Lambda \|v_m\|_2$$

is absolutely summable, and further then

$$\mathbb{E}[g_m(z,\mathcal{I})] = \sum_{n=m}^\infty \frac{z^{n-m}}{n!} \mathbb{E}[\mathcal{I}^n] = \sum_{n=m}^\infty z^{n-m} \langle \mathbf{1} | K^{n-m+1} v_m \rangle$$

$$= \langle \mathbf{1} | \sum_{\ell=0}^\infty (zK)^\ell K v_m \rangle = \langle \mathbf{1} | (\mathbf{I} - zK)^{-1} K v_m \rangle \, ,$$

where the last equality is the standard Newman series for the resolvent.

Turning to the proof of (6.5), since $\Re(z) < 0$, the integral on the right is finite. By definition we have that

$$\mathbb{E}\left(g_m(z,\mathcal{I})\right) = \int_0^\infty g_m(z,u) \, d\mathbb{P}(\mathcal{I} > u) \, ,$$

and (6.5) is obtained by integrating by parts m times. To justify this, we note that, on the one hand by Lemma 6.3 all moments of $\mathcal{I}$ are finite, so that $\mathbb{P}(\mathcal{I} > u) \leq c_N / u^N$ for any N, and therefore $f_m(u)$ is bounded and goes to 0 as u tends to ∞. On the other hand, $d^m g(z,u)/d^m u = e^{zu}$ with $d^k g(z,u)/d^k u = 0$ at $(z,0)$ for $k = 0, \ldots, m-1$ which controls the boundary terms at $u = 0$, and writing $g(z,u) = z^{-m}(e^{zu} - \sum_{n=0}^{m-1} z^n u^n / n!)$ and using the fact that $\Re(z) < 0$ controls the boundary terms at $u = \infty$. $\square$

We next diagonalize K. Since K is a convolution operator on $B(0,1)$ with locally $L^1(\mathbb{R}^d, dx)$ kernel, it follows easily as in [HS78, Corollary 12.3] that K is a (symmetric) compact operator. Moreover, the Fourier transform relation $\int e^{i(x \cdot p)} u^0(x) \, dx = |p|^{-\beta} > 0$ implies that K is strictly positive definite. By the standard theory for symmetric compact operators, K has discrete spectrum (except for a possible accumulation point at 0) with all eigenvalues positive, of finite multiplicity, and the corresponding eigenvectors of K, denoted $\{\phi_j\}$ form a complete orthonormal basis of $L^2(B(0,1), dx)$ (see [RS78a, Theorems VI.15, VI.16]). Moreover, $\langle f | K g \rangle > 0$ for any non-negative, non-zero, f, g, so

K is ergodic and by the generalized Perron-Frobenius Theorem, see [RS78b, Theorem XIII.43], the eigenspace corresponding to $\Lambda = \lambda_1$ is one dimensional, and we may and shall choose ϕ_1 such that $\phi_1(y) > 0$ for all $y \in B(0,1)$. Noting that ϕ_j are also eigenvectors of $(\mathbf{I} - zK)^{-1}$ with corresponding eigenvalues $(1 - z\lambda_j)^{-1}$, we get, for $|z| < 1/\Lambda$ and $v_m \in L^2(B(0,1))$, that

$$\langle \mathbf{1}|(\mathbf{I} - zK)^{-1}Kv_m \rangle = \sum_{j=1}^{\infty} \frac{c_j \lambda_j}{1 - z\lambda_j} = \int_0^{\infty} e^{zu} \left(\sum_{j=1}^{\infty} c_j e^{-u/\lambda_j} \right) du,$$

where we could change the order of summation and integration since, by Cauchy-Schwartz, $c_j := \langle \mathbf{1}|\phi_j \rangle \langle v_m|\phi_j \rangle$ is absolutely summable. Hence, comparing (6.4) and (6.5), we have, for any $z \in [-3/(4\Lambda), -1/(4\Lambda)]$ that

$$\int_0^{\infty} e^{zu} \left(\sum_{j=1}^{\infty} c_j e^{-u/\lambda_j} \right) du = \int_0^{\infty} e^{zu} f_m(u) du \qquad (6.6)$$

(see also Fig. 6.1). With c_j absolutely summable both integrals in (6.6) are analytic in the strip $\Re(z) \in [-3/(4\Lambda), -1/(4\Lambda)]$, and since they agree for z real inside the strip, they agree throughout this strip. Considering $z = -1/(2\Lambda) + it$, $t \in \mathbb{R}$ we have that the Fourier transforms of $f_n(u)e^{-u/(2\Lambda)}$ and $\sum_{j=1}^{\infty} c_j e^{-u/\lambda_j} e^{-u/(2\Lambda)}$ coincide, hence by uniqueness of the Fourier transform, we have that for almost every $u \geq 0$,

$$f_m(u) = \sum_{j=1}^{\infty} c_j e^{-u/\lambda_j}. \qquad (6.7)$$

In fact, $f_m(u)$ is continuous when $m \geq 2$ and the sum converges uniformly in $[0, \infty)$, so (6.7) holds for all $u \geq 0$.

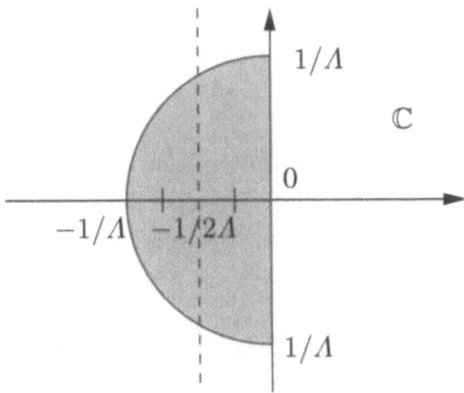

Fig. 6.1. Both identities of Lemma 6.4 hold in the shaded region

Considering the $(m-1)$-st derivative of f_m, the uniform convergence of the sums

$$\sum_{j=1}^{\infty} c_j \lambda_j^{-k} e^{-u/\lambda_j}$$

for $u > 0$ bounded away from 0 and $k = 1, \ldots, m-1$, shows that, for all $u > 0$,

$$\mathbb{P}\left(\mathcal{I} > u\right) = \sum_{j=1}^{\infty} c_j \lambda_j^{-(m-1)} e^{-u/\lambda_j} ,$$

which is (6.2) with $\psi_j = \lambda_j^{-(m-1)} c_j$. Recall Lemma 6.1 that the eigenvectors ϕ_j of K are continuous, hence bounded, and so are $K^{m-1}\phi_j = \lambda_j^{m-1}\phi_j$. Note that $\langle v_m | \phi_j \rangle = (K^{m-1}\phi_j)(0) = \lambda_j^{m-1}\phi_j(0)$, so $\lambda_j^{-(m-1)} c_j = \phi_j(0)\langle 1 | \phi_j \rangle$ as stated. Since $\phi_1(y) > 0$ for all $y \in B(0,1)$ as are the functions $\mathbf{1}$ and $v_m = K^{m-2}u^0$ it follows that $c_1 = \langle 1 | \phi_1 \rangle \langle v_m | \phi_1 \rangle > 0$. Obviously, $\lambda_1 = \Lambda$ and $\psi_1 = \lambda_1^{-(m-1)} c_1 > 0$. Since $\lambda_1 > \sup_{j \geq 2} \lambda_j$ and $c_j = \lambda_j^{m-1}\psi_j$ is absolutely summable, we immediately get (6.3) as well.   □

6.2 Kac's Moment Formula

The derivation of the moments of $\mathcal{I} = \mu_\infty^X(B(0,1))$ applies more generally to additive functionals of Markov processes. We outline here part of the general exposition of this method in [FP99].

Let τ be a *killing time* for a time homogeneous Markov process $\{X_t\}$. That is, $\{X_t : 0 \leq t < \tau\}$ is a Markov process with sub-Markovian semi-group, $K_t f(x) = \mathbb{E}_x(f(X_t)\mathbf{1}_{\{t<\tau\}})$. While $\tau < \infty$ non-random is typically not a killing time, $\tau = \infty$, a random τ independent of $\{X_t\}$ and of Exponential law, or τ a first hitting time, are some of the classical examples of killing times. Let E be the state space of $\{X_t\}$ and v denote a given non-negative, measurable function on E. To simplify our exposition, we do not spell out the precise measurability setting (as provided in [FP99]). In any case, this is of little concern as our main interest is the classical setting where X_t is the Brownian motion in $\mathbb{R}^d$. Define the additive functional

$$A_v = \int_0^\tau v(X_t)dt .$$

Example 6.5. *We have in mind the Brownian motion w_t in $\mathbb{R}^d$, $v = \mathbf{1}_{B(0,1)}$ and two special examples:*

i. $\tau = \infty$ *for which* $A_v = \mu_\infty^w(B(0,1))$);
ii. $\tau = \sigma_d(0,1) = \inf\{t \geq 0 : |w_t| \geq 1\}$, *for which* $A_v = \sigma_d(0,1)$.

As noted in [Kac51], the moments of A_v can be computed in the same manner we had done it in Lemma 6.3, that is,

Proposition 6.6 (Kac's Moment Formula). *Let* $G_v(x, dy) = v(y)G(x, dy)$ *and* $G(x, dy)$ *be the Green's operator associated with* X_t *and* τ, *i.e.*

$$Gf(x) = \mathbb{E}_x \left[\int_0^\tau f(X_t)dt \right] = \int_0^\infty K_t f(x)dt ,$$

for any non-negative measurable f. *Then, for any* $x \in E$ *and* $n \geq 1$,

$$\mathbb{E}_x[A_v^n] = n! \, [G_v^n \mathbf{1}](x) \tag{6.8}$$

Proof. Let ∂ denote the cemetery state of the killed process Y_t on the enlarged state space $E' = E \cup \partial$, that is, $Y_t = X_t$ for $t < \tau$ and $Y_t = \partial$ for $t \geq \tau$. Since τ is a killing time, $\{Y_t\}$ is a Markov process whose semi-group is derived from K_t, and τ is just the hitting time of ∂ by this process. Setting $v(\partial) = 0$ we get by ordering that

$$A_v^n = \int_0^\infty \cdots \int_0^\infty \prod_{i=1}^n v(Y_{t_i})dt_1 \ldots dt_n = n! \, I_n ,$$

with

$$I_n := \int_{0 < t_1 < \cdots < t_n} \prod_{i=1}^n v(Y_{t_i})dt_1 \ldots dt_n .$$

The Markov property of $\{Y_t\}$ at time t_1 gives for any $y \in E'$ the induction formula

$$\mathbb{E}_y[I_{n+1}] = \mathbb{E}_y \left[\int_0^\infty v(Y_{t_1}) \mathbb{E}_{Y_{t_1}}[\tilde{I}_n]dt_1 \right] , \tag{6.9}$$

where $\tilde{I}_n$ denotes a copy of I_n that is independent of $\{Y_t\}$. Note that

$$G_v f(y) = \mathbb{E}_y \left[\int_0^\infty f(Y_t)v(Y_t)dt \right] , \tag{6.10}$$

for any $y \in E'$ and any non-negative measurable function f on E', resulting with $G_v f(\partial) = 0$ for all f. Taking the induction hypothesis to be

$$\mathbb{E}_y[I_n] = (G_v^n \mathbf{1})(y) \quad \text{for all } y \in E' ,$$

which is true for $n = 1$ by the definition of A_v and G_v, we get by (6.9) and (6.10) that

$$\mathbb{E}_y[I_{n+1}] = \mathbb{E}_y \left[\int_0^\infty v(Y_{t_1})(G_v^n \mathbf{1})(Y_{t_1})dt_1 \right] = (G_v^{n+1} \mathbf{1})(y) ,$$

and the proof is complete by induction on n. $\square$

In Lemma 6.3 we considered the special case where

$$G_v f(x) = \int_{B(0,1)} g_v(x - y)f(y)dy ,$$

with g_v a symmetric function, for which it is easy to verify that $(G_v^n 1)(0) = \langle 1|G_v^{n-1}g_v\rangle$.

The weighted sum of the moment formula yields the Feynman-Kac formula,

Theorem 6.7 (Feynman-Kac Formula). *For $v \geq 0$ and $\theta \geq 0$, the moment generating function $f_{v,\theta}(x) = \mathbb{E}_x[e^{\theta A_v}]$ is the minimal positive solution of $f = \theta G_v f + 1$. Further, if $|\theta| < 1/\|G_v 1\|_\infty$ then the latter equation has a unique finite solution which is $f_{v,\theta}$.*

The second part of the theorem, due to Kha'sminskii [Kha59], is what we have used in the Brownian context when proving (5.16). It is also what is used by [DPR03] in conjunction with Theorem 2.5 and Lemma 2.7 for going beyond the result of Theorem 2.8. See [FP99] and the references therein for some of the many other applications of this theorem.

Proof. By Proposition 6.6 we have that for $v \geq 0$ and $\theta \geq 0$,

$$f_{v,\theta}(x) = \mathbb{E}_x[e^{\theta A_v}] = \sum_{n=0}^{\infty} \frac{\theta^n}{n!} \mathbb{E}_x[A_v^n] = \sum_{n=0}^{\infty} \theta^n (G_v^n 1)(x) . \tag{6.11}$$

Note that $f := \sum_{n=0}^{\infty} \theta^n G_v^n 1$ appearing in the right side of (6.11) is a solution of $f = \theta G_v f + 1$, though it may well diverge for some or even all $x \in E$. Since G_v is a positive operator, $\|G_v^n 1\|_\infty \leq (\|G_v 1\|_\infty)^n$, hence if $|\theta| < 1/\|G_v 1\|_\infty$ then the series $\sum_n \theta^n G_v^n 1$ converges uniformly to a finite, positive solution of $f = \theta G_v f + 1$, which by dominated convergence is $f_{v,\theta}$. □

We conclude this section by briefly explaining part of the derivation in [CT62] of the identity in law of $\mu_\infty^w(B(0,1))$ and that of $\sigma_{d-2}(0,1)$. In part (ii) of Example 6.5 we have that $G_v 1(x) = \mathbb{E}_x[\sigma_d(0,1)]$ when $|x| \leq 1$, and $G_v 1(x) = 0$ otherwise. Hence, in this case

$$\|G_v 1\|_\infty = \sup_{x \in B(0,1)} \mathbb{E}_x[\sigma_d(0,1)] = \mathbb{E}_0[\sigma_d(0,1)] = \frac{1}{d} < \infty$$

(see for example Lemma 2.10). Applying Kha'sminskii's condition we see that $f(x) = \mathbb{E}_x[e^{\theta \sigma_d(0,1)}] < \infty$ for all $\theta < d$ and $x \in B(0,1)$. Similar to the proof of Lemma 2.11 it is easy to check that then $f(x)$ is a solution of the Dirichlet problem

$$\frac{1}{2}\Delta f + \theta f = 0 \quad \text{in } B(0,1) , \tag{6.12}$$

subject to $f = 1$ in $\partial B(0,1)$ (this can also be derived from Ito's formula, as for example in [Bas99, Page 109], or recognizing that the solution of Poisson's equation of Lemma 2.10 for $\kappa = 0$ and $D = B(0,1)$ is by definition $u = G_v f$ and using the identity $f = \theta u + 1$ to get the PDE (6.12) for f). The solution to (6.12) is of the form $f(x) = u(|x|)$, with u thus solving the ODE

$$\begin{cases} \frac{1}{2}u''(r) + \frac{d-1}{2r}u'(r) + \theta u(r) = 0 \\ u(1) = 1 \end{cases}$$

whose solution can be expressed in terms of a Bessel function of the first kind. The positive operator G_v for part (i) of Example 6.5 is $K_{2,d}$ of Sect. 6.1 which is unbounded. However, recall Lemma 6.1 that it is a bounded operator from $L^2(B(0,1))$ to itself and as done in deriving (6.4), the moment generating function $f_{v,\theta}(x)$ of $A_v = \mu_\infty^w(B(0,1))$ is finite and smooth in x for any $\theta \leq 0$ with $|\theta|$ small enough. Applying the Laplacian to both sides of the identity $f = \theta G_v f + 1$ results with f being the radial solution of (6.12) which is bounded in $B(0,1)$, with the scaling constant determined by the identity $f = \theta G_v f + 1$. This is expressed also in terms of a Bessel function of the first kind. Considering $x = 0$, the analyticity in $\theta \in \mathbb{C}$ of the relevant functions allows [CT62] to find explicit power series expressions for the corresponding laws, and using identities for Bessel functions, show that these coincide when one takes $d - 2$ in part (ii) of the example versus d in part (i) of the example.

6.3 The Ciesielski-Taylor Identities

Denote by $\big(R_\delta(t)\big)_{t \geq 0}$ a Bessel process of dimension $\delta > 0$ starting at 0. For positive integer δ this is just the Euclidean norm of a (standard) δ-dimensional Brownian motion. A definition for all $\delta \geq 0$ is given in [RY99, Chapter XI]. The Ciesielski-Taylor identities are thus the statement that for all $\delta > 0$,

$$\int_0^\infty \mathbf{1}_{\{R_{\delta+2}(s) \leq 1\}} ds \stackrel{\mathcal{L}}{=} T_1(R_\delta) \tag{6.13}$$

where $T_1(R_\delta) := \inf\{t : R_\delta(t) = 1\}$. For positive integer δ this is the statement we made in connection with (3.18), that the total occupation measure of the unit ball in $\mathbb{R}^{\delta+2}$ for a standard $(\delta + 2)$-dimensional Brownian motion has the same law as the time it takes a standard δ-dimensional Brownian motion to exit the unit ball in $\mathbb{R}^\delta$. In this section we denote by $l_t^a(X)$ the local time of the process $(X_s)_{s \geq 0}$ in a at time t. Since the local time process $(l_\infty^a)_{a \geq 0}$ is the density of the occupation measure (see [RY99, Corollary VI.1.6]) and the quadratic variation process for any Bessel process equals $(t)_{t \geq 0}$, the identity (6.13) is equivalent to

$$\int_0^1 l_\infty^a(R_{\delta+2})da \stackrel{\mathcal{L}}{=} \int_0^1 l_{T_1}^a(R_\delta)da . \tag{6.14}$$

We present here Yor's proof of (6.14), following [Yor92, Chapter 4]. This proof has two ingredients, first using the Ray-Knight theorem to substitute the local time processes in (6.14) by more manageable processes (see Theorem 6.8), then applying an "integration by parts" identity in law for quadratic functionals of one-dimensional Brownian motion (see Proposition 6.9). The first ingredient of the proof is thus,

Theorem 6.8. *Let $(w_t)_{t\geq 0}$ and $(b_t)_{t\geq 0}$ denote the planar standard Brownian motion and the standard planar Brownian bridge, respectively. Then,*

i. For $\gamma > 0$,

$$\left(l_\infty^a(R_{2+\gamma})\right)_{a>0} \stackrel{\mathcal{L}}{=} \left(\frac{1}{\gamma a^{\gamma-1}}|w_{a^\gamma}|^2\right)_{a>0} \tag{6.15}$$

$$\left(l_{T_1}^a(R_{2+\gamma})\right)_{a\in(0,1]} \stackrel{\mathcal{L}}{=} \left(\frac{1}{\gamma a^{\gamma-1}}|b_{a^\gamma}|^2\right)_{a\in(0,1]} \tag{6.16}$$

ii. For $\gamma = 0$,

$$\left(l_{T_1}^a(R_2)\right)_{a\in(0,1]} \stackrel{\mathcal{L}}{=} \left(a|w_{\log(1/a)}|^2\right)_{a\in(0,1]} \tag{6.17}$$

iii. For $0 < \gamma \leq 2$,

$$\left(l_{T_1}^a(R_{2-\gamma})\right)_{a\in(0,1]} \stackrel{\mathcal{L}}{=} \left(\frac{1}{\gamma a^{\gamma-1}}|w_{1-a^\gamma}|^2\right)_{a\in(0,1]} \tag{6.18}$$

Throughout this section, the standard planar Brownian bridge is the process $b_t = \left((b_t^1, b_t^2)\right)_{t\in[0,1]}$ with b_t^1 and b_t^2 two independent standard one-dimensional Brownian bridges on $[0,1]$.

In view of Theorem 6.8, the identities of (6.14) are equivalent to

$$\frac{1}{\delta}\int_0^1 \frac{|w_{a^\delta}|^2}{a^{\delta-1}}da \stackrel{\mathcal{L}}{=} \frac{1}{\delta-2}\int_0^1 \frac{|b_{a^{\delta-2}}|^2}{a^{\delta-3}}da, \qquad \text{for } \delta > 2 \quad (6.19)$$

$$\frac{1}{2}\int_0^1 \frac{|w_{a^2}|^2}{a}da \stackrel{\mathcal{L}}{=} \int_0^1 a|w_{\log(1/a)}|^2 da, \qquad (\delta = 2) \quad (6.20)$$

$$\frac{1}{\delta}\int_0^1 \frac{|w_{a^\delta}|^2}{a^{\delta-1}}da \stackrel{\mathcal{L}}{=} \frac{1}{2-\delta}\int_0^1 \frac{|w_{1-a^{2-\delta}}|^2}{a^{1-\delta}}da, \qquad \text{for } \delta < 2. \quad (6.21)$$

Each side of the identities (6.19)–(6.21) is the sum of two i.i.d. random variables, representing the corresponding contribution for a one-dimensional Brownian motion w or Brownian bridge b. It thus suffices to prove these identities in the one-dimensional case, which we proceed to do while deferring the proof of Theorem 6.8. To this end, we rely on the following "integration by parts" formula.

Proposition 6.9. *Let $(w_t)_{t\geq 0}$ be a one-dimensional Brownian motion starting at $w_0 = 0$. For all $0 \leq \alpha \leq \beta < \infty$ and any non-random continuous functions, $f, g : [\alpha, \beta] \to [0, \infty)$ such that f is decreasing, and g is increasing,*

$$f(\beta)w_{g(\beta)}^2 - \int_\alpha^\beta w_{g(x)}^2 df(x) \stackrel{\mathcal{L}}{=} g(\alpha)w_{f(\alpha)}^2 + \int_\alpha^\beta w_{f(x)}^2 dg(x).$$

Proof. Using the convention $f(t_{n+1}) = g(t_0) = 0$, it is enough to show that the identity in law

$$\sum_{i=1}^{n} \big(f(t_i) - f(t_{i+1})\big)w_{g(t_i)}^2 \overset{\mathcal{L}}{=} \sum_{j=1}^{n} \big(g(t_j) - g(t_{j-1})\big)w_{f(t_j)}^2 \qquad (6.22)$$

holds for any partition $\{\alpha = t_1 < t_2 < \cdots < t_n = \beta\}$ of $[\alpha, \beta]$, where we may and shall assume without loss of generality that $f(t_i) \neq f(t_{i+1})$ and $g(t_i) \neq g(t_{i-1})$ for $i = 1, \ldots, n$. To this end, fixing such partition, define $v_i = \sqrt{f(t_i) - f(t_{i+1})} > 0$ and $u_j = \sqrt{g(t_j) - g(t_{j-1})} > 0$ for $i, j \in \{1, 2, \ldots, n\}$. Note that $X_i = v_i^{-1}(w_{f(t_i)} - w_{f(t_{i+1})})$, $i = 1, \ldots, n$ are i.i.d. Normal random variables of zero mean and unit variance, as are $Y_j = u_j^{-1}(w_{g(t_j)} - w_{g(t_{j-1})})$ for $j = 1, \ldots, n$. Taking $\mathbf{X} = (X_1, \ldots, X_n)$ and $\mathbf{Y} = (Y_1, \ldots, Y_n)$ note that the left-hand side of (6.22) is $|\mathbf{AY}|^2$ for the $n \times n$-dimensional matrix $\mathbf{A}$ with entries $A_{i,j} = v_i u_j \mathbf{1}_{i \geq j}$, whereas the right-hand side of (6.22) is $|\mathbf{A}^t \mathbf{X}|^2$ (here $\mathbf{A}^t$ denote the transpose of the matrix $\mathbf{A}$ and $|\cdot|$ is the Euclidean norm in $\mathbb{R}^n$). With $\mathbf{Y}$ having the same law as $\mathbf{X}$, it thus suffices to show that $|\mathbf{AX}| \overset{\mathcal{L}}{=} |\mathbf{A}^t \mathbf{X}|$ for any non-random $n \times n$-dimensional matrix $\mathbf{A}$. Writing the singular value decomposition $\mathbf{A} = \mathbf{UDV}$ with $\mathbf{U}$ and $\mathbf{V}$ orthogonal and $\mathbf{D}$ diagonal, we have that the law of $\mathbf{X}$ is invariant under the non-random orthogonal transformation $\mathbf{V}$ and the Euclidean norm is invariant under the isometry $\mathbf{U}$ of $\mathbb{R}^n$. Hence, $|\mathbf{AX}| = |\mathbf{DVX}| \overset{\mathcal{L}}{=} |\mathbf{DX}|$. By similar reasoning $|\mathbf{A}^t \mathbf{X}| = |\mathbf{DU}^t \mathbf{X}| \overset{\mathcal{L}}{=} |\mathbf{DX}|$ completing the proof of (6.22), and with it, that of the proposition. $\qquad \square$

Remark 6.10. *The preceding proof of Proposition 6.9 is via a spectral argument. With $\widetilde{\mathbf{X}}$ denoting an independent copy of $\mathbf{X}$, an alternative proof based on combining Fubini's theorem with the Fourier-Laplace identity*

$$\mathbb{E}\left[\exp\left(-\frac{\lambda^2}{2}|\mathbf{AX}|^2\right)\right] = \mathbb{E}\left[\exp\left(i\lambda \widetilde{\mathbf{X}}^t \mathbf{AX}\right)\right], \qquad (6.23)$$

is provided in [DSY94]. The infinite-dimensional version of (6.23) provides there also identities in law for quadratic functionals of a one-dimensional Brownian motion (w_t), such as

$$\int_0^{\infty} ds \left(\int_0^{\infty} k(s,t)dw_t\right)^2 \overset{\mathcal{L}}{=} \int_0^{\infty} ds \left(\int_0^{\infty} k(t,s)dw_t\right)^2$$

(for any non-random $k(\cdot, \cdot)$), as well as its extension to a relation between the corresponding β and α powers of integrals of symmetric α and β stable processes. For example, in case $k(s,t) = (\mathbf{1}_{t \leq s} - s)\mathbf{1}_{s,t \leq 1}$ this amounts to $\int_0^1 ds(w_s - sw_1)^2$ having the same law as $\int_0^1 ds(w_s - \int_0^1 w_t dt)^2$.

See also [DY97] for a variety of other identities in law for functionals of Brownian motion, the derivation of many of whom relies upon the Fourier-Laplace identity (6.23).

We establish (6.20) by applying Proposition 6.9 for $f(x) = \log(1/x)$ and $g(x) = x^2$, with $\alpha = 0$ and $\beta = 1$. Similarly, using $f(x) = 1 - x^{2-\delta}$ and $g(x) = x^\delta$ results with (6.21). To get (6.19) we first apply Proposition 6.9 for $f(x) = x^{2-\delta} - 1$ and $g(x) = x^\delta$ to get that

$$(\delta - 2) \int_0^1 x^{1-\delta} w_{x^\delta}^2 dx \stackrel{\mathcal{L}}{=} \delta \int_0^1 x^{\delta-1} w_{x^{2-\delta}-1}^2 dx .$$

Since $(w_{t^{-1}-1})_{t\in(0,1]} \stackrel{\mathcal{L}}{=} (\frac{1}{t} b_t)_{t\in(0,1]}$ (see [RY99, Exercise I.3.10]) the right-hand side of this relation has the same law as $\delta \int_0^1 \frac{1}{x^{\delta-3}} b_{x^{\delta-2}}^2 dx$, thus completing the proof of (6.19).

We conclude the section with an outline of the proof of Theorem 6.8.

Proof of Theorem 6.8:. The case $\gamma = 1$ of the relation (6.15) follows from the first Ray-Knight theorem (see [RY99, Theorem XI.2.2]) and the time reversal result of Williams (see [RY99, Corollary VII.4.6]). For the case $\gamma \neq 1$ we use [RY99, Proposition XI.1.11] with the choices $q = 1/\gamma$, $\nu = \gamma/2$ to see that there is a three dimensional Bessel process R_3 starting from 0, such that

$$R_{2+\gamma}^\gamma = R_3(\tau.) ,$$

where $\tau_t = \int_0^t (\gamma R_{2+\gamma}^{\gamma-1}(s))^2 ds$ for $t \geq 0$. Then, by [RY99, Exercise VI.1.23], for all $a > 0$ we have

$$\gamma a^{\gamma-1} l_\infty^a (R_{2+\gamma}) = l_\infty^{a^\gamma} (R_3(\tau.))$$

and by [RY99, Exercise VI.1.27] the right-hand side equals $l_\infty^{a^\gamma} (R_3)$ (by continuity, this holds almost surely, for all values of a). As we have already seen by (6.15) with $\gamma = 1$ this equals in law to $|w_{a^\gamma}|^2$, thus finishing the proof of (6.15).

Relation (6.16) is proved similarly. First we prove the case $\gamma = 1$. Let $R_{3,r}$ denote the three dimensional Bessel process starting at $r > 0$. By [RY99, Exercise VI.1.23] applied to the function $f(x) = -x^{-1}$ we have that for any $a \in (0,1)$

$$l_{T_1}^a (R_{3,r}) = a^2 l_{T_1}^{\frac{1}{a}} \left(\frac{1}{R_{3,r}} \right)$$

By [RY99, Lemma VI.3.12], the process $1/R_{3,r}$ is a time-changed one dimensional Brownian motion $\beta.$ starting at $\beta_0 = 1/r$, restricted to $[0, T_0[$. Hence, by [RY99, Exercise VI.1.27] we have that

$$l_{T_1}^{\frac{1}{a}} \left(\frac{1}{R_{3,r}} \right) = l_{T_1}^{\frac{1}{a}} (\beta) .$$

By the time reversal result of Williams (see [RY99, Corollary VII.4.6]), there exists a three dimensional Bessel process R_3 starting at zero, such that

$$\left(l_{T_1}^{\frac{1}{a}} (\beta) \right)_{a\in(0,1]} \stackrel{\mathcal{L}}{=} \left(l_{\mathcal{T}_r}^{\frac{1}{a}} (R_3) \right)_{a\in(0,1]} ,$$

where $\mathcal{I}_r$ denotes the time interval between $\sigma_1 = \sup\{t : R_3(t) = 1\}$ and $\sigma_r = \sup\{t : R_3(t) = 1/r\}$. Thus,

$$\left(l_{T_1}^{ra}(R_{3,r})\right)_{a\in(0,1]} \overset{\mathcal{L}}{=} \left(l_{\mathcal{I}_r}^{\frac{1}{a}}(R_3)\right)_{a\in(0,1]}.$$

As $r \to 0$ we have $\sigma_r \uparrow \infty$ and by Tanaka's formula (see [RY99, Theorem VI.1.2]), the left-hand side converges as $r \to 0$ to the left-hand side of (6.16). Further, the path $R_3(\sigma_1 + \cdot) - 1$ has the same law as $R_3(\cdot)$ (see [RY99, Proposition VI.3.9]), implying, using (6.15), that the local time we want equals in distribution to $|w_{\frac{1}{a}-1}|^2$. Therefore

$$\left(l_{T_1}(R_3)\right)_{a\in(0,1]} \overset{\mathcal{L}}{=} \left(a^2|w_{\frac{1}{a}-1}|^2\right)_{a\in(0,1]}$$

and the last equals in distribution to the right-hand side of (6.16) due to [RY99, Exercise I.3.10]. The case $\gamma \neq 1$ is then proved following the same procedure as in the proof of (6.15).

Relation (6.17) is part 1 of [RY99, Exercise XI.2.6].

The proof of (6.18) is similar to that of (6.15). For $\gamma = 1$ it reads

$$\left(l_{T_1}^{ra}(R_1)\right)_{a\in(0,1]} \overset{\mathcal{L}}{=} \left(|w_{1-a}|^2\right)_{a\in(0,1]}.$$

This is a consequence of the first Ray-Knight theorem and the fact that the law of the path obtained by gluing together the positive excursions of a standard one-dimensional Brownian motion is a reflecting Brownian motion (see [KS91, Theorem 6.3.1]). For $\gamma \neq 1$ we apply again [RY99, Proposition XI.1.11] with the choices $q = 1/\gamma, \nu = -\gamma/2$ to see that there is a Bessel process of dimension one, R_1 starting from zero, such that

$$R_{2-\gamma}^\gamma = R_1(\tau.),$$

where $\tau_t = \int_0^t (\gamma R_{2-\gamma}^{\gamma-1}(s))^2 ds$ for $t \geq 0$. Then, by [RY99, Exercise VI.1.23], for all $a > 0$ we have

$$\gamma a^{\gamma-1} l_{T_1}^{a}(R_{2-\gamma}) = l_{T_1}^{a^\gamma}(R_1(\tau.))$$

and by [RY99, Exercise VI.1.27] the right-hand side of the last equation equals $l_{T_1}^{a^\gamma}(R_1)$. By the case $\gamma = 1$ this also equals in law to $|w_{1-a^\gamma}|^2$ finishing the proof. $\square$

References

[AF01]   D. J. Aldous and J. Fill, *Reversible Markov chains and random walks on graphs*, http://stat-www.berkeley.edu/users/aldous/RWG/book.html, 2001.

[Ald89]   D. J. Aldous, *Probability approximations via the Poisson clumping heuristic*, vol. 77, Springer-Verlag, 1989.

[Ald91a]   D. Aldous, *Random walk covering of some special trees*, J. Math. Analysis Appl. **157** (1991), no. 1, 271–283.

[Ald91b]   _____, *Threshold limits for cover times*, J. Theoret. Probab. **4** (1991), no. 1, 197–211.

[Aub82]   T. Aubin, *Nonlinear analysis on manifolds. Monge-Ampére equations*, Springer-Verlag, New York, 1982.

[Aub98]   _____, *Some nonlinear problems in Riemannian geometry*, Springer Monographs in Mathematics, Springer-Verlag, Berlin, 1998.

[Bas99]   R. F. Bass, *Diffusions and elliptic operators. probability and its applications*, Springer-Verlag, New York, 199.

[BBT97]   A. Bruckner, J. Bruckner, and B. Thompson, *Real Analysis*, Prentice Hall, NJ, 1997.

[BDG01]   E. Bolthausen, J.-D. Deuschel, and G. Giacomin, *Entropic repulsion and the maximum of the two dimensional free field*, Ann. Probab. **29** (2001), no. 4, 1670–92.

[BG85]   R. F. Bass and P. S. Griffin, *The most visited site of Brownian motion and simple random walk*, Z. Wahrsch. Verw. Gebiete **70** (1985), 417–436.

[BGT87]   N. H. Bingham, C. M. Goldie, and J. L. Teugels, *Regular variation*, Cambridge University Press, 1987.

[BH91]   M. Brummelhuis and H. Hilhorst, *Covering of a finite lattice by a random walk*, Physica A. **176** (1991), 387–408.

[Bra83]   M. Bramson, *Convergence of solutions of the Kolmogorov equation to travelling waves*, Mem. Amer. Math. Soc. **44** (1983), no. 285, 1–190.

[Bra86]   _____, *Location of the traveling wave for the Kolmogorov equation*, Probab. Theory Related Fields **73** (1986), no. 4, 481–515.

[BZ05]   M. Bramson and O. Zeitouni, *Recursions and tightness*, Preprint (2005).

[CT62]   Z. Ciesielski and J. Taylor, *First passage and sojourn time and the exact Hausdorff measure of the sample path*, Transactions of the American Mathematical Society **103** (1962), 434–52.

[Dav52]   R. O. Davies, *Subsets of finite measure in analytic sets*, Indagationes Math. **14** (1952), 488–489.

[Dav04]   O. Daviaud, *Extremes of the discrete two-dimensional gaussian free field*, Preprint, ArXiv math.PR/0406609 (2004).

[Dav05]   _____, *Thick points for the Cauchy process*, Annals Inst. Henri Poincaré, Prob. and Stat. (2005).

[DPR03]   A. Dembo, Y. Peres, and J. Rosen, *Brownian motion on compact manifolds: cover time and late points*, Electronic Journal of Probability **8** (2003).

[DPR05]   _____, *How large a disc is covered by a random walk in n steps?*, Preprint (2005).

[DPRZ99]   A. Dembo, Y. Peres, J. Rosen, and O. Zeitouni, *Thick points for transient symmetric stable processes*, Electronic Journal of Probability **4** (1999), no. 10, 1–13.

[DPRZ00a] _____, *Thick points for spatial Brownian motion: Multifractal analysis of occupation measure*, The Annals of Probability **28** (2000), 1–35.

[DPRZ00b] _____, *Thin points for Brownian motion*, Annals Inst. Henri Poincaré, Prob. and Stat. **36** (2000), 749–74.

[DPRZ01]   _____, *Thick points for planar Brownian motion and the Erdős-Taylor conjecture on random walk*, Acta Mathematica **186** (2001), 239–70.

[DPRZ02]   ———, *Thick points for intersections of planar Brownian paths*, Transactions of the American Mathematical Society **354** (2002), 4969–5003.

[DPRZ04]   ———, *Cover times for Brownian motion and random walks in two dimensions*, Annals Mathematics **160** (2004), 433–64.

[DPRZ05]   ———, *Late points for random walks in two dimensions*, The Annals of Probability **33** (2005).

[DSY94]   C. Donati-Martin, S. Song, and M. Yor, *Symmetric stable processes, fubini's theorem and some extensions of the ciesielski-taylor identities in law*, Stochastics and Stochastics Reports **50** (1994), 1–33.

[DY97]   C. Donati-Martin and M. Yor, *Some brownian functionals and their laws*, The Annals of Probability **25** (1997), no. 3, 1011–58.

[Ein89]   U. Einmahl, *Extensions of results of Komlós, Major, and Tusnády to the multivariate case*, Journal of Multivariate Analysis **28** (1989), 20–68.

[ET60]   P. Erdős and S. J. Taylor, *Some problems concerning the structure of random walk paths*, Acta Scientifica Hungarica **11** (1960), 137–62.

[Fal90]   K. J. Falconer, *Fractal geometry: mathematical foundations and applications*, Wiley, Chichester, New York, 1990.

[FP99]   P. J. Fitzsimmons and J. Pitman, *Kac's moment formula and the Feynman-Kac formula for additive functionals of a Markov process*, Stochastic Processes and its Applications **79** (1999), no. 1, 117–34.

[Fro35]   O. Frostman, *Potential d'équilibre et capacité des ensembles avec quelques applications a la théorie des fonctions*, Meddel. Lunds Univ. Math. Sem. **3** (1935), 1–118.

[Haw81]   John Hawkes, *Trees generated by a simple branching process*, J. London Math. Soc. (2) **24** (1981), no. 2, 373–384.

[HS78]   P. Halmos and V. Sunder, *Bounded integral operators on l^2 spaces*, Springer-Verlag, New York, 1978.

[JP95]   H. Joyce and D. Preiss, *On the existence of subsets of finite positive packing measure*, Mathematika **42** (1995), 15–42.

[JS00]   J. Jonasson and O. Schramm, *On the cover time of planar graphs*, Electronic Communications in Probability **5** (2000), no. 10, 85–90.

[Kac51]   M. Kac, *On some connections between probability theory and differential and integral equations*, Proc. Second Berkeley Symp. Math. Stat. Prob. (1951), 189–215.

[Kah85]   J.-P. Kahane, *Some random series of functions: Second edition*, Cambridge University Press, 1985.

[Kau69]   R. Kaufman, *Une propriété metriqué du mouvement Brownien*, C. R. Acad. Sci. Paris **268** (1969), 727–28.

[Kha59]   R. Z. Khas'minskii, *On positive solutions of the equation au+vu=0*, Theory Probab. Appl. **4** (1959), 309–18.

[KMT75]   J. Komlós, P. Major, and G. Tusnády, *An approximation of partial sums of independent RVs, and the sample DF. I*, Zeitschrift für Verwandte Gebiete **32** (1975), 111–31.

[KPX00]   D. Khoshnevisan, Y. Peres, and Y. Xiao, *lim sup random fractals*, Electronic Journal of Probability **5** (2000), no. 4, 1–24.

[KS91]   I. Karatzas and S. E. Shreve, *Brownian motion and stochastic calculus*, second ed., no. Graduate Text in Mathematics, Springer-Verlag, New York, 1991.

[Law91]   G. Lawler, *Intersections of random walks*, Birkhauser, 1991.

[Law92]   _____, *On the covering time of a disc by a random walk in two dimensions*, Seminar in Stochastic Processes (1992), 189–208.

[lG92]   J.-F. le Gall, *Some properties of planar Brownian motion, école d'ete de probabilities de St. Flour XX, 1990 (Berlin)*, Lecture Notes in Mathematics, vol. 1527, Springer-Verlag, Berlin, 1992.

[LS04]   M. A. Lifshits and Z. Shi, *The escape rate of favorite sites of simple random walk and brownian motion*, The Annals of Probability **32** (2004), 159–152.

[Lyo90]   R. Lyons, *Random walks and percolation on trees*, The Annals of Probability **18** (1990), 931–958.

[Mat88a]   P. Matthews, *Covering problems for Brownian motion on spheres*, The Annals of Probability **16** (1988), 189–99.

[Mat88b]   _____, *Covering problems for Markov chains*, The Annals of Probability **16** (1988), 1215–28.

[Mat95]   P. Mattila, *Geometry of sets and measures in Euclidean spaces*, Cambridge University Press, 1995.

[McK75]   H. P. McKean, *Application of Brownian motion to the equation of Kolmogorov-Petrovskii-Piskunov*, Communications on Pure and Applied Mathematics **28** (1975), 323–331.

[Per96a]   Y. Peres, *Intersection equivalance of Brownian path and certain branching processes*, Commun. Math. Phys. **177** (1996), 417–434.

[Per96b]   _____, *Remarks on intersection-equivalence and capacity-equivalence.*, Ann. Inst. H. Poincaré, Phys. Theo. **64** (1996), no. 3, 339–347.

[Per99]   _____, *Probability on trees: An introductory climb, école d'ete de probabilities de St. Flour, 1997 (Berlin)*, Lecture Notes in Mathematics, vol. 1717, Springer-Verlag, Berlin, 1999.

[Per01]   _____, *An invitation to sample paths of the Brownian motion*, http://stat-www.berkeley.edu/users/peres/bmall.pdf, 2001.

[Per03]   _____, *Brownian intersections, cover times and thick points via trees*, Proceedings of the ICM, Beijing 2002 **3** (2003), 73–78.

[PT87]   E. A. Perkins and S. J. Taylor, *Uniform measure results for the image of subsets under Brownian motion*, Probability Theory and Related Fields **76** (1987), 257–89.

[PY03]   J. Pitman and M. Yor, *Hitting, occupation, and inverse local times of one-dimensional diffusions: martingale and excursion approaches*, Bernoulli **9** (2003), 1–24.

[Rév90]   P. Révész, *Random walk in random and non-random environments*, World Scientific, Singapore, 1990.

[Roy88]   H. L. Royden, *Real analysis*, Macmillan, 1988.

[RS78a]   M. Reed and B. Simon, *Methods of modern mathematical physics I: Functional analysis*, Academic Press, 1978.

[RS78b]   _____, *Methods of modern mathematical physics IV: Analysis of operators*, Academic Press, 1978.

[RY99]   D. Revuz and M. Yor, *Continuous martingales and Brownian motion: Third edition*, Grundlehren der Mathematischen Wissenschaften, vol. 293, Springer-Verlag, New York, 1999.

[Spi64]   F. Spitzer, *Principles of random walk*, Van Nostrand, Princeton, New Jersey, 1964.

[Tay66]   S. J. Taylor, *Multiple points for the sample paths of the symmetric stable process*, Z. Wahrscheinlichkeitstheorie und Verw. Gebiete **6** (1966), 170–180.

[Tay86]   J. Taylor, *The measure theory of random fractals*, Math. Proceed. Cambr. Phil. Soc. **100** (1986), 383–486.

[Tet91]   P. Tetali, *Random walks and the effective resistance of networks*, J. Theoret. Probab. **4** (1991), 101–09.

[Wil89]   H. S. Wilf, *The editor's corner: the white screen problem*, Amer. Math. Monthly **96** (1989), 704–707.

[Yor91]   M. Yor, *Une explication du théorème de Ciesielski-Taylor*, Ann. Inst. H. Poincaré, Probab. Statist. **27** (1991), no. 2, 201–213.

[Yor92]   _____, *Some aspects of Brownian motion. Part I. Some special functionals*, birkhauser verlag ed., Lectures in Mathematics ETH Zurich., Springer-Verlag, Basel, 1992.

[Zai03]   A. Y. Zaitsev, *Estimates for the strong approximation in multidimensional central limit theorem*, Proceedings of the ICM, 2002, World scientific, 2003, pp. 107–116.

Part II

Tadahisa Funaki: Stochastic Interface Models

Stochastic Interface Models

Tadahisa Funaki

Graduate School of Mathematical Sciences, The University of Tokyo, Komaba, Tokyo 153-8914, JAPAN, funaki@ms.u-tokyo.ac.jp

Contents

Abstract. In these notes we try to review developments in the last decade of the theory on stochastic models for interfaces arising in two phase system, mostly on the so-called $\nabla\varphi$ interface model. We are, in particular, interested in the scaling limits which pass from the microscopic models to macroscopic level. Such limit procedures are formulated as classical limit theorems in probability theory such as the law of large numbers, the central limit theorem and the large deviation principles.

Key words: Random interfaces, Effective interfaces, Phase coexistence and separation, Ginzburg-Landau model, Massless model, Random walk representation, Surface tension, Wulff shape, Hydrodynamic limit, Motion by mean curvature, Evolutionary variational inequality, Fluctuations, Large deviations, Free boundaries.

2000 Mathematics Subject Classification: 60-02 (60K35, 60H30, 60H15), 82-02 (82B24, 82B31, 82B41, 82C24, 82C31, 82C41), 35J20, 35K55, 35R35

1 Introduction

1.1 Background

The water changes its state to ice or vapor together with variations in temperature. Each of these three states (liquid/solid/gas) is macroscopically homogeneous and called a phase (or a pure phase) in physics. The water and the ice can coexist at temperature $0°\text{C}$. In fact, under various physical situations especially at low temperature, more than one distinct pure phases coexist in space and different phases are separated by fairly sharp hypersurfaces called **interfaces**. Snow crystals in the vapor or alloys consisting of two types of metals are typical examples. Crystals are macroscopic objects, which have ordered arrangements of atoms or molecules in microscopic scale.

Wulff [254] in 1901 proposed a variational principle, at thermodynamic level or from the phenomenological point of view, for determining the shape of interfaces for crystals. Let $E \subset \mathbb{R}^d$ be a crystal shape. Its boundary ∂E is then an interface and an energy called the **total surface tension** is associated with each interface by

$$\mathcal{W}(E) = \int_{\partial E} \sigma(\mathbf{n}(x)) \, dx \,, \tag{1.1}$$

where $\sigma = \sigma(\mathbf{n}) \geq 0$ is the **surface tension** of flat hyperplane in $\mathbb{R}^d$ with unit normal vector $\mathbf{n} \in S^{d-1}$ and dx represents the volume element on ∂E. The interface has locally an energy $\sigma(\mathbf{n}(x))$ depending on its tilt $\mathbf{n} = \mathbf{n}(x)$ and, integrating it over the surface ∂E, the **Wulff functional** $\mathcal{W}(E)$ is defined. For an alloy consisting of two types of metals A and B, E is the region occupied by A-type's metal so that its volume is always kept invariant if the amount of each metal is fixed.

It is expected that the interface, which is in equilibrium and stable, minimizes its total energy and this naturally leads us to the variational problem:

$$\min_{\text{vol}\,(E)=v} \mathcal{W}(E) \tag{1.2}$$

under the condition that the total volume of the crystal E (e.g., the region occupied by A-type's metal) is fixed to be $v > 0$. The minimizer E of (1.2) and its explicit geometric expression are called the **Wulff shape** and the Wulff construction, respectively. Especially when the surface tension σ is independent of the direction $\mathbf{n}$, $\mathcal{W}(E)$ coincides with the surface area of ∂E (except constant multipliers) and (1.2) is equivalent to the well-known isoperimetric problem. It is one of quite general and fundamental principles in physics that physically realizable phenomena might be characterized by variational principles. Wulff's variational problem is one of the typical examples.

Crystals are, as we have already pointed out, macroscopic objects. It is a principal goal of statistical mechanics to understand such macroscopic phenomena in nature from microscopic level of atoms or molecules. Dobrushin,

Kotecký and Shlosman [86] studied the Wulff's problem from microscopic point of view for the first time. They have employed the ferromagnetic Ising model as a microscopic model and established, at sufficiently low temperatures, the large deviation principle for the sequence of corresponding Gibbs measures on finite domains when the volumes of these domains diverge to infinity. It was shown that the large deviation rate functional is exactly the Wulff functional $\mathcal{W}(E)$ with the surface tension $\sigma(\mathbf{n})$ determined thermodynamically from the underlying Gibbs measures. As a consequence, under the canonical Gibbs measures obtained by conditioning the macroscopic volume occupied by + spins to be constant, a law of large numbers is proved and the Wulff shape is obtained in the limit. The results of Dobrushin et al. were afterward generalized by Ioffe and Schonmann [152], Bodineau [20], Cerf and Pisztora [52] and others; see a review paper [22].

Once an equilibrium situation is understood to some extent, the next target is obviously the analysis of the corresponding dynamics. The situation that two distinct pure phases coexist and are separated by a sharp interface will persist under the time evolution and the interface will relax slowly. The goal is to investigate the motion of interface on a properly chosen coarse space-time scale. The time evolution corresponding to the Ising model is a reversible spin-flip dynamics, the so-called Glauber or Kawasaki dynamics which may be the prime examples. Spin at each site randomly flips and changes its sign under the dynamics without or with conservation law. At sufficiently low temperatures, the interactions between spins on two neighboring sites become strong enough to incline them to have the common signs with high probability and most changes occur near the interface. The shape of interface is however rather complicated; for instance, it has overhangs or bubbles.

A class of effective interface models is introduced by avoiding such complications and directly modeling the interface degree of freedom at microscopic level; see Sect. 1.3. These models are, at one side, compromises between the description of physical phenomena and mathematical requirements but, on the other side, explain the phenomena in satisfactory good way. The aim of these notes is to try to give an overview of results mostly on the $\nabla\varphi$ interface model, which is one of such effective interface models.

As we have observed, in statistical mechanics, there are at least two different scales: macroscopic and microscopic ones. The procedures connecting microscopic models with the macroscopic phenomena are realized by taking the scaling limits. The scaling parameter $N \in \mathbb{N}$ represents the ratio of the macroscopically typical length (e.g., 1 cm) to the microscopic one (e.g., 1 nm) and it is finite, but turns out to be quite large ($N = 10^7$ in this example). The physical phenomena can be mathematically understood only by taking the limit $N \to \infty$. The dynamics involves the scalings also in time. Within a macroscopic unit length of time, the molecules collide with each other with tremendous frequency. Since the microscopic models such as the Ising model and the $\nabla\varphi$ interface model involve randomness, the limit procedure $N \to \infty$

can be formulated in the framework of classical limit theorems in probability theory.

The principal ideas behind these limit theorems are that, by the ergodic or mixing properties of the microscopic systems, the microscopic (physical) quantities are locally in macroscopic scale averaged or homogenized under the scaling limits. The macroscopic observables are obtained under such averaging effects. However, the $\nabla\varphi$ interface model which we shall discuss in the present notes has only an extremely weak mixing property and this sometimes makes the analysis of the model difficult. For instance, the thermodynamic quantity may diverge under the usual scaling. This suggests the necessity of introducing scalings different from the usual one to obtain a nontrivial limit.

1.2 Quick Overview of the Results

In Sect. 2, the $\nabla\varphi$ interface model is precisely introduced. The basic microscopic objects are height variables ϕ of interfaces. Assigning an energy $H(\phi)$ to each height variable, its statistical ensemble in equilibrium is defined by the Gibbs measures. Then, the corresponding time evolution called the Ginzburg-Landau $\nabla\varphi$ interface model is constructed in such a way that it is reversible under the Gibbs measures, in other words, the detailed balance is fulfilled. The scaling limits connecting microscopic and macroscopic levels will be explained.

The $\nabla\varphi$ interface model with quadratic potentials is discussed in Sect. 3 as a warming up before studying general case with convex potentials. In Sect. 4, fundamental tools like Helffer-Sjöstrand (random walk) representation, FKG inequality and Brascamp-Lieb inequality are presented.

A basic role in various limit theorems is played by the so-called surface tension $\sigma(u), u \in \mathbb{R}^d$. The function σ is a macroscopic or thermodynamic function and will be introduced in Sect. 5. The limit theorems under the scalings can be formulated in the terminology of probability theory as follows:

Law of large numbers (LLN): Macroscopic quantity obtained under the scaling limit from randomly fluctuating microscopic objects, i.e., height variables of interfaces in our model, becomes deterministic due to certain averaging effects.

Central limit theorem (CLT): Fluctuations around the deterministic limit are studied.

Large deviation principle (LDP): LDPs for macroscopically scaled height variables are sometimes useful to show the LLNs.

From the physical point of view, these limit theorems are classified into two types: static results on the equilibrium Gibbs measures and dynamic results:

(1) Static results, Sects. 6-9.

LDP, LLN and derivation of variational principles (VP), Sect. 6: LDP was studied for Gaussian case by Ben Arous and Deuschel [12] and for general Gibbsian case by Deuschel, Giacomin and Ioffe [77]. For height variables conditioned to be positive and to have definite total volume, the shape of most probable droplet called the **Wulff shape** is determined as a minimizer of the total surface tension as a consequence of LDP. Adding an effect of weak self potentials to the system, Funaki and Sakagawa [123] derived the VPs of Alt and Caffarelli [5] or Alt, Caffarelli and Friedman [6]. Bolthausen and Ioffe [31] discussed under additional pinning effect at a wall for 2+1 dimensional system and obtained the **Winterbottom shape** in the limit.

Entropic repulsion (wall effect), Sect. 7.1: The entropic repulsion is the problem to study, when a hard wall is settled at the height level 0, how high the interfaces are pushed up by the randomness (i.e., the entropic effect) naturally existing in the Gibbs measures. The problem was posed by Lebowitz and Maes [186] and then investigated by Bolthausen, Deuschel and Zeitouni [29], Deuschel [74], Deuschel and Giacomin [75] for Gaussian case and by Deuschel and Giacomin [76] for general Gibbsian case.

Pinning and wetting transition, Sects. 7.2, 7.3: The pinning is the problem to study, under the effect of weak force attracting interfaces to the height level 0, whether the field is really localized or not. The problem was discussed by Dunlop, Magnen, Rivasseau and Roche [93], Deuschel and Velenik [81], Ioffe and Velenik [153] and Bolthausen and Velenik [32]. The two effects of entropic repulsion and pinning conflict with each other, and a natural question to be addressed is which effect is dominant in the system. In one and two dimensions, a phase transition called wetting transition occurs. This fact was first observed by Fisher [101] in one dimension, followed by Bolthausen, Deuschel and Zeitouni [30] and Caputo and Velenik [49].

CLT, Sect. 8: Naddaf and Spencer [202] investigated CLT for Gibbs measures. The result is nontrivial since the Gibbs measures have long correlations.

Characterization of $\nabla\varphi$-Gibbs measures, Definition 2.2, Sect. 9: The family of all (tempered and shift invariant) $\nabla\varphi$-Gibbs measures is characterized based on the coupling argument for the corresponding dynamics. This result plays a key role in the proof of the hydrodynamic limit.

(2) Dynamic results I, Sects. 10-12.

Hydrodynamic limit (LLN) and derivation of motion by mean curvature with anisotropy, Sect. 10: LLN is shown under the time evolution. This procedure is called the hydrodynamic limit and established by Funaki and Spohn [124]. Motion by mean curvature (MMC) except for some anisotropy is derived in the limit. The diffusion matrix of the limit equation is formally given by Hessian of the surface tension.

Equilibrium fluctuation (CLT), Sect. 11: Dynamic CLT in equilibrium is studied and an infinite dimensional Ornstein-Uhlenbeck process is derived in the limit by Giacomin, Olla and Spohn [135]. The identification of the covariance matrix with Hessian of the surface tension, however, still remains open.

LDP, Sect. 12: Dynamic LDP was discussed by Funaki and Nishikawa [121].

(3) Dynamic results II, Sects. 13-15.

The dynamics under the effects of wall or additional weak self potentials is studied.

Hydrodynamic limit on a wall, Sect. 13: The limit is MMC with reflection and described by an **evolutionary variational inequality**, Funaki [117].

Equilibrium fluctuation (CLT) on a wall, Sects. 14.1, 14.2: A **stochastic PDE with reflection** is obtained under the scaling limit, Funaki and Olla [122].

Dynamic entropic repulsion, Sect. 14.3: The problem of entropic repulsion is investigated under the dynamics, Deuschel and Nishikawa [80] and others.

Dynamics in two media, Sect. 15.1: The dynamics associated with the Hamiltonian added a weak self potential is discussed.

Pinning dynamics on a wall, Sect. 15.2: Dynamics under the effects of both pinning and repulsion is constructed.

(4) Other dynamic models for interfaces, Sect. 16.

The following five topics are discussed in the last supplementary section.

Stochastic lattice gas and free boundary problems
Interacting Brownian particles at zero temperature
Singular limits for stochastic reaction-diffusion equations
Limit shape of random Young diagrams
Growing interfaces

Funaki [116] and Giacomin [130, 132, 133] are survey papers on the $\nabla\varphi$ interface model. See also [125, 210, 224] for problems on interfaces and crystals.

1.3 Derivation of Effective Interface Models from Ising Model

Let us briefly and rather formally explain how one can derive the effective interface models from the ferromagnetic Ising model at sufficiently low temperature. In the Ising model, the energy is associated to each $\pm$ spin configuration $s = \{s(x); x \in \Lambda_\ell\} \in \{+1, -1\}^{\Lambda_\ell}$ on a large box $\Lambda_\ell := [-\ell, \ell]^d \cap \mathbb{Z}^d$ as the sum over all bonds $\langle x, y \rangle$ in Λ_ℓ (i.e., $x, y \in \Lambda_\ell : |x - y| = 1$)

$$H(s) = - \sum_{\langle x,y \rangle \subset \Lambda_\ell} s(x)s(y) .$$

The sum is usually defined under certain boundary conditions. We shall consider, for simplicity, only when $d = 2$. The function $H(s)$ can be rewritten as

$$H(s) = 2|\gamma| \ (+ \text{ constant})$$

in terms of the set of contours $\gamma = \gamma(s)$ on the dual lattice corresponding to s, which separate two regions consisting of sites occupied by $+$ and $-$ spins, respectively, where $|\gamma|$ denotes the number of bonds in γ (the total length of γ) and an additional constant in $H(s)$ is independent of the configurations s. Under the Gibbs measure

$$\mu_\ell(s) = \frac{1}{Z_\ell} e^{-\beta H(s)}, \quad s \in \{+1, -1\}^{\Lambda_\ell} ,$$

with the normalization constant Z_ℓ, if the temperature T ($\beta = 1/kT$, $k > 0$ is the Boltzmann constant) is sufficiently low, the configurations of spins which have the same values on neighboring sites overwhelm the probability, since such configurations have smaller energies. In other words, when there is a single large contour γ, the probability that the configurations in Fig. 1.2 having bubbles arise is very little and almost negligible. We can therefore disregard (with high probability) the configurations with bubbles and assume that the configurations like in Fig. 1.1 can only appear. Such spin configurations s are equivalently represented by the height variables $\phi = \{\phi(x) \in [-\ell, \ell] \cap \mathbb{Z}; x \in [-\ell, \ell]^{d-1} \cap \mathbb{Z}^{d-1}\}$ (in fact, we are considering the case of $d = 2$) which measure the distances of γ from the x-axis, one fixed hyperplane. Then, the energy $H(s)$ has another form

$$H(\phi) = 2 \sum_{\langle x,y \rangle \subset [-\ell, \ell]^{d-1} \cap \mathbb{Z}^{d-1}} |\phi(x) - \phi(y)| \qquad (1.3)$$

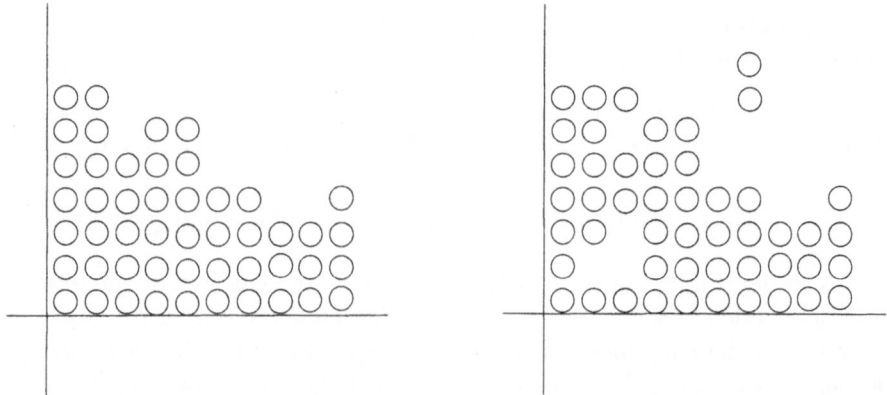

Fig. 1.1. Possible configurations   **Fig. 1.2.** Neglected configurations

up to an additional constant; notice that the number of horizontal bonds in γ is always fixed. The model for random interfaces $\phi : [-\ell, \ell]^{d-1} \cap \mathbb{Z}^{d-1} \to \mathbb{Z}$ with the energy (1.3) is called the **SOS (Solid on Solid) model**. One can further replace the space $\mathbb{Z}$ for values of height variables with continuum $\mathbb{R}$ and $|\phi(x) - \phi(y)|$ with $V(\phi(x) - \phi(y))$, and this leads us to the $\nabla\varphi$ **interface model**. As a generalization of the function $V(\eta) = |\eta|$, it is natural to suppose that the potential function V is convex and symmetric (even) so that the energy is small when the differences of heights $\phi : [-\ell, \ell]^{d-1} \cap \mathbb{Z}^{d-1} \to \mathbb{R}$ on neighboring sites are small, in other words, when the interfaces are more flat.

1.4 Basic Notation

- For $\Lambda \subset \mathbb{Z}^d$ (d dimensional square lattice),

$$\partial^+ \Lambda = \{x \notin \Lambda; |x - y| = 1 \text{ for some } y \in \Lambda\}$$

is the outer boundary of Λ and $\overline{\Lambda} = \Lambda \cup \partial^+ \Lambda$ is the closure of Λ, respectively, where $x \notin \Lambda$ means $x \in \Lambda^c = \mathbb{Z}^d \setminus \Lambda$. The inner boundary of Λ is

$$\partial^- \Lambda = \{x \in \Lambda; |x - y| = 1 \text{ for some } y \notin \Lambda\} .$$

- $\Lambda \Subset \mathbb{Z}^d$ means that Λ is a finite subset of $\mathbb{Z}^d$: $|\Lambda|(= \sharp\Lambda) < \infty$.
- $O \in \mathbb{Z}^d$ stands for the origin and, for $\ell \in \mathbb{N}$, $\Lambda_\ell = [-\ell, \ell]^d \cap \mathbb{Z}^d$ denotes the lattice cube with center O and side length $2\ell + 1$.
- $|x| = \max_{1 \le i \le d} |x_i|$ for $x = (x_i)_{i=1}^d \in \mathbb{Z}^d$ and $|u| = \sqrt{\sum_{i=1}^d u_i^2}$ for $u = (u_i)_{i=1}^d \in \mathbb{R}^d$ (There will be some exceptional usages in Sect. 3). The inner product of $\mathbb{R}^d$ is denoted by $u \cdot x$ or sometimes by (u, x) for $u, x \in \mathbb{R}^d$.
- For a bounded domain D in $\mathbb{R}^d$, we denote $D_N = ND \cap \mathbb{Z}^d$, where $ND = \{N\theta; \theta \in D\} \subset \mathbb{R}^d$ and $N \in \mathbb{N}$ stands for the scaling parameter. The set D_N is a microscopic correspondence, which is discretized, to the macroscopic domain D.
- The set $\mathbb{T}^d = (\mathbb{R}/\mathbb{Z})^d \equiv (0, 1]^d$ denotes a d dimensional unit torus (identifying 0 with 1) and $\mathbb{T}_N^d = (\mathbb{Z}/N\mathbb{Z})^d \equiv \{1, 2, \ldots, N\}^d$ is the corresponding microscopic lattice torus (identifying 0 with N). We also use the notation $\tilde{\mathbb{T}}^d = (-\pi, \pi]^d$.
- For a topological space S, $\mathcal{P}(S)$ stands for the family of all Borel probability measures on S.

Acknowledgment

The results stated in Sects. 9 and 10, one of the cores in these notes, have grown out of the visit of H. Spohn to Japan in the spring of 1995. I am much indebted to him, who actually got me started to work on the problems related to the $\nabla\varphi$ interface model. I was stimulated by discussions with many people, in particular, with J.-D. Deuschel, G. Giacomin, D. Ioffe, S. Olla, G.S. Weiss and N. Yoshida. H. Sakagawa read an early version in part and gave me several suggestions for improvement. Professor J. Picard invited me to deliver a series of lectures at the International Probability School at Saint-Flour, 2003. I deeply thank all of these people.

2 $\nabla\varphi$ Interface Model

The $\nabla\varphi$ interface model has a rather simplified feature, for example, when it is compared with the Ising model, as we have pointed out. It is, however, equipped with a sufficiently wide variety of nontrivial aspects and serves as a useful model to explain physical behavior of interfaces from microscopic point of view. In this section we introduce the model.

2.1 Height Variables

We are concerned with a hypersurface (interface) embedded in $d+1$ dimensional space $\mathbb{R}^{d+1}$, which separates two distinct pure phases. Notice that, in Sect. 1.3, we discussed in d dimensional space; however, here and after d is replaced with $d+1$. To avoid complications, we assume that the interface has no overhangs nor bubbles and accordingly that it is represented as a graph viewed from a certain d dimensional fixed reference hyperplane Γ located in the space $\mathbb{R}^{d+1}$. In other words, the location of the interface is described by the height variables $\phi = \{\phi(x) \in \mathbb{R}; x \in \Gamma\}$, which measure the vertical distances between the interface and Γ. The variables ϕ are microscopic objects, and the space Γ is discretized and taken as $\Gamma = \Lambda(\Subset \mathbb{Z}^d)$, in particular, $\Gamma = D_N$ with a (macroscopic) bounded domain D in $\mathbb{R}^d$ or lattice torus $\mathbb{T}_N^d$ or $\mathbb{Z}^d$. Here N represents the size of the microscopic system, and our main interest will be in analyzing the asymptotic behavior of the system under the scaling limit $N \to \infty$.

2.2 Hamiltonian

An **energy** is associated with each height variable $\phi : \Gamma \to \mathbb{R}$ by assigning penalty according to its tilts (slopes). Namely, we define the Hamiltonian $H(\phi)$ as the sum over all bonds (i.e., pairs of nearest neighbor sites) $\langle x, y \rangle$ in Γ when $\Gamma = \mathbb{T}_N^d$ or $\mathbb{Z}^d$, and in $\overline{\Gamma}$ when $\Gamma = D_N$ or $\Gamma = \Lambda \Subset \mathbb{Z}^d$ in general

$$H(\phi) \equiv H_\Gamma^\psi(\phi) = \sum_{\langle x,y \rangle \subset \Gamma(\text{or } \overline{\Gamma})} V(\phi(x) - \phi(y)) . \tag{2.1}$$

Note that the boundary conditions $\psi = \{\psi(x); x \in \partial^+\Gamma\}$ are required to define the sum (2.1) for $\Gamma = D_N$, i.e., we assume

$$\phi(x) = \psi(x), \quad x \in \partial^+ \Gamma \,,$$

in the sum. When $\Gamma = \mathbb{Z}^d$, (2.1) is a formal infinite sum. The (interaction) **potential** V is smooth, symmetric and strictly convex. More precisely, throughout the present notes we require the following three conditions on the potential $V = V(\eta)$:

(V1) (smoothness) $V \in C^2(\mathbb{R})$,

(V2) (symmetry) $V(-\eta) = V(\eta), \quad \eta \in \mathbb{R}$, $\qquad\qquad\qquad\qquad$ (2.2)

(V3) (strict convexity) $c_- \leq V''(\eta) \leq c_+, \quad \eta \in \mathbb{R}, \quad$ for some $c_-, c_+ > 0$.

The surface ϕ has low energy if the tilts $|\phi(x) - \phi(y)|$ are small. The energy (2.1) of the interface ϕ is constructed in such a manner that it is invariant under a uniform translation $\phi(x) \to \phi(x) + h$ for all $x \in \mathbb{Z}^d$ (or $x \in \overline{\Gamma}$) and $h \in \mathbb{R}$. A typical example of V satisfying the conditions (2.2) is a quadratic potential $V(\eta) = \frac{c}{2}\eta^2, c > 0$.

For every $\Lambda \subset \mathbb{Z}^d$, Λ^* denotes the set of all directed bonds $b = \langle x, y \rangle$ in Λ, which are directed from y to x. We write $x_b = x$, $y_b = y$ for $b = \langle x, y \rangle$. For each $b \in (\mathbb{Z}^d)^*$ and $\phi = \{\phi(x); x \in \mathbb{Z}^d\} \in \mathbb{R}^{\mathbb{Z}^d}$, define

$$\nabla \phi(b) = \phi(x_b) - \phi(y_b) \,.$$

We also define $\nabla_i \phi(x) = \phi(x + e_i) - \phi(x), \; 1 \leq i \leq d$ for $x \in \mathbb{Z}^d$ where $e_i \in \mathbb{Z}^d$ is the i-th unit vector given by $(e_i)_j = \delta_{ij}$. The variables $\nabla \phi(x) = \{\nabla_i \phi(x)\}_{1 \leq i \leq d} \in \mathbb{R}^d$ represent vector field of **height differences** or sometimes called **tilt** (or **gradients**) of ϕ. The Hamiltonian $H(\phi)$ is then rewritten as

$$H(\phi) = \frac{1}{2} \sum_{b \in \Gamma^* (\text{or } \overline{\Gamma}^*)} V(\nabla \phi(b)) \,. \qquad\qquad (2.3)$$

The factor $1/2$ is needed because each undirected bond $b = \langle x, y \rangle$ is counted twice in the sum. Since the energy is determined from the height differences $\nabla \phi$, the model is called the $\nabla \varphi$ interface model.

Remark 2.1. (1) *The sum (2.1) is meaningful only when the potential V is symmetric, while the expression (2.3) makes sense for asymmetric V. However, note that the sum (2.3) is essentially invariant (except for the boundary contributions) if V is replaced with its symmetrization $\frac{1}{2}\{V(\eta) + V(-\eta)\}$.*
(2) *The potential V can be generalized to the bond-dependent case: $\{V_b = V_b(\eta); b \in (\mathbb{Z}^d)^*\}$ so that the corresponding Hamiltonian is defined by (2.3) with V replaced by V_b; see Example 5.3, Problem 10.1 below and [230]. This formulation truly covers the asymmetric potentials.*

Remark 2.2. (1) *In the quantum field theory, H is called massless Hamiltonian and well studied in '80s. Massive Hamiltonian is given by $H_m(\phi) = H(\phi) + \frac{1}{2}m^2 \sum_x \phi(x)^2, m > 0$. The Hamiltonian with weak self potentials or pinning potentials will be introduced in Sect. 6.1 or in Sect. 6.4 (see also*

Sect. 7.2), respectively. (2) In our model, height variables $\phi(x)$ themselves are not discretized. The SOS (solid on solid) model is a model obtained discretizing the height variables simultaneously: $\phi(x) \in \mathbb{Z}_+$ and with $V(\eta) = |\eta|$, cf. Sect. 1.3 and [54], [55], [106].

(3) ($\Delta\varphi$ interface model) In the $\nabla\varphi$ interface model, the energy $H(\phi)$ is roughly the surface area of the microscopic interface ϕ. In fact, this is true for $V(\eta) = \sqrt{1+\eta^2}$. However, if we are concerned for example with the membrane as the object of our study, its surface area is preserved and always constant. Therefore the energy should be determined by taking into account the next order term like $\sum_x (\Delta\phi(x))^2$, which may be regarded as the curvature of ϕ, see [145].

2.3 Equilibrium States (Gibbs Measures)

Once the Hamiltonian H is specified, in the formulation of statistical mechanics, equilibrium states called Gibbs measures can be naturally associated taking the effect of random fluctuations into account.

φ-Gibbs Measures

For a finite region $\Lambda \Subset \mathbb{Z}^d$, the **Gibbs measure** (more exactly, φ-Gibbs measure, finite volume φ-Gibbs measure or local specification) for the field of height variables $\phi \in \mathbb{R}^\Lambda$ over Λ is defined by

$$\mu(d\phi) \equiv \mu_\Lambda^\psi(d\phi) := \frac{1}{Z_\Lambda^\psi} \exp\left\{-H_\Lambda^\psi(\phi)\right\} d\phi_\Lambda , \qquad (2.4)$$

with the boundary conditions $\psi \in \mathbb{R}^{\partial^+\Lambda}$. The term $e^{-H_\Lambda^\psi(\phi)}$ is the Boltzmann factor, while

$$d\phi_\Lambda = \prod_{x \in \Lambda} d\phi(x)$$

is the Lebesgue measure on $\mathbb{R}^\Lambda$ which represents uniform fluctuations of the interface. The constant Z_Λ^ψ is for normalization defined by

$$Z_\Lambda^\psi = \int_{\mathbb{R}^\Lambda} \exp\left\{-H_\Lambda^\psi(\phi)\right\} d\phi_\Lambda . \qquad (2.5)$$

Note that the conditions (2.2) imply $Z_\Lambda^\psi < \infty$ for every $\Lambda \Subset \mathbb{Z}^d$ and therefore $\mu_\Lambda^\psi \in \mathcal{P}(\mathbb{R}^\Lambda)$.

The reason for introducing these measures is based on a physical argument. The uniform measure $d\phi_\Lambda$ arises from the postulate in equilibrium statistical mechanics called **principle of equal a priori probabilities**, while the Boltzmann factor naturally appears from the Gibbs' principle which is sometimes called **equivalence of ensembles**: a subsystem in a very large closed system distributed under the microcanonical ensemble (= equal probabilities on a system with conservation law) is described by the Gibbs measure, [72, 128, 223].

We shall often regard $\mu_\Lambda^\psi \in \mathcal{P}(\mathbb{R}^{\overline{\Lambda}})$ by considering $\phi(x) = \psi(x)$ for $x \in \partial^+\Lambda$ under μ_Λ^ψ. The boundary condition ψ is sometimes taken from $\mathbb{R}^{\Lambda^c}$, and we regard $\mu_\Lambda^\psi \in \mathcal{P}(\mathbb{R}^{\mathbb{Z}^d})$ in such case. When $\Gamma = \mathbb{T}_N^d$, the Gibbs measure is unnormalizable, since $H_\Lambda^\psi(\phi)$ is translation invariant and this makes the normalization $Z_{\mathbb{T}_N^d} = \infty$.

For an infinite region $\Lambda : |\Lambda| = \infty$, the expression (2.4) has no meaning since the Hamiltonian $H_\Lambda(\phi)$ is a formal sum. Nevertheless, one can define the notion of Gibbs measures on $\mathbb{Z}^d$ based on the well-known DLR formulations. For $A \subset \mathbb{Z}^d$, we shall denote $\mathcal{F}_A$ the σ-field of $\mathbb{R}^{\mathbb{Z}^d}$ generated by $\{\phi(x); x \in A\}$.

Definition 2.1. *The probability measure $\mu \in \mathcal{P}(\mathbb{R}^{\mathbb{Z}^d})$ is called a Gibbs measure for φ-field (φ-**Gibbs measure** for short), if its conditional probability on $\mathcal{F}_{\Lambda^c}$ satisfies the DLR equation*

$$\mu(\,\cdot\,|\mathcal{F}_{\Lambda^c})(\psi) = \mu_\Lambda^\psi(\,\cdot\,), \quad \mu\text{-a.e. } \psi,$$

for every $\Lambda \Subset \mathbb{Z}^d$.

It is known that the φ-Gibbs measures exist when the dimension $d \geq 3$, but not for $d = 1, 2$. An infinite volume limit (thermodynamic limit) for μ_Λ^0 as $\Lambda \nearrow \mathbb{Z}^d$ exists only when $d \geq 3$ (cf. Sect. 4.5).

$\nabla\varphi$-Gibbs Measures

The height variables $\phi = \{\phi(x); x \in \mathbb{Z}^d\}$ on $\mathbb{Z}^d$ automatically determines a field of height differences $\nabla\phi = \{\nabla\phi(b); b \in (\mathbb{Z}^d)^*\}$. One can therefore consider the distribution μ^∇ of $\nabla\varphi$-field under the φ-Gibbs measure μ. We shall call μ^∇ the $\nabla\varphi$-**Gibbs measure**. In fact, it is possible to define the $\nabla\varphi$-Gibbs measures directly by means of the DLR equations and, in this sense, $\nabla\varphi$-Gibbs measures exist for all dimensions $d \geq 1$ (cf. Sect. 4.4).

In order to describe the DLR equation for $\nabla\varphi$-Gibbs measures, we first clarify the structure of the state space for the $\nabla\varphi$-field. It is obvious that the height variable $\phi \in \mathbb{R}^{\mathbb{Z}^d}$ determines $\nabla\phi \in \mathbb{R}^{(\mathbb{Z}^d)^*}$; however, all $\eta = \{\eta(b)\} \in \mathbb{R}^{(\mathbb{Z}^d)^*}$ can not be the $\nabla\varphi$-field, i.e., it may not be possible to find ϕ such that $\eta = \nabla\phi$ in general. Indeed, $\nabla\phi$ always satisfies the loop condition: every sum of $\nabla\phi$ along a closed loop must vanish. To state more precisely, we introduce some notion.

A sequence of bonds $\mathfrak{C} = \{b^{(1)}, b^{(2)}, \ldots, b^{(n)}\}$ is called a chain connecting y and x $(y, x \in \mathbb{Z}^d)$ if $y_{b^{(1)}} = y, x_{b^{(i)}} = y_{b^{(i+1)}}$ for $1 \leq i \leq n-1$ and $x_{b^{(n)}} = x$. The chain $\mathfrak{C}$ is called a closed loop if $x_{b^{(n)}} = y_{b^{(1)}}$. A plaquette is a closed loop $\mathfrak{P} = \{b^{(1)}, b^{(2)}, b^{(3)}, b^{(4)}\}$ such that $\{x_{b^{(i)}}, i = 1, .., 4\}$ consists of four different points. The field $\eta = \{\eta(b)\} \in \mathbb{R}^{(\mathbb{Z}^d)^*}$ is said to satisfy the **plaquette condition** if

(P1) $\eta(b) = -\eta(-b)$ for all $b \in (\mathbb{Z}^d)^*$,

(P2) $\displaystyle\sum_{b \in \mathfrak{P}} \eta(b) = 0$ for all plaquettes $\mathfrak{P}$ in $\mathbb{Z}^d$,

where $-b$ denotes the reversed bond of b. Note that, if $\phi = \{\phi(x)\} \in \mathbb{R}^{\mathbb{Z}^d}$, then $\nabla\phi = \{\nabla\phi(b)\} \in \mathbb{R}^{(\mathbb{Z}^d)^*}$ automatically satisfies the plaquette condition. The plaquette condition is equivalent to the **loop condition**:

(L) $\displaystyle\sum_{b \in \mathfrak{C}} \eta(b) = 0$ for all closed loops $\mathfrak{C}$ in $\mathbb{Z}^d$.

Notice that the condition (P1) follows from (L) by taking the closed loop $\mathfrak{C} = \{b, -b\}$. We set

$$\mathcal{X} = \{\eta \in \mathbb{R}^{(\mathbb{Z}^d)^*}; \eta \text{ satisfies the loop condition}\},$$

then $\mathcal{X}$ is the state space for the $\nabla\varphi$-field endowed with the topology induced from the space $\mathbb{R}^{(\mathbb{Z}^d)^*}$ having product topology. In fact, the height differences $\eta^\phi \in \mathcal{X}$ are associated with the heights $\phi \in \mathbb{R}^{\mathbb{Z}^d}$ by

$$\eta^\phi(b) := \nabla\phi(b), \quad b \in (\mathbb{Z}^d)^*, \tag{2.6}$$

and, conversely, the heights $\phi^{\eta,\phi(O)} \in \mathbb{R}^{\mathbb{Z}^d}$ can be constructed from height differences η and the height variable $\phi(O)$ at $x = O$ as

$$\phi^{\eta,\phi(O)}(x) := \sum_{b \in \mathfrak{C}_{O,x}} \eta(b) + \phi(O), \tag{2.7}$$

where $\mathfrak{C}_{O,x}$ is an arbitrary chain connecting O and x. Note that $\phi^{\eta,\phi(O)}$ is well-defined if $\eta = \{\eta(b)\} \in \mathcal{X}$.

We next define the finite volume $\nabla\varphi$-Gibbs measures. For every $\xi \in \mathcal{X}$ and $\Lambda \Subset \mathbb{Z}^d$ the space of all possible configurations of height differences on $\overline{\Lambda^*} := \{b = \langle x, y\rangle \in (\mathbb{Z}^d)^*; x \text{ or } y \in \Lambda\}$ for given boundary condition ξ is defined as

$$\mathcal{X}_{\overline{\Lambda^*},\xi} = \{\eta = (\eta(b))_{b \in \overline{\Lambda^*}}; \eta \vee \xi \in \mathcal{X}\},$$

where $\eta \vee \xi \in \mathcal{X}$ is determined by $(\eta \vee \xi)(b) = \eta(b)$ for $b \in \overline{\Lambda^*}$ and $= \xi(b)$ for $b \notin \overline{\Lambda^*}$. The finite volume $\nabla\varphi$-Gibbs measure in Λ (or, more precisely, in $\overline{\Lambda^*}$) with boundary condition ξ is defined by

$$\mu_{\Lambda,\xi}^\nabla(d\eta) = \frac{1}{Z_{\Lambda,\xi}} \exp\left\{-\frac{1}{2}\sum_{b \in \overline{\Lambda^*}} V(\eta(b))\right\} d\eta_{\Lambda,\xi} \in \mathcal{P}(\mathcal{X}_{\overline{\Lambda^*},\xi}),$$

where $d\eta_{\Lambda,\xi}$ denotes a uniform measure on the affine space $\mathcal{X}_{\overline{\Lambda^*},\xi}$ and $Z_{\Lambda,\xi}$ is the normalization constant. We shall sometimes regard $\mu_{\Lambda,\xi}^\nabla \in \mathcal{P}(\mathcal{X})$ by considering $\eta(b) = \xi(b)$ for $b \notin \overline{\Lambda^*}$ under $\mu_{\Lambda,\xi}^\nabla$ as before. Note that the dimension of the space $\mathcal{X}_{\overline{\Lambda^*},\xi}$ is $|\Lambda|$ at least if $\mathbb{Z}^d \setminus \Lambda$ is connected, since one can associate η with $\phi = \phi_\Lambda$ by

$$\phi(x) = \sum_{b \in \mathfrak{C}_{x_0,x}} (\eta \vee \xi)(b), \quad x \in \Lambda, \tag{2.8}$$

where $x_0 \notin \Lambda$ is fixed and $\mathfrak{C}_{x_0,x}$ is a chain connecting x_0 and x.

The finite volume φ-Gibbs measures and the finite volume $\nabla\varphi$-Gibbs measures are associated with each other as we have pointed out above. Namely, given $\xi \in \mathcal{X}$ and $h \in \mathbb{R}$, define $\psi \in \mathbb{R}^{\mathbb{Z}^d}$ as $\psi = \phi^{\xi,h}$ by (2.7). Then, if ϕ is μ_Λ^ψ-distributed with the boundary condition ψ constructed in this way, $\nabla\phi$ is $\mu_{\Lambda,\xi}^\nabla$-distributed. The distribution of $\nabla\phi$ is certainly independent of the choice of h.

Now, similarly to the definition of the φ-Gibbs measures on $\mathbb{Z}^d$, one can introduce the $\nabla\varphi$-Gibbs measures on $(\mathbb{Z}^d)^*$.

Definition 2.2. *The probability measure $\mu^\nabla \in \mathcal{P}(\mathcal{X})$ is called a Gibbs measure for the height differences ($\nabla\varphi$-Gibbs measure for short), if it satisfies the DLR equation*

$$\mu^\nabla(\,\cdot\,|\mathcal{F}_{(\mathbb{Z}^d)^* \setminus \overline{\Lambda^*}})(\xi) = \mu_{\Lambda,\xi}^\nabla(\,\cdot\,), \quad \mu^\nabla\text{-a.e. } \xi,$$

for every $\Lambda \Subset \mathbb{Z}^d$, where $\mathcal{F}_{(\mathbb{Z}^d)^* \setminus \overline{\Lambda^*}}$ stands for the σ-field of $\mathcal{X}$ generated by $\{\eta(b); b \in (\mathbb{Z}^d)^* \setminus \overline{\Lambda^*}\}$.

Markov Property

In the Hamiltonian $H(\phi)$, the interactions among the height variables are only counted through the neighboring sites. This structure is reflected as the **Markov property** of the field of height variables $\phi = \{\phi(x)\}$ under the (finite or infinite volume) φ-Gibbs measures μ_Λ^ψ and μ:

Proposition 2.1. (1) *Let $\Lambda \Subset \mathbb{Z}^d$ and the boundary condition $\psi \in \mathbb{R}^{\partial^+\Lambda}$ be given. Suppose that Λ is decomposed into three regions A_1, A_2, B and B separates A_1 and A_2; namely, $\Lambda = A_1 \cup A_2 \cup B$, $A_1 \cap A_2 = A_1 \cap B = A_2 \cap B = \emptyset$ and $|x_1 - x_2| > 1$ holds for every $x_1 \in A_1$ and $x_2 \in A_2$. Then, under the conditional probability $\mu_\Lambda^\psi(\,\cdot\,|\mathcal{F}_B)$, the random variables ϕ_{A_1} and ϕ_{A_2} are mutually independent, where we denote $\phi_{A_1} = \{\phi(x); x \in A_1\}$ etc.*
(2) *Let $\mu \in \mathcal{P}(\mathbb{R}^{\mathbb{Z}^d})$ be a φ-Gibbs measure. Then, for every $A \Subset \mathbb{Z}^d$, the random variables ϕ_A and $\phi_{\overline{A}^c}$ are mutually independent under the conditional probability $\mu(\,\cdot\,|\mathcal{F}_{\partial^+ A})$.*

In particular, in one dimension, $\phi = \{\phi(x)\}$ is a **pinned random walk** under μ_Λ^ψ regarding x as time variables. Let $\{\eta(y); y = 1, 2, \ldots\}$ be an $\mathbb{R}$-valued *i.i.d.* defined on a certain probability space (Ω, P) having the distribution $p(a)da$, where

$$p(a) = \frac{1}{z}e^{-V(a)}, \quad a \in \mathbb{R}$$

and $z = \int_{\mathbb{R}} e^{-V(a)} \, da$ is the normalization constant. Then, we have the following.

Proposition 2.2. *Let* $\Lambda = \{1, 2, \ldots, N-1\} \subset \mathbb{Z}^1$ *and assume that the boundary conditions are given by* $\psi(0) = h_0, \psi(N) = h_1$. *Define the height variables* $\phi = \{\phi(x); x \in \overline{\Lambda}\}, \overline{\Lambda} = \{0, 1, 2, \ldots, N\}$ *by*

$$\phi(x) = h_0 + \sum_{y=1}^{x} \eta(y), \quad x \in \overline{\Lambda},$$

and consider them under the conditional probability $P(\cdot | \phi(N) = h_1)$. *Then,* $\phi_\Lambda = \{\phi(x); x \in \Lambda\}$ *is* μ_Λ^ψ-*distributed.*

Shift Invariance and Ergodicity

Here, we recall the notion of shift invariance and ergodicity under the shifts for φ-fields and $\nabla\varphi$-fields, respectively, see, e.g., [128]. For $x \in \mathbb{Z}^d$, we define the shift operators $\tau_x : \mathbb{R}^{\mathbb{Z}^d} \to \mathbb{R}^{\mathbb{Z}^d}$ for heights by $(\tau_x\phi)(y) = \phi(y-x)$ for $y \in \mathbb{Z}^d$ and $\phi \in \mathbb{R}^{\mathbb{Z}^d}$. The shifts for height differences are also denoted by τ_x. Namely, $\tau_x : \mathcal{X} \to \mathcal{X}$ $\left(\text{or } \tau_x : \mathbb{R}^{(\mathbb{Z}^d)^*} \to \mathbb{R}^{(\mathbb{Z}^d)^*}\right)$ are defined by $(\tau_x\eta)(b) = \eta(b-x)$ for $b \in (\mathbb{Z}^d)^*$ and $\eta \in \mathcal{X}$ $\left(\text{or } \eta \in \mathbb{R}^{(\mathbb{Z}^d)^*}\right)$, where $b - x = \langle x_b - x, y_b - x \rangle \in (\mathbb{Z}^d)^*$.

Definition 2.3. *A probability measure* $\mu \in \mathcal{P}(\mathbb{R}^{\mathbb{Z}^d})$ *is called shift invariant if* $\mu \circ \tau_x^{-1} = \mu$ *for every* $x \in \mathbb{Z}^d$. *A shift invariant* $\mu \in \mathcal{P}(\mathbb{R}^{\mathbb{Z}^d})$ *is called ergodic (under the shifts) if* $\{\tau_x\}$-*invariant functions* $F = F(\phi)$ *on* $\mathbb{R}^{\mathbb{Z}^d}$ *(i.e., functions satisfying* $F(\tau_x\phi) = F(\phi)$ μ-*a.e. for every* $x \in \mathbb{Z}^d$*) are constant (*μ-*a.e.). Similarly, the shift invariance and ergodicity for a probability measure* $\mu \in \mathcal{P}(\mathcal{X})$ $\left(\text{or } \mu \in \mathcal{P}(\mathbb{R}^{(\mathbb{Z}^d)^*})\right)$ *are defined.*

2.4 Dynamics

Corresponding to the Hamiltonian $H(\phi)$, one can naturally introduce a random time evolution of microscopic height variables ϕ of the interface. Indeed, we consider the stochastic differential equations (SDEs) for $\phi_t = \{\phi_t(x); x \in \Gamma\} \in \mathbb{R}^\Gamma, t > 0$

$$d\phi_t(x) = -\frac{\partial H}{\partial \phi(x)}(\phi_t)dt + \sqrt{2}dw_t(x), \quad x \in \Gamma, \tag{2.9}$$

where $w_t = \{w_t(x); x \in \Gamma\}$ is a family of independent one dimensional standard Brownian motions. The derivative of $H(\phi)$ in the variable $\phi(x)$ is given by

$$\frac{\partial H}{\partial \phi(x)}(\phi) = \sum_{y \in \Gamma \,(\text{or } \overline{\Gamma}) : |x-y|=1} V'(\phi(x) - \phi(y)) , \qquad (2.10)$$

for $x \in \Gamma$. When $\Gamma \Subset \mathbb{Z}^d$, the SDEs (2.9) have the form

$$d\phi_t(x) = - \sum_{y \in \overline{\Gamma} : |x-y|=1} V'(\phi_t(x) - \phi_t(y))dt + \sqrt{2}dw_t(x), \quad x \in \Gamma , \qquad (2.11)$$

subject to the boundary conditions

$$\phi_t(y) = \psi(y), \quad y \in \partial^+ \Gamma . \qquad (2.12)$$

When $\Gamma = \mathbb{Z}^d$, although the Hamiltonian H is a formal sum, its derivative (2.10) has an affirmative meaning and we can write down the SDEs for $\phi_t = \{\phi_t(x); x \in \mathbb{Z}^d\} \in \mathbb{R}^{\mathbb{Z}^d}, t > 0$

$$d\phi_t(x) = - \sum_{y \in \mathbb{Z}^d : |x-y|=1} V'(\phi_t(x) - \phi_t(y))dt + \sqrt{2}dw_t(x), \quad x \in \mathbb{Z}^d . \qquad (2.13)$$

The SDEs (2.11) with (2.12) or the SDEs (2.13) have unique solutions, since the coefficient V' in the drift term is Lipschitz continuous by our assumptions (2.2). For (2.13), since it is an infinite system, one need to introduce a proper function space for solutions, cf. Lemmas 9.1 and 9.2. The evolution of ϕ_t is designed in such a manner that it is stationary and, moreover, reversible under the Gibbs measures μ_Λ^ψ or μ, cf. Proposition 9.4 for the associated $\nabla\varphi$-dynamics. In physical terminology, the equation fulfills the detailed balance condition. Such evolution or the SDEs are called **Ginzburg-Landau dynamics**, distorted Brownian motion or the **Langevin equation** associated with $H(\phi)$.

The drift term in the SDEs (2.9) determines the gradient flow along which the energy $H(\phi)$ decreases. In fact, since the function V is symmetric and convex, $\phi_t(x) > \phi_t(y)$ implies that $-V'(\phi_t(x) - \phi_t(y)) < 0$ so that the drift term of (2.11) or (2.13) is negative and therefore $\phi_t(x)$ decreases. Conversely if $\phi_t(x) < \phi_t(y)$, the drift is positive and $\phi_t(x)$ increases. Therefore, in both cases, the drift has an effect to make the interface ϕ flat. The term $\sqrt{2}w_t(x)$ gives a random fluctuation which competes against the drift.

The **Dirichlet form** corresponding to the SDEs (2.13) is

$$\mathcal{E}(F, G) \equiv -E^\mu[FLG] = \sum_{x \in \mathbb{Z}^d} E^\mu[\partial F(x, \phi)\partial G(x, \phi)] , \qquad (2.14)$$

for $F = F(\phi), G = G(\phi)$, where $E^\mu[\cdot]$ denotes the expectation under the Gibbs measure μ, L is the generator of the process ϕ_t and $\partial F(x, \phi) := \partial F/\partial \phi(x)$, cf. Sects. 4.1 and 10.3. Indeed, at least when $\Gamma \Subset \mathbb{Z}^d$, the generator L of the process $\phi_t \in \mathbb{R}^\Gamma$ determined by the SDEs (2.9) is the differential operator of second order

$$L = \sum_{x \in \Gamma} \left(\frac{\partial}{\partial \phi(x)} \right)^2 - \sum_{x \in \Gamma} \frac{\partial H}{\partial \phi(x)} \frac{\partial}{\partial \phi(x)} \tag{2.15}$$

and, by integration by parts formula, we have

$$\int_{\mathbb{R}^\Gamma} FLG \cdot e^{-H} \, d\phi_\Gamma = \int_{\mathbb{R}^\Gamma} F \sum_{x \in \Gamma} \frac{\partial}{\partial \phi(x)} \left\{ \frac{\partial G}{\partial \phi(x)} e^{-H} \right\} d\phi_\Gamma$$

$$= - \sum_{x \in \Gamma} \int_{\mathbb{R}^\Gamma} \frac{\partial F}{\partial \phi(x)} \frac{\partial G}{\partial \phi(x)} \cdot e^{-H} \, d\phi_\Gamma \,,$$

for every $F = F(\phi), G = G(\phi) \in C_b^2(\mathbb{R}^\Gamma)$. The Hamiltonians H may be more general than (2.1), for instance, those with self potentials, see (6.3)

Remark 2.3. (1) *The dynamics corresponding to the massive Hamiltonian H_m (recall Remark 2.2) can be introduced similarly. It forces the heights $\phi = \{\phi(x)\}$ to stay bounded.*
(2) *Interface dynamics of SOS type was studied by several authors, e.g., Dunlop [90] considered the dynamics for the corresponding gradient fields in one dimension; see also Remark 13.1 and Sect. 16.5.*

2.5 Scaling Limits

Our main interest is in the analysis of the **scaling limit**, which passes from microscopic to macroscopic levels. For the microscopic height variables $\phi = \{\phi(x); x \in \Gamma\}$ with $\Gamma = D_N, \mathbb{T}_N^d$ or $\mathbb{Z}^d$, the macroscopic height variables $h^N = \{h^N(\theta)\}$ are associated by

$$h^N(\theta) = \frac{1}{N} \phi([N\theta]), \quad \theta \in D, \mathbb{T}^d \text{ or } \mathbb{R}^d \,, \tag{2.16}$$

where $[N\theta]$ stands for the integer part of $N\theta (\in \mathbb{R}^d)$ taken componentwise. Note that both x- and ϕ-axes are rescaled by a factor $1/N$. This is because the φ-field represents a hypersurface embedded in $d + 1$ dimensional space. The functions h^N are step functions. Sometimes interpolations by polilinear functions (or polygonal approximations) are also considered, see (6.9) and (6.21) below.

For the time evolution $\phi_t = \{\phi_t(x); x \in \Gamma\}, t > 0$ of the interface, we shall mostly work under the **space-time diffusive scaling**

$$h^N(t, \theta) = \frac{1}{N} \phi_{N^2 t}([N\theta]) \,. \tag{2.17}$$

2.6 Quadratic Potentials

Here we take a quadratic function $V(\eta) = \frac{1}{2}\eta^2$ as a typical example of the potential satisfying our basic conditions (2.2). To rewrite the Hamiltonian

$H(\phi)$ for such V, let us introduce the discrete Laplacian $\Delta \equiv \Delta_{\Lambda,\psi}$ for $\Lambda \Subset \mathbb{Z}^d$ with boundary conditions $\psi \in \mathbb{R}^{\partial^+ \Lambda}$

$$\Delta\phi(x) = \sum_{y \in \overline{\Lambda}:|x-y|=1} ((\phi \vee \psi)(y) - \phi(x)), \quad x \in \Lambda, \qquad (2.18)$$

where $\phi \vee \psi \in \mathbb{R}^{\overline{\Lambda}}$ stands for the height variables which coincide with ϕ on Λ and with ψ on $\partial^+ \Lambda$, respectively; i.e., $\phi \vee \psi(x) = \phi(x)$ for $x \in \Lambda$ and $= \psi(x)$ for $x \in \partial^+ \Lambda$. The summation by parts formula proves that

$$H_\Lambda^0(\phi) = -\frac{1}{2}(\phi, \Delta_{\Lambda,0}\phi)_\Lambda \qquad (2.19)$$

where $(\phi_1, \phi_2)_\Lambda = \sum_{x \in \Lambda} \phi_1(x)\phi_2(x)$ denotes an inner product of ϕ_1 and $\phi_2 \in \mathbb{R}^\Lambda$. The boundary condition is taken $\psi = 0$ for simplicity. In particular, the finite volume Gibbs measure μ_Λ^0 can be expressed as

$$\mu_\Lambda^0(d\phi_\Lambda) = \frac{1}{Z_\Lambda^0} e^{\frac{1}{2}(\phi, \Delta_{\Lambda,0}\phi)_\Lambda} d\phi_\Lambda,$$

and accordingly, ϕ_Λ forms a Gaussian field under the distribution μ_Λ^0 with mean 0 and covariance $(-\Delta_{\Lambda,0})^{-1}$, the inverse operator of $-\Delta_{\Lambda,0}$, see Sect. 3.1 for more details.

For $V(\eta) = \frac{1}{2}\eta^2$, the corresponding dynamics (2.9) is a simple discrete stochastic heat equation

$$d\phi_t(x) = \Delta\phi_t(x)dt + \sqrt{2}dw_t(x), \quad x \in \Gamma. \qquad (2.20)$$

3 Gaussian Equilibrium Systems

As a warming up before studying general systems, let us consider the $\nabla\varphi$ interface model in the case where the potential is quadratic: $V(\eta) = \frac{1}{2}\eta^2$. The corresponding system formed by the height variables ϕ is then Gaussian and sometimes called a free lattice field or a harmonic oscillator in physical literatures. For a Gaussian system, one can explicitly compute the mean, covariance (two-point correlation function) and characteristic functions. In particular, as we shall see, the covariance of our field ϕ can be represented by means of the simple random walks on the lattice, Proposition 3.2. This will be extended to general potentials V and called the Helffer-Sjöstrand representation, see Sect. 4.1 below.

We begin with systems on finite and connected regions $\Lambda(\Subset \mathbb{Z}^d)$ in Sect. 3.1 and then, by taking the thermodynamic limit (i.e., $\Lambda \nearrow \mathbb{Z}^d$), infinite systems on $\mathbb{Z}^d$ will be constructed in Sect. 3.2. We shall also discuss massive system and see significant differences in massive and massless systems, for instance, in the speed of decay of correlation functions or the dependence of the system on the boundary conditions, see Sect. 3.3. Sect. 3.4 deals with the macroscopic scaling limits for φ and $\nabla\varphi$-fields.

3.1 Gaussian Systems in a Finite Region

We assume that $\Lambda \in \mathbb{Z}^d$ is connected. When $V(\eta) = \frac{1}{2}\eta^2$ and the boundary conditions $\psi \in \mathbb{R}^{\partial^+\Lambda}$ (or $\psi \in \mathbb{R}^{\mathbb{Z}^d}$ or $\psi \in \mathbb{R}^{\Lambda^c}$) are given, the corresponding Hamiltonian $H(\phi) \equiv H_\Lambda^\psi(\phi)$ defined by (2.1) is a quadratic form of ϕ so that the finite volume φ-Gibbs measure $\mu_\Lambda^\psi \in \mathcal{P}(\mathbb{R}^{\overline{\Lambda}})$ (or $\in \mathcal{P}(\mathbb{R}^{\mathbb{Z}^d})$) determined by (2.4) is Gaussian.

Harmonic Functions and Green Functions

The mean and covariance of the height variables $\phi = \{\phi(x); x \in \Lambda\}$ under μ_Λ^ψ are computable by solving the Dirichlet boundary value problem on Λ for the discrete Laplacian Δ. Indeed, we consider the difference equation on Λ with the boundary condition ψ

$$
\begin{cases}
\Delta \overline{\phi}(x) := \displaystyle\sum_{y \in \mathbb{Z}^d : |x-y|=1} (\overline{\phi}(y) - \overline{\phi}(x)) = 0, & x \in \Lambda, \\
\overline{\phi}(x) = \psi(x), & x \in \partial^+\Lambda,
\end{cases}
\tag{3.1}
$$

which is equivalent to

$$
\Delta_{\Lambda,\psi} \overline{\phi}(x) = 0, \qquad x \in \Lambda,
$$

where $\Delta_{\Lambda,\psi}$ is the discrete Laplacian determined by (2.18). The solution $\overline{\phi} \equiv \overline{\phi}_{\Lambda,\psi} = \{\overline{\phi}(x); x \in \Lambda\}$ of (3.1) is unique and called a (discrete) **harmonic function** on Λ.

Let $G_\Lambda(x,y)$, $x \in \overline{\Lambda}, y \in \Lambda$ be the **Green function** (potential kernel) for the discrete Laplacian $\Delta_{\Lambda,0}$ with boundary condition 0, i.e., the solution of equations

$$
\begin{cases}
-\Delta G_\Lambda(x,y) = \delta(x,y), & x \in \Lambda, \\
G_\Lambda(x,y) = 0, & x \in \partial^+\Lambda,
\end{cases}
\tag{3.2}
$$

where $\delta(x,y)$ is the Kronecker's δ, and Δ acts on the variable x and y is thought of as a parameter. In fact, $\{G_\Lambda(x,y); x, y \in \Lambda\}$ is the inverse matrix of $\{-\Delta_\Lambda(x,y); x, y \in \Lambda\}$ so that we shall denote

$$
G_\Lambda(x,y) = (-\Delta_\Lambda)^{-1}(x,y),
$$

note that $\Delta_\Lambda(x,y)$ is the kernel of $\Delta_\Lambda \equiv \Delta_{\Lambda,0}$: $\Delta_\Lambda\phi(x) = \sum_{y \in \Lambda} \Delta_\Lambda(x,y)\phi(y)$.

Mean, Covariance and Characteristic Functions

The next proposition is an extension of the fact stated in Sect. 2.6 when the boundary conditions are $\psi \equiv 0$.

Proposition 3.1. (1) *Under μ_Λ^ψ, $\phi = \{\phi(x); x \in \Lambda\}$ is Gaussian with mean $\overline{\phi}_{\Lambda,\psi} = \{\overline{\phi}_{\Lambda,\psi}(x); x \in \Lambda\}$ and covariance $G_\Lambda(x,y)$, i.e., $\mu_\Lambda^\psi = N(\overline{\phi}_{\Lambda,\psi}, G_\Lambda)$. In particular, for $x, y \in \Lambda$,*

$$E^{\mu_\Lambda^\psi}[\phi(x)] = \overline{\phi}_{\Lambda,\psi}(x), \qquad (3.3)$$

$$E^{\mu_\Lambda^\psi}[\phi(x); \phi(y)] = G_\Lambda(x,y), \qquad (3.4)$$

where

$$E^{\mu}[\phi(x); \phi(y)] := E^\mu\left[\{\phi(x) - E^\mu[\phi(x)]\}\{\phi(y) - E^\mu[\phi(y)]\}\right]$$

stands for the covariance of $\phi(x)$ and $\phi(y)$ under μ.
(2) *The characteristic function of μ_Λ^ψ is given by*

$$E^{\mu_\Lambda^\psi}\left[e^{\sqrt{-1}(\xi,\phi)_\Lambda}\right] = \exp\left\{\sqrt{-1}(\xi, \overline{\phi}_{\Lambda,\psi})_\Lambda - \frac{1}{2}(\xi, (-\Delta_\Lambda)^{-1}\xi)_\Lambda\right\}$$

for $\xi \in \mathbb{R}^\Lambda$.
(3) *If ϕ is μ_Λ^0-distributed, then $\phi + \overline{\phi}_{\Lambda,\psi}$ is μ_Λ^ψ-distributed.*

Proof. A careful rearrangement of the sum in the Hamiltonian $H_\Lambda^\psi(\phi)$ applying the summation by parts formula leads us to

$$H_\Lambda^\psi(\phi) = -\frac{1}{2}\left((\phi - \overline{\phi}_{\Lambda,\psi}), \Delta_\Lambda(\phi - \overline{\phi}_{\Lambda,\psi})\right)_\Lambda$$
$$+ \frac{1}{2}\sum_{\substack{x \in \Lambda, y \notin \Lambda \\ |x-y|=1}} \overline{\phi}_{\Lambda,\psi}(y)\nabla\overline{\phi}_{\Lambda,\psi}(\langle y, x\rangle),$$

for every $\phi \in \mathbb{R}^\Lambda$. This is an extension of (2.19) for 0-boundary conditions and a discrete analogue of Green-Stokes' formula. Note that the second term in the right hand side depends only on the boundary conditions ψ and not on ϕ. Therefore, we have that

$$\mu_\Lambda^\psi(d\phi_\Lambda) = \frac{1}{\tilde{Z}_\Lambda^\psi}\exp\left\{\frac{1}{2}((\phi - \overline{\phi}_{\Lambda,\psi}), \Delta_\Lambda(\phi - \overline{\phi}_{\Lambda,\psi}))_\Lambda\right\}d\phi_\Lambda$$

with a proper normalization constant $\tilde{Z}_\Lambda^\psi$. This immediately shows the assertions (1) and (2). The third assertion (3) follows from (1) or (2).

It might be useful to give another proof for (1). Actually, to show (3.3), set its left hand side as $h(x)$. Then, $h(x)$ satisfies the equation (3.1). In fact, the boundary condition is obvious and, for $x \in \Lambda$,

$$\Delta h(x) = E^{\mu_\Lambda^\psi}[\Delta\phi(x)] = -E^{\mu_\Lambda^\psi}\left[\frac{\partial H_\Lambda^\psi}{\partial\phi(x)}\right] = 0$$

by the integration by parts under μ_A^ψ. The uniqueness of solutions of (3.1) proves (3.3). The proof of (3.4) is similar; one may check its left hand side solves (3.2) in place of $G_A(x, y)$. This can be shown again by the integration by parts.

It is standard to calculate the mean, covariance and other higher moments from the characteristic function. Indeed, for instance, (3.4) has the third proof: We may assume $\psi \equiv 0$ by translating the field ϕ by $\bar\phi_{A,\psi}$ and in this case

$$
E^{\mu_A^0}[(\xi, \phi)_A^2] = -\frac{d^2}{d\alpha^2} E^{\mu_A^0}\left[e^{\sqrt{-1}\alpha(\xi,\phi)_A}\right]\Bigg|_{\alpha=0}
$$

$$
= -\frac{d^2}{d\alpha^2} e^{-\frac{\alpha^2}{2}(\xi, (-\Delta_A)^{-1}\xi)_A}\Bigg|_{\alpha=0} = (\xi, (-\Delta_A)^{-1}\xi)_A .
$$

Then, the identity (3.4) follows by taking $\xi = \delta_x, \delta_y$ or $\delta_x + \delta_y$ in this formula and computing their differences, where $\delta_x(\cdot) = \delta(x, \cdot)$.   □

In particular, for $\mu_N \equiv \mu_{D_N}^0$ with $\Lambda = D_N$ taking $D = (-1, 1)^d$ and with 0-boundary conditions, we have $E^{\mu_N}[\phi(O)] = 0$ and the variance behaves as $N \to \infty$

$$
E^{\mu_N}[\phi(O)^2] = (-\Delta_{D_N})^{-1}(O, O) \approx \begin{cases} 1, & d \geq 3 , \\ \log N, & d = 2 , \\ N, & d = 1 , \end{cases} \tag{3.5}
$$

where $\approx$ means that the ratio of the both sides stays uniformly positive and bounded. The number of the sites neighboring to each site is $2d$ and therefore one can expect that, as the lattice dimension d increases, the fluctuations of the interfaces become smaller, in other words, they gain more stiffness. The behavior (3.5) of the variance agrees with this observation. When $d \geq 3$, the second moment stays bounded as $N \to \infty$ and accordingly φ-Gibbs measure is normalizable in the sense that it admits the thermodynamic limit, see Sect. 3.2. For general convex potentials V, Brascamp-Lieb inequality gives at least the corresponding upper bound in (3.5), see Sect. 4.2. When $d = 1$, $\phi(x)$ is essentially the pinned Brownian motion with discrete time parameter $x \in (-N, N) \cap \mathbb{Z}$ and therefore (3.5) is standard.

Random Walk Representation

Let $X = \{X_t\}_{t \geq 0}$ be the simple random walk on $\mathbb{Z}^d$ with continuous time parameter t, i.e., the generator of X is the discrete Laplacian Δ and the jump of X to the adjacent sites is accomplished by choosing one of them with equal probabilities after an exponentially distributed waiting time with mean $\frac{1}{2d}$. Let τ_Λ be the exit time of X from the region Λ:

$$
\tau_\Lambda := \inf\{t \geq 0; X_t \in \Lambda^c\} .
$$

The transition probability of the simple random walk on Λ with absorbing boundary $\partial^+\Lambda$ is denoted by $p_\Lambda(t,x,y) \equiv E_x[1_{\{y\}}(X_t), t < \tau_\Lambda], t \geq 0, x,y \in \Lambda$, where $E_x[\cdot]$ stands for the expectation for X starting at x: $X_0 = x$. Then, the following representations are easy.

Proposition 3.2. *For every* $x,y \in \Lambda$, *we have*

$$\overline{\phi}_{\Lambda,\psi}(x) = E_x\left[\psi\left(X_{\tau_\Lambda}\right)\right], \tag{3.6}$$

$$G_\Lambda(x,y) = E_x\left[\int_0^{\tau_\Lambda} 1_{\{y\}}\left(X_t\right) dt\right] = \int_0^\infty p_\Lambda(t,x,y)\,dt. \tag{3.7}$$

The middle term of (3.7) is the average of the occupation time of X_t at y before leaving Λ.

3.2 Gaussian Systems on $\mathbb{Z}^d$

Let us assume that a harmonic function $\psi = \{\psi(x); x \in \mathbb{Z}^d\} \in \mathbb{R}^{\mathbb{Z}^d}$ is given on the whole lattice $\mathbb{Z}^d$ and consider the Gaussian finite volume φ-Gibbs measures $\mu_\Lambda^\psi \in \mathcal{P}(\mathbb{R}^{\mathbb{Z}^d})$ for all connected $\Lambda \Subset \mathbb{Z}^d$. We shall see that, if $d \geq 3$, μ_Λ^ψ admits a weak limit $\mu^\psi \in \mathcal{P}(\mathbb{R}^{\mathbb{Z}^d})$ as $\Lambda \nearrow \mathbb{Z}^d$ (i.e., along an increasing sequence $\{\Lambda^{(n)}\}_{n=1,2,\ldots}$ satisfying $\cup_{n=1}^\infty \Lambda^{(n)} = \mathbb{Z}^d$) and the limit μ^ψ is a φ-Gibbs measure (on $\mathbb{Z}^d$) corresponding to the potential $V(\eta) = \frac{1}{2}\eta^2$. A simple but important class of the harmonic functions on $\mathbb{Z}^d$ is given by $\psi(x) = u\cdot x + h$ for $u \in \mathbb{R}^d$ and $h \in \mathbb{R}$, where $u \cdot x$ denotes the inner product in $\mathbb{R}^d$. The two-point correlation function of μ^ψ decays slowly in algebraic (i.e., polynomial) order. The $\nabla\varphi$-Gibbs measures exist for arbitrary dimension d.

Thermodynamic Limit

Since $\overline{\phi}_{\Lambda,\psi} = \psi$ on Λ for every harmonic function ψ, from (3.3), the mean of ϕ under μ_Λ^ψ is ψ. The covariance of μ_Λ^ψ is $G_\Lambda(x,y)$, recall (3.4). Let $G(x,y) \equiv (-\Delta)^{-1}(x,y) = G(x-y)$ be the Green function (of 0th order) of the operator Δ on $\mathbb{Z}^d$, i.e.,

$$G(x,y) = \int_0^\infty p(t,x,y)\,dt, \quad x,y \in \mathbb{Z}^d,$$

where $p(t,x,y)$ denotes the transition probability of the simple random walk X on $\mathbb{Z}^d$. It is well-known that $G(x,y) < \infty$ if and only if $d \geq 3$ (i.e., if X is transient). This can be also seen from an explicit formula for $G(x)$ by the Fourier transform:

$$G(x) = \frac{1}{2(2\pi)^d} \int_{\tilde{\mathbb{T}}^d} \frac{e^{\sqrt{-1}x\cdot\theta}}{\sum_{j=1}^d(1 - \cos\theta_j)}\,d\theta, \tag{3.8}$$

where $\tilde{\mathbb{T}}^d = (-\pi,\pi]^d$ and $d\theta = \prod_{j=1}^d d\theta_j$. Since $p_\Lambda(t,x,y) \uparrow p(t,x,y)$ as $\Lambda \nearrow \mathbb{Z}^d$, we have

$$\lim_{\Lambda \nearrow \mathbb{Z}^d} G_\Lambda(x, y) = G(x, y), \quad x, y \in \mathbb{Z}^d .$$

To study the limit of μ_Λ^ψ as $\Lambda \nearrow \mathbb{Z}^d$, recalling Proposition 3.1-(3), we may assume $\psi \equiv 0$. Let $\mu \in \mathcal{P}(\mathbb{R}^{\mathbb{Z}^d})$ be the distribution of a Gaussian system $\phi = \{\phi(x); x \in \mathbb{Z}^d\}$ with mean 0 and covariance $G(x, y)$, whose characteristic function is given by

$$E^\mu \left[e^{\sqrt{-1}(\xi, \phi)} \right] = e^{-\frac{1}{2}(\xi, (-\Delta)^{-1}\xi)}, \quad \xi \in C_0(\mathbb{Z}^d) ,$$

where $(\xi, \phi) = \sum_{x \in \mathbb{Z}^d} \xi(x)\phi(x)$ is the inner product (on the whole lattice $\mathbb{Z}^d$) and $C_0(\mathbb{Z}^d)$ denotes the family of all $\xi : \mathbb{Z}^d \to \mathbb{R}$ satisfying $\xi(x) = 0, x \notin \Lambda$ for some $\Lambda \Subset \mathbb{Z}^d$. The convergence of the covariances

$$(\xi, (-\Delta_\Lambda)^{-1}\xi) = \sum_{x, y \in \mathbb{Z}^d} G_\Lambda(x, y)\xi(x)\xi(y)$$

$$\longrightarrow \sum_{x, y \in \mathbb{Z}^d} G(x, y)\xi(x)\xi(y) = (\xi, (-\Delta)^{-1}\xi)$$

for $\xi \in C_0(\mathbb{Z}^d)$ (note that both sums are finite) implies the convergence of the characteristic functions so that μ_Λ^0 weakly converges to μ as $\Lambda \nearrow \mathbb{Z}^d$ on the space $\mathbb{R}^{\mathbb{Z}^d}$ endowed with the product topology.

In fact, the convergence holds under stronger topologies. To see that, let us introduce weighted ℓ^2-spaces on $\mathbb{Z}^d$

$$\ell^2(\mathbb{Z}^d, z) := \{\phi \in \mathbb{R}^{\mathbb{Z}^d}; \|\phi\|_z^2 := \sum_{x \in \mathbb{Z}^d} \phi(x)^2 z(x) < \infty\}$$

for weight functions $z = \{z(x) > 0; x \in \mathbb{Z}^d\}$. We shall especially concern with two classes of spaces $(\ell_\alpha^2, \| \cdot \|_\alpha)$ and $(\ell_r^2, \| \cdot \|_r)$ for $\alpha, r > 0$ taking $z(x) = (1 + |x|)^{-\alpha}$ and $z(x) = e^{-2r|x|}$, respectively.

Proposition 3.3. *Assume $d \geq 3$. Then μ_Λ^0 weakly converges to μ as $\Lambda \nearrow \mathbb{Z}^d$ on the spaces $\ell_\alpha^2, \alpha > d$ or $\ell_r^2, r > 0$.*

Proof. The proof is concluded once the tightness of $\{\mu_\Lambda^0\}_\Lambda$ on these spaces is shown. However, since

$$0 \leq E^{\mu_\Lambda^0} \left[\phi(x)^2 \right] = G_\Lambda(x, x) \leq G(x, x) = G(O) < \infty ,$$

we have that $\sup_\Lambda E^{\mu_\Lambda^0}[\|\phi\|_\alpha^2] < \infty$ (if $\alpha > d$) and $\sup_\Lambda E^{\mu_\Lambda^0}[\|\phi\|_r^2] < \infty$. These uniform estimates imply the tightness noting that the imbeddings $\ell_{\alpha_1}^2 \subset \ell_{\alpha_2}^2$ or $\ell_{r_1}^2 \subset \ell_{r_2}^2$ are compact when $0 < \alpha_1 < \alpha_2$ or $0 < r_1 < r_2$, respectively. □

Remark 3.1. *Fernique's theorem for Gaussian random variables shows that*

$$\sup_\Lambda E^{\mu_\Lambda^0}[e^{\epsilon\|\phi\|_\alpha^2}] < \infty, \quad \alpha > d ,$$

for some $\epsilon > 0$, e.g. [232]. In particular, this implies $\sup_\Lambda E^{\mu_\Lambda^0}[\|\|\phi\|\|_\alpha^k] < \infty$ for $\alpha > d$ and $k \in \mathbb{N}$. The same uniform estimates hold for the norm $\|\phi\|_r$.

Finally, let us show that the limit μ of μ_Λ^0 is actually a φ-Gibbs measure. The argument below is applicable also when the potentials V are general. We call a function $g = g(\phi)$ on $\mathbb{R}^{\mathbb{Z}^d}$ local if it is $\mathcal{F}_\Lambda$-measurable for some $\Lambda \Subset \mathbb{Z}^d$ and the smallest Λ is denoted by $\mathrm{supp}\,(g)$.

Proposition 3.4. *μ is a φ-Gibbs measure.*

Proof. Let $\Sigma \Subset \mathbb{Z}^d$, $\mathcal{F}_\Sigma$-measurable bounded function f and $\mathcal{F}_{\Sigma^c}$-measurable bounded local function g be given. Then, if Λ is sufficiently large such that $\Sigma \cup \mathrm{supp}\,(g) \subset \Lambda$, we have

$$E^{\mu_\Lambda^0}[fg] = E^{\mu_\Lambda^0}[g(\phi)E^{\mu_\Sigma^\phi}[f]] \,.$$

However, under the limit $\Lambda \nearrow \mathbb{Z}^d$, the left and the right hand sides converge to $E^\mu[fg]$ and $E^\mu[g(\phi)E^{\mu_\Sigma^\phi}[f]]$, respectively. Thus we obtain the DLR equation for μ. $\qquad\square$

In summary, we see that for every harmonic function ψ on $\mathbb{Z}^d$ a weak limit $\mu^\psi \in \mathcal{P}(\mathbb{R}^{\mathbb{Z}^d})$ of μ_Λ^ψ as $\Lambda \nearrow \mathbb{Z}^d$ exists and it is a φ-Gibbs measure if $d \geq 3$.

Remark 3.2. (1) *Take $\psi(x) = u \cdot x + h$ for the boundary conditions. Then the limit μ^ψ is not the same for different h. In this sense the massless field is quite sensitive on the boundary conditions.*
(2) *As the results on a massive model ([15], see Remark 3.3 below) suggest, the extremal sets of the class of φ-Gibbs measures might be much wider than $\{\mu^\psi; \psi \text{ are harmonic}\}$. However, $\nabla\varphi$-Gibbs measures are exhausted, under the assumption of shift invariance (and temperedness), by the convex hull of the gradient fields associated to these fields (see Sect. 9).*

Long Correlations

As we have seen, the two-point correlation function of ϕ under the φ-Gibbs measure μ^ψ coincides with the Green function $G(x, y)$ of the simple random walk on $\mathbb{Z}^d$, and it decays only algebraically (or in polynomial order) and not exponentially fast. In this sense the field has long dependence.

Proposition 3.5. *Assume $d \geq 3$. Then the two-point correlation function of μ^ψ is always positive and behaves like*

$$E^{\mu^\psi}[\phi(x); \phi(y)] \sim \frac{k_1}{|x - y|^{d-2}}$$

as $|x - y| \to \infty$, where $|x - y|$ stands for the Euclidean distance and $\sim$ means that the ratio of both sides converges to 1. The constant k_1 is determined by

$$k_1 = \frac{1}{2} \int_0^\infty (2\pi t)^{-\frac{d}{2}} e^{-\frac{1}{2t}} \, dt \,.$$

Proof. The conclusion follows from the behavior $G(x) \sim k_1/|x|^{d-2}, |x| \to \infty$ of the Green function established by Itô-McKean [155] (2.7, p.121); see also Spitzer [238], p.308, P1 for $d = 3$ and Lawler [185]. Note that, in these references, Δ is normalized by dividing it by $2d$.   □

This proposition, in particular, implies that one of the important thermodynamic quantities called the compressibility diverges in massless model:

$$\sum_{x \in \mathbb{Z}^d} E^{\mu^\psi}[\phi(x); \phi(y)] = \infty .$$

Note that $k_1/|x - y|^{d-2}, x, y \in \mathbb{R}^d$ is the Green function (of the continuum Laplacian) on $\mathbb{R}^d$ and the constant k_1 has another expression

$$k_1 = (4\pi^{d/2})^{-1} \Gamma\left(\frac{d}{2} - 1\right) = \frac{1}{(d-2)\Omega_d} ,$$

where Ω_d is the surface area of the $d-1$ dimensional unit sphere. For general potential V, similar asymptotics for the two-point correlation function are obtained by [202], see Sect. 4.3.

$\nabla\varphi$-Gaussian Field

We have required the assumption $d \geq 3$ to construct φ-field on the infinite volume lattice $\mathbb{Z}^d$, but for its gradient the thermodynamic limit exists in arbitrary dimensions d including $d = 1, 2$. To see this, we first notice the next lemma which is immediate from Proposition 3.1-(1). Recall that

$$\nabla_i \phi(x) := \phi(x + e_i) - \phi(x)(\equiv \nabla\phi(x + e_i)), \quad x \in \mathbb{Z}^d, 1 \leq i \leq d .$$

The bond $\langle x+e_i, x\rangle$ is simply denoted by $x+e_i$ and, in particular, e_i sometimes represents the bond $\langle e_i, O\rangle$.

Lemma 3.6. *Let $\Lambda \Subset \mathbb{Z}^d$ and $\psi \in \mathbb{R}^{\mathbb{Z}^d}$ be given. Then we have*

$$E^{\mu_\Lambda^\psi}[\nabla_i \phi(x)] = \nabla_i \overline{\phi}_{\Lambda,\psi}(x) , \tag{3.9}$$

$$E^{\mu_\Lambda^\psi}[\nabla_i \phi(x); \nabla_j \phi(y)] = \nabla_{i,x} \nabla_{j,y} G_\Lambda(x, y) , \tag{3.10}$$

for every $x, y \in \Lambda, 1 \leq i, j \leq d$, where $\nabla_{i,x}$ and $\nabla_{j,y}$ indicate that these operators act on the variables x and y, respectively.

When $d = 1, 2$, although $G_\Lambda(x, y)$ itself is not convergent as $\Lambda \nearrow \mathbb{Z}^d$, its normalization

$$\tilde{G}_\Lambda(x, y) := \int_0^\infty \{p_\Lambda(t, x, y) - p_\Lambda(t, 0, 0)\} dt$$

admits the finite limit

$$G(x, y) := \int_0^\infty \{p(t, x, y) - p(t, 0, 0)\} \, dt, \quad x, y \in \mathbb{Z}^d \,,$$

which is called the (normalized 0th order) Green function. One can replace G_Λ in the right hand side of (3.10) with $\tilde{G}_\Lambda$ so that the covariance of the $\nabla\varphi$-field has the limit as $\Lambda \nearrow \mathbb{Z}^d$. We therefore obtain the next proposition.

Proposition 3.7. *For a harmonic function ψ on $\mathbb{Z}^d$, let $\mu^{\psi, \nabla} \in \mathcal{P}(\mathbb{R}^{(\mathbb{Z}^d)^*})$ be the distribution of the Gaussian field on $(\mathbb{Z}^d)^*$ with mean and covariance*

$$E^{\mu^{\psi, \nabla}} [\nabla_i \phi(x)] = \nabla_i \psi(x) \,,$$

$$E^{\mu^{\psi, \nabla}} [\nabla_i \phi(x); \nabla_j \phi(y)] = \nabla_{i,x} \nabla_{j,y} G(x, y) \,,$$

respectively. Then $\mu^{\psi, \nabla}$ is a $\nabla\varphi$-Gibbs measure (see Definition 2.2 and Sect. 9).

We have a family of $\nabla\varphi$-Gibbs measures $\{\mu^{\psi_u, \nabla}; u \in \mathbb{R}^d\}$ by taking $\psi(x) \equiv \psi_u(x) := u \cdot x$. When $d \geq 3$, if $\phi = \{\phi(x); x \in \mathbb{Z}^d\}$ is μ^ψ-distributed, then its gradient field $\nabla\phi = \{\nabla\phi(b); b \in (\mathbb{Z}^d)^*\}$ is $\mu^{\psi, \nabla}$-distributed. When $d = 1$, the Green function is given by $G(x) = -\frac{1}{2}|x|$, which proves $E^{\mu^{\psi, \nabla}} [\nabla\phi(x); \nabla\phi(y)] = \delta(x - y)(\equiv \delta(x, y))$. This, in particular, shows that $\{\nabla\phi(b); b \in (\mathbb{Z})^*\}$ is an independent Gaussian system in one dimension. When $d = 2$, the Green function behaves like

$$G(x) = -\frac{1}{2\pi} \log |x| + c_0 + O(|x|^{-2}), \quad |x| \to \infty \,,$$

see Stöhr [241], Spitzer [238].

3.3 Massive Gaussian Systems

In the present subsection, we study the φ-field associated with the massive Hamiltonian $H_m(\phi)$ introduced in Remark 2.2-(1). The mass term of H_m actually has a strong influence on the field. It is localized and exhibits very different features from the massless case. In particular, (1) the φ-Gibbs measure exists for arbitrary dimensions $d \geq 1$, (2) the effect of the boundary conditions is weak (see Corollary 3.9 below) and (3) the two-point correlation function decays exponentially fast; in other words, the field has a strong mixing property.

Massive Gaussian φ-Gibbs Measures

For $\Lambda \Subset \mathbb{Z}^d$ and the boundary condition $\psi \in \mathbb{R}^{\mathbb{Z}^d}$, the finite volume φ-Gibbs measure $\mu^\psi_{\Lambda;m} \in \mathcal{P}(\mathbb{R}^\Lambda)$ having mass $m > 0$ is defined by

$$\mu^\psi_{\Lambda;m}(d\phi_\Lambda) := \frac{1}{Z^\psi_{\Lambda;m}} e^{-H^\psi_{\Lambda;m}(\phi)} \, d\phi_\Lambda$$

where

$$H_{\Lambda;m}^{\psi}(\phi) = H_{\Lambda}^{\psi}(\phi) + \frac{m^2}{2}\sum_{x\in\Lambda}\phi(x)^2$$

is the massive Hamiltonian and $Z_{\Lambda;m}^{\psi}$ is the normalization constant. As before, we sometimes regard $\mu_{\Lambda;m}^{\psi} \in \mathcal{P}(\mathbb{R}^{\mathbb{Z}^d})$. The φ-Gibbs measure $\mu \equiv \mu_m \in \mathcal{P}(\mathbb{R}^{\mathbb{Z}^d})$ (on $\mathbb{Z}^d$) having mass m is defined by means of the DLR equation with the local specifications $\mu_{\Lambda;m}^{\psi}$ in place of μ_{Λ}^{ψ} in Definition 2.1. We are always concerning the case where $V(\eta) = \frac{1}{2}\eta^2$ throughout this section.

Finite Systems

Similarly to the massless case, the mean and covariance of the field ϕ under $\mu_{\Lambda;m}^{\psi}$ can be expressed as solutions of certain difference equations and admit the random walk representation. Indeed, consider the equations (3.1) and (3.2) with Δ replaced by $\Delta - m^2$, respectively, i.e.,

$$\begin{cases} (\Delta-m^2)\overline{\phi}(x) = 0, & x\in\Lambda , \\ \overline{\phi}(x) = \psi(x), & x\in\partial^+\Lambda , \end{cases} \quad (3.11)$$

and

$$\begin{cases} -(\Delta-m^2)G_{\Lambda;m}(x,y) = \delta(x,y), & x\in\Lambda , \\ G_{\Lambda;m}(x,y) = 0, & x\in\partial^+\Lambda , \end{cases} \quad (3.12)$$

for $y\in\Lambda$. The solution of (3.11) is denoted by $\overline{\phi} = \overline{\phi}_{\Lambda,\psi;m}$, while $G_{\Lambda;m}(x,y)$ is sometimes written as

$$G_{\Lambda;m}(x,y) = (-\Delta_\Lambda + m^2)^{-1}(x,y) .$$

Consider the simple random walk $X = \{X_t\}_{t\geq 0}$ on $\mathbb{Z}^d$ as before and let σ be an exponentially distributed random variable with mean $\frac{1}{m^2}$ being independent of X. The random walk X is killed at the time σ, in other words, it jumps to a point $\Delta(\notin \mathbb{Z}^d)$ at σ and stays there afterward. Every function ψ on $\mathbb{Z}^d$ is extended to $\mathbb{Z}^d \cup \{\Delta\}$ setting $\psi(\Delta) = 0$. The next proposition is an extension of Propositions 3.1 and 3.2 to the massive case. The proof is similar.

Proposition 3.8. *Under $\mu_{\Lambda;m}^{\psi}$, $\phi = \{\phi(x); x\in\Lambda\}$ is Gaussian with mean $\overline{\phi}_{\Lambda,\psi;m}(x)$ and covariance $G_{\Lambda;m}(x,y)$. In particular, we have for $x,y\in\Lambda$*

$$E^{\mu_{\Lambda;m}^{\psi}}[\phi(x)] = \overline{\phi}_{\Lambda,\psi;m}(x) = E_x[\psi(X_{\tau_\Lambda\wedge\sigma})], \quad (3.13)$$

$$E^{\mu_{\Lambda;m}^{\psi}}[\phi(x);\phi(y)] = G_{\Lambda;m}(x,y) = E_x\left[\int_0^{\tau_\Lambda\wedge\sigma} 1_{\{y\}}(X_t)\,dt\right]. \quad (3.14)$$

Thermodynamic Limit

The random walk representation is useful to observe that the limit of $\mu^{\psi}_{\Lambda;m}$ as $\Lambda \nearrow \mathbb{Z}^d$ does not depend on the boundary condition ψ if it grows at most in polynomial order as $|x| \to \infty$. This property for massive field is essentially different from the massless case. If ψ grows exponentially fast, its effect may remain in the limit of $\mu^{\psi}_{\Lambda;m}$, see Remark 3.3 below.

Corollary 3.9. *If the function ψ on $\mathbb{Z}^d$ satisfies $|\psi(x)| \leq C(1 + |x|^n)$ for some $C, n > 0$, then we have for every $x \in \mathbb{Z}^d$*

$$\lim_{\Lambda \nearrow \mathbb{Z}^d} \overline{\phi}_{\Lambda, \psi;m}(x) = 0 .$$

Proof. To prove the conclusion, from (3.13), it suffices to show that $P_x(\tau_{\Lambda_\ell} < \sigma) = o(\ell^{-n})$ as $\ell \to \infty$, where $\Lambda_\ell = [-\ell, \ell]^d \cap \mathbb{Z}^d$. However, since $P(\sigma > \sqrt{\ell}) = e^{-m^2\sqrt{\ell}}$, this follows from the large deviation type estimate on τ_Λ: $P_x(\tau_{\Lambda_\ell} < \sqrt{\ell}) \leq e^{-C\ell}$ for some $C > 0$. $\qquad\square$

The covariance $G_{\Lambda;m}(x, y)$ of $\mu^{\psi}_{\Lambda;m}$ converges as $\Lambda \nearrow \mathbb{Z}^d$ to $G_m(x, y) = G_m(x - y)$, where $G_m(x)$ is defined by

$$G_m(x) := \int_0^\infty e^{-m^2 t} p(t, x) \, dt$$

$$= \frac{1}{(2\pi)^d} \int_{\tilde{\mathbb{T}}^d} \frac{e^{\sqrt{-1}x \cdot \theta}}{2 \sum_{j=1}^d (1 - \cos\theta_j) + m^2} \, d\theta ,$$

where $p(t, x) = p(t, x, O)$. Note that, since $m > 0$, $G_m(x) < \infty$ for all $d \geq 1$. The function $G_m(x, y)$ is sometimes written as $(-\Delta + m^2)^{-1}(x, y)$ and called the Green function of m^2th order of the operator Δ on $\mathbb{Z}^d$. When the boundary condition ψ satisfies the condition in Corollary 3.9, the thermodynamic limit $\mu_m \in \mathcal{P}(\mathbb{R}^{\mathbb{Z}^d})$ of $\mu^{\psi}_{\Lambda;m}$ exists and it is the Gaussian measure with the mean 0, covariance $G_m(x, y)$ and characteristic function

$$E^{\mu_m}\left[e^{\sqrt{-1}(\xi, \phi)}\right] = e^{-\frac{1}{2}(\xi, (-\Delta + m^2)^{-1}\xi)} ,$$

for $\xi \in C_0(\mathbb{Z}^d)$. The limit measure is independent of the choice of ψ as long as it satisfies the condition in Corollary 3.9.

Remark 3.3. *Benfatto et al. [15] characterized the structure of the class of all massive Gaussian φ-Gibbs measures on $\mathbb{R}^{\mathbb{Z}}$ when $d = 1$. Their result shows that its extremal set $\mathcal{E}$ is given by*

$$\mathcal{E} = \{\mu_{\alpha_-, \alpha_+}; (\alpha_-, \alpha_+) \in \mathbb{R}^2\} ,$$

where μ_{α_-, α_+} is the limit of the sequence of finite φ-Gibbs measures $\mu^{\psi}_{[-\ell, \ell];m}$ with boundary condition ψ satisfying

$$\alpha_\pm = (1 - \rho^2) \lim_{\ell \to \infty} \rho^\ell \psi(\pm(\ell + 1))$$

for certain $\rho \in (0, 1)$. For instance, if ψ is replaced by $\psi + u \cdot x + h$, the constants $\alpha_\pm$ are the same. In this respect, φ-Gibbs measure is not much sensitive to the boundary conditions. The mass term $\frac{m^2}{2} \sum \phi(x)^2$ has an effect to localize the field. In fact, the above mentioned result implies that the shift invariant massive Gaussian φ-Gibbs measure is unique in one dimension.

Short Correlations

The exponential decay of the two-point correlation function

$$E^{\mu_m}[\phi(x); \phi(y)] = E^{\mu_m}[\phi(x)\phi(y)] = G_m(x - y)$$

under μ_m is precisely stated in the next proposition. The proof of (2) is given by a simple calculation based on the residue theorem.

Proposition 3.10. (1) *When $d \geq 2$, for each $\omega \in S^{d-1}$ (i.e., $|\omega| = 1$), determine $b(\omega) = b_m(\omega) \in \mathbb{R}^d$ and $\gamma \in \mathbb{R} \setminus \{0\}$ by*

$$
\begin{cases}
\dfrac{1}{2d} \displaystyle\sum_{|y|=1} e^{b \cdot y} = \dfrac{m^2}{2d} + 1 \,, \\[4mm]
\dfrac{1}{2d} \displaystyle\sum_{|y|=1} y e^{b \cdot y} = \gamma \omega \,.
\end{cases}
$$

Then, we have

$$G_m(x) \sim C_d |x|^{-\frac{d-1}{2}} e^{-b_m(x/|x|) \cdot x}$$

as $|x| \to \infty$ for some $C_d > 0$.
(2) When $d = 1$, $G_m(x)$ has an explicit formula:

$$G_m(x) = \frac{1}{2\pi} \int_{-\pi}^{\pi} \frac{e^{\sqrt{-1} x \theta}}{2(1 - \cos \theta) + m^2} \, d\theta = \frac{e^{-\tilde{m}|x|}}{2 \sinh \tilde{m}}, \quad x \in \mathbb{Z}$$

where $\tilde{m} > 0$ is the solution of an algebraic equation $\cosh \tilde{m} = \frac{m^2}{2} + 1$. In particular, $\tilde{m}$ behaves such that $\tilde{m} = m + O(m^2)$ as $m \downarrow 0$.

Remark 3.4. *Let $C_m(x - y) = C_m(x, y), x, y \in \mathbb{R}^d$ be the Green function of m^2th order for the (continuum) Laplacian on $\mathbb{R}^d$, i.e.,*

$$C_m(x) := (-\Delta + m^2)^{-1} \delta(x) = \frac{1}{(2\pi)^d} \int_{\mathbb{R}^d} \frac{e^{\sqrt{-1} x \cdot p}}{p^2 + m^2} \, dp, \quad x \in \mathbb{R}^d \,.$$

The function $C_m(x)$ has an expression by means of the modified Bessel functions. For example, when $d = 3$, we have

$$C_m(x) = \frac{1}{4\pi|x|}e^{-m|x|} ,$$

and, for general $d \geq 1$, it behaves

$$C_m(x) \sim \text{const } m^{\frac{d-3}{2}}|x|^{-\frac{d-1}{2}}e^{-m|x|} ,$$

as $m|x| \to \infty$, see [138] p.126. Note that the exponential decay rates for G_m and C_m are different, see also [234] p.257 for $d = 2$.

Proposition 3.10 gives the exact exponential decay rates of the Green function G_m for $m > 0$. However, in order just to see the exponentially decaying property of G_m, one can apply the **Aronson's type estimate** on the transition probability $p(t, x, y) = p(t, x - y)$ of the simple random walk on $\mathbb{Z}^d$:

$$p(t, x) \leq \min\left\{\frac{C}{t^{d/2}}e^{-|x|^2/Ct}, 1\right\}, \quad t > 0, \ x \in \mathbb{Z}^d , \qquad (3.15)$$

for some $C > 0$; see [202] §2, [50] for general random walks. In fact, we divide the integral

$$G_m(x) = \int_0^\infty e^{-m^2 t}p(t, x)\, dt$$

into the sum of those on two intervals $[0, |x|)$ and $[|x|, \infty)$. Then, on the first interval, if $x \neq 0$, we estimate the integrand as

$$e^{-m^2 t}p(t, x) \leq \frac{C}{t^{d/2}}e^{-|x|^2/2Ct}e^{-|x|^2/2Ct}$$

$$\leq \frac{C}{t^{d/2}}e^{-1/2Ct}e^{-|x|^2/2C|x|} \leq \text{const } e^{-|x|/2C} ,$$

while on the second

$$e^{-m^2 t}p(t, x) \leq e^{-\frac{m^2 t}{2}}e^{-\frac{m^2}{2}|x|} .$$

This proves that

$$0 < G_m(x) \leq Ce^{-c|x|}, \quad x \in \mathbb{Z}^d ,$$

for some $c, C > 0$. See [202] Theorem B for non-Gaussian case.

The Aronson's type estimate is applicable to the massless case as well and, though it is weaker than Proposition 3.5, we have the following:

$$0 < G(x) \leq \frac{C}{|x|^{d-2}} ,$$

for some $C > 0$ when $d \geq 3$. In fact, the change of the variables $t = |x|^2 s$ in the integral implies

$$G(x) = \int_0^\infty p(t, x)\, dt \leq \int_0^\infty \frac{C}{t^{d/2}}e^{-|x|^2/Ct}\, dt = \frac{C}{|x|^{d-2}}\int_0^\infty \frac{1}{s^{d/2}}e^{-1/Cs}\, ds .$$

Note that the last integral converges when $d \geq 3$. See [202] Theorem C or Theorem 4.13 in Sect. 4.3 for non-Gaussian case.

3.4 Macroscopic Scaling Limits

The random field $\phi = \{\phi(x); x \in \mathbb{Z}^d\}$ is a microscopic object and our goal is to study its macroscopic behavior. In this subsection, we discuss such problem under the Gaussian measures $\mu =: \mu_0$ (massless case, $d \geq 3$) and $\mu_m, m > 0$ (massive case, $d \geq 1$). Recall that μ and μ_m are φ-Gibbs measures on $\mathbb{Z}^d$ obtained by the thermodynamic limit with boundary conditions $\psi \equiv 0$; see Sects. 3.2 and 3.3, respectively.

Scaling Limits

Let N be the ratio of typical lengths at macroscopic and microscopic levels. Then the point $\theta = (\theta_i)_{i=1}^d \in \mathbb{R}^d$ at macroscopic level corresponds to the lattice point $[N\theta] := ([N\theta_i])_{i=1}^d \in \mathbb{Z}^d$ at microscopic level, recall Sect. 2.5. If $x \in \mathbb{Z}^d$ is close to $[N\theta]$ in such sense that $|x - [N\theta]| \ll N$, then x also macroscopically corresponds to θ. This means that observing the random field ϕ at macroscopic point θ is equivalent to taking its sample mean around the microscopic point $[N\theta]$. Such averaging yields a cancellation in the fluctuations of ϕ.

Motivated by these observations, let us consider the sample mean of ϕ over the microscopic region $\Lambda_N = (-N, N]^d \cap \mathbb{Z}^d$, which corresponds to the macroscopic box $D = (-1, 1]^d$:

$$\overline{\phi}^N := \frac{1}{(2N)^d} \sum_{x \in \Lambda_N} \phi(x) \,,$$

note that $(2N)^d = |\Lambda_N|$. The field ϕ is distributed under μ_m for $m \geq 0$.

Lemma 3.11. *As $N \to \infty$, $\overline{\phi}^N$ converges to 0 in L^2 under μ_m for all $m \geq 0$.*

Proof. If we denote $G(x)$ by $G_0(x)$, we have for all $m \geq 0$

$$E^{\mu_m}\left[\left(\overline{\phi}^N\right)^2\right] = \frac{1}{(2N)^{2d}} \sum_{x,y \in \Lambda_N} E^{\mu_m}[\phi(x)\phi(y)] = \frac{1}{(2N)^{2d}} \sum_{x,y \in \Lambda_N} G_m(x - y) \,.$$

However, the Green functions admit bounds for some $C, c > 0$

$$0 < G_m(x) \leq \begin{cases} \dfrac{C}{|x|^{d-2}}, & m = 0, \, d \geq 3 \,, \\ Ce^{-c|x|}, & m > 0, \, d \geq 1 \,, \end{cases}$$

which prove the conclusion.   □

This lemma is the law of large numbers for φ-field and the next natural question is to study the fluctuation of $\overline{\phi}^N$ around its limit 0 under a proper rescaling. As we shall see, the necessary scalings will change according as $m = 0$ (i.e., massless case) or $m > 0$ (i.e., massive case) due to the difference in the mixing property of the field ϕ.

Fluctuations in Massive φ-Gaussian Field

First, let us consider the massive case: $m > 0$. Then the right scaling for the fluctuation of $\overline{\phi}^N$ will be

$$\tilde{\Phi}^N := (2N)^{\frac{d}{2}}\overline{\phi}^N \equiv \frac{1}{(2N)^{\frac{d}{2}}}\sum_{x\in\Lambda_N}\phi(x)\,. \qquad (3.16)$$

Since $(2N)^{\frac{d}{2}} = |\Lambda_N|^{\frac{1}{2}}$, this is the usual scaling for the central limit theorem; recall that $\phi = \{\phi(x); x \in \mathbb{Z}^d\}$ distributed under μ_m has a "nice" exponential mixing property when $m > 0$.

Proposition 3.12. *The fluctuation $\tilde{\Phi}^N$ weakly converges to the Gaussian distribution $N(0, m^{-2})$ with mean 0 and variance m^{-2} as $N \to \infty$.*

Proof. Since $\tilde{\Phi}^N$ is Gaussian distributed with mean 0, the conclusion follows from the convergence of its variance:

$$E^{\mu_m}\left[\left(\tilde{\Phi}^N\right)^2\right] = \frac{1}{(2N)^d}\sum_{x,y\in\Lambda_N}G_m(x-y)\xrightarrow[N\to\infty]{}\frac{1}{m^2}\,,$$

note (1) in the next remark. □

Remark 3.5. *From Bricmont et al. [39] Proposition A1 (p. 294), we have for $\mu_m, m > 0$*
(1) $\sum_{x\in\mathbb{Z}^d} E^{\mu_m}[\phi(O)\phi(x)] = m^{-2}$,
(2) $\sum_{x\in\mathbb{Z}^d} E^{\mu_m}[\phi(O)\nabla_i\phi(x)] \sim \text{const}\, m^{-1}$ $(m\downarrow 0)$,
(3) $\sum_{x\in\mathbb{Z}^d} E^{\mu_m}[\nabla_i\phi(O)\nabla_i\phi(x)]$ *is absolutely converging for each m and stays bounded as $m\downarrow 0$ (see Lemma 3.13 below).*
However, if $i\neq j$,
(4) $\sum_{x\in\mathbb{Z}^d} E^{\mu_m}[\nabla_i\phi(O)\nabla_j\phi(x)] \sim \text{const}\,|\log m|$ $(m\downarrow 0)$.
Loosely speaking, as $m\downarrow 0$, ϕ is expected to converge to the massless field so that its covariances (or those of its gradients) might behave like $|x-y|^{2-d}$ (or making its gradients in x), and this may prove that

$$\sum_{x\in\mathbb{Z}^d} E^{\mu_m}[\nabla_i\phi(O)\nabla_j\phi(x)] \approx \int^R r^{(2-d)-2}\cdot r^{d-1}\,dr \approx \log R\,,$$

where $R \approx m^{-1}$ is the correlation length.

Fluctuations in Massless φ-Gaussian Field

Next, let us consider the massless case: $m = 0$ and $d \geq 3$. Let ϕ be μ_0-distributed. Since the variance m^{-2} of the limit distribution of $\tilde{\Phi}^N$ under μ_m diverges as $m \downarrow 0$, the scaling (3.16) must not be correct in the massless case. However, if we further scale-down the value of $\tilde{\Phi}^N$ dividing it by N and introduce

$$\Phi^N := \frac{1}{N}\tilde{\Phi}^N \equiv \frac{1}{(2N)^{\frac{d}{2}} \cdot N} \sum_{x \in \Lambda_N} \phi(x), \qquad (3.17)$$

then it has the limit under μ_0. In fact, the variance of Φ^N behaves

$$E^{\mu_0}\left[\left(\Phi^N\right)^2\right] = 2^{-d}N^{-d-2} \sum_{x,y \in \Lambda_N} G(x - y)$$

$$\sim k_1 N^{-2} \sum_{x \in \Lambda_N} |x|^{2-d} \sim k_1 N^{-2} \int_{|\theta| \leq N} |\theta|^{2-d}\,d\theta$$

$$= k_1 N^{-2} \int_0^N r^{(2-d)+(d-1)}\,dr = O(1).$$

Therefore, (3.17) is the right scaling when $m = 0$. This actually coincides with the interpretation stated in Sect. 2.5: $\phi = \{\phi(x); x \in \mathbb{Z}^d\}$ represents the height of an interface embedded in $d + 1$ dimensional space so that both x- and ϕ-axes should be rescaled by the factor $1/N$ at the same time.

If we introduce random signed measures on $\mathbb{R}^d$ by

$$\Phi^N(d\theta) := \frac{1}{N^{\frac{d}{2}+1}} \sum_{x \in \mathbb{Z}^d} \phi(x)\delta_{x/N}(d\theta), \qquad \theta \in \mathbb{R}^d, \qquad (3.18)$$

then Φ^N in (3.17) is represented as $\Phi^N = 2^{-d/2}\langle \Phi^N(\cdot), 1_D \rangle$, where $\langle \Phi^N(\cdot), f \rangle$ stands for the integral of $f = f(\theta)$ under the measure $\Phi^N(\cdot)$. In this way, studying the limit of Φ^N is reduced to investigating more general problem for the properly scaled empirical measures of ϕ.

Fluctuations in Massless $\nabla\varphi$-Gaussian Field

When $f = f(\theta)$ has the form $f = -\frac{\partial g}{\partial \theta_i}$ with certain $g = g(\theta)$, we can rewrite $\langle \Phi^N(\cdot), f \rangle$ as

$$\langle \Phi^N(\cdot), f \rangle = \left\langle \Phi^N(\cdot), -\frac{\partial g}{\partial \theta_i} \right\rangle = -N^{-\frac{d}{2}-1} \sum_{x \in \mathbb{Z}^d} \phi(x)\frac{\partial g}{\partial \theta_i}(x/N)$$

$$\sim -N^{-\frac{d}{2}-1} \sum_{x \in \mathbb{Z}^d} \phi(x) \cdot N\{g((x + e_i)/N) - g(x/N)\}$$

$$= -N^{-\frac{d}{2}} \sum_{x \in \mathbb{Z}^d} (\phi(x - e_i) - \phi(x))g(x/N)$$

$$= N^{-\frac{d}{2}} \sum_{x \in \mathbb{Z}^d} \nabla_i\phi(x)g((x + e_i)/N).$$

The second line is the approximation of $\frac{\partial g}{\partial \theta_i}$ by its discrete derivatives. This rearrangement, in particular, implies that the scaling in $\Phi^N(d\theta)$ coincides with the usual one of the central limit theorem, if one deals with the corresponding gradient fields $\nabla\phi = \{\nabla\phi(x); x \in \mathbb{Z}^d\}$ instead of ϕ.

Thus it is natural to introduce the scaled empirical measures of $\nabla\phi = \{\nabla_i\phi; 1 \le i \le d\}$:

$$\Psi_i^N(d\theta) \equiv \Psi_i^N(d\theta; u) := \frac{1}{N^{\frac{d}{2}}} \sum_{x \in \mathbb{Z}^d} \{\nabla_i\phi(x) - u_i\}\delta_{x/N}(d\theta) . \qquad (3.19)$$

The field $\nabla\phi$ is μ_u^∇-distributed, where $\mu_u^\nabla, u = (u_i)_{i=1}^d \in \mathbb{R}^d$ is the $\nabla\varphi$-Gibbs measure $\mu^{\psi_u, \nabla}$ having boundary conditions $\psi(x) = \psi_u(x) \equiv u \cdot x$ obtained in Proposition 3.7. Note that $u_i = E^{\mu_u^\nabla}[\nabla_i\phi(x)]$ and, since $\nabla_i\phi(x) - u_i = \nabla_i(\phi - \psi_u)(x)$, the distribution of $\Psi_i^N(d\theta; u)$ under μ_u^∇ coincides with that of $\Psi_i^N(d\theta; 0)$ under μ_0^∇. We may therefore assume $u = 0$ to study the limit. The limit of the variance $E^{\mu_0^\nabla}\left[\langle\Psi_i^N, g\rangle^2\right]$ as $N \to \infty$ can be computed based on the next lemma.

Lemma 3.13.

$$\sum_{y \in \mathbb{Z}^d} E^{\mu_0^\nabla}[\nabla_i\phi(O)\nabla_i\phi(y)] = \frac{1}{d} .$$

Proof. Each term in the sum can be rewritten as

$$E^\mu\left[(\phi(e_i) - \phi(O))(\phi(y + e_i) - \phi(y))\right]$$
$$= \frac{1}{(2\pi)^d}\int_{\widetilde{\mathbb{T}}^d}\frac{2e^{\sqrt{-1}y\cdot\theta} - e^{\sqrt{-1}(y-e_i)\cdot\theta} - e^{\sqrt{-1}(y+e_i)\cdot\theta}}{2\sum_{j=1}^d(1 - \cos\theta_j)}\,d\theta$$
$$= \frac{1}{(2\pi)^d}\int_{\widetilde{\mathbb{T}}^d}e^{\sqrt{-1}y\cdot\theta}\frac{1 - \cos\theta_i}{\sum_{j=1}^d(1 - \cos\theta_j)}\,d\theta ,$$

which implies

$$\sum_{i=1}^d E^{\mu_0^\nabla}[\nabla_i\phi(O)\nabla_i\phi(y)] = \frac{1}{(2\pi)^d}\int_{\widetilde{\mathbb{T}}^d}e^{\sqrt{-1}y\cdot\theta}\,d\theta = \delta(y) .$$

The conclusion is shown by taking the sum in $y \in \mathbb{Z}^d$ of the both sides of this identity. $\square$

4 Random Walk Representation and Fundamental Inequalities

We are now at the position to enter into the study of the $\nabla\varphi$ interface model for general convex potentials V satisfying the three basic conditions (V1)-(V3) in (2.2). We shall first establish in this section three fundamental tools for analyzing the model, i.e., **Helffer-Sjöstrand representation**, **FKG (Fortuin-Kasteleyn-Ginibre) inequality** and **Brascamp-Lieb inequality**. Helffer-Sjöstrand representation describes for the correlation functions under the Gibbs measures by means of a certain random walk in random environments. Its original idea comes from [144], [237]. This representation readily implies FKG and Brascamp-Lieb inequalities. The latter is an inequality between the variances of non-Gaussian fields and those of Gaussian fields, which we can explicitly compute as we have seen in Sects. 3.1 and 3.2. In particular, uniform moment estimates on the non-Gaussian fields are obtained and these make us possible to construct $\nabla\varphi$-Gibbs measures on $(\mathbb{Z}^d)^*$ (for every $d \geq 1$) and φ-Gibbs measures on $\mathbb{Z}^d$ (for $d \geq 3$) by passing to the thermodynamic limit. The arguments in this section heavily rely on the convexity of the potential V, i.e., the attractiveness of the interaction.

4.1 Helffer-Sjöstrand Representation and FKG Inequality

Idea Behind the Representation

Let us shortly explain the idea behind the Helffer-Sjöstrand representation. It gives the following identity for the covariance of $F = F(\phi)$ and $G = G(\phi)$ under the Gibbs measure μ:

$$E^\mu[F; G] = \sum_{x \in \mathbb{Z}^d} \int_0^\infty E^\mu[\partial F(x, \phi_0) \partial G(X_t, \phi_t)]\, dt . \tag{4.1}$$

In the right hand side, $\phi_t = \{\phi_t(x); x \in \mathbb{Z}^d\}$ is the φ-dynamics defined by the SDEs (2.13) with μ-distributed initial data ϕ_0, X_t is the random walk on $\mathbb{Z}^d$ starting at x with (temporary inhomogeneous) generator Q^{ϕ_t} defined by

$$Q^\phi f(x) = \sum_{y:|x-y|=1} V''(\phi(x) - \phi(y))\{f(y) - f(x)\} ,$$

for $f : \mathbb{Z}^d \to \mathbb{R}$. Indeed, assuming $E^\mu[G] = 0$, let H be the solution of the Poisson equation $-LH = G$, where L is the generator of ϕ_t determined by (2.15) with $\Gamma = \mathbb{Z}^d$. Then, from (2.14)

$$E^\mu[F; G] = E^\mu[F(-LH)] = \sum_{x \in \mathbb{Z}^d} E^\mu[\partial F(x, \phi) \partial H(x, \phi)] . \tag{4.2}$$

However, a simple computation (cf. (4.7) below) shows

$$\partial(LH)(x, \phi) = L\partial H(x, \phi) + (Q\partial H(\cdot, \phi))(x) \equiv \{(L + Q)\partial H\}(x, \phi)$$

and therefore

$$\partial H(x, \phi) = (L + Q)^{-1} \partial (LH) = E_{(x,\phi)} \left[\int_0^\infty \partial G(X_t, \phi_t) \, dt \right] .$$

This implies the identity (4.1). The above argument is rather formal and, in particular, one should replace the measure μ with the finite volume Gibbs measure [77] or with the Gibbs measure for $\nabla\varphi$-field [135]. Note that the convexity condition on V (i.e., $V'' \geq 0$) is essential for the existence of the random walk X_t.

Precise Formulation

Let the finite region $\Lambda \Subset \mathbb{Z}^d$ and the boundary condition $\psi \in \mathbb{R}^{\mathbb{Z}^d}$ be given. We shall consider slightly general Hamiltonian having **external field (chemical potential)** $\rho = \{\rho(x); x \in \Lambda\} \in \mathbb{R}^\Lambda$:

$$H_\Lambda^{\psi,\rho}(\phi) = H_\Lambda^\psi(\phi) - (\rho, \phi)_\Lambda \tag{4.3}$$

and the corresponding finite volume φ-Gibbs measure

$$\mu_\Lambda^{\psi,\rho}(d\phi_\Lambda) = \frac{1}{Z_\Lambda^{\psi,\rho}} e^{-H_\Lambda^{\psi,\rho}(\phi)} \, d\phi_\Lambda \in \mathcal{P}(\mathbb{R}^\Lambda) , \tag{4.4}$$

where $Z_\Lambda^{\psi,\rho}$ is the normalization constant. This generalization will be useful for the proof of Brascamp-Lieb inequality, cf. Lemma 4.6 and Theorem 4.9. The operator $L_\Lambda^{\psi,\rho}$ defined by

$$\begin{aligned} L_\Lambda^{\psi,\rho} F(\phi) :=& e^{H_\Lambda^{\psi,\rho}(\phi)} \sum_{x \in \Lambda} \frac{\partial}{\partial \phi(x)} \left\{ e^{-H_\Lambda^{\psi,\rho}(\phi)} \frac{\partial F}{\partial \phi(x)} \right\} \\ =& \sum_{x \in \Lambda} \left\{ \frac{\partial^2 F}{\partial \phi(x)^2} - \frac{\partial H_\Lambda^{\psi,\rho}}{\partial \phi(x)} \frac{\partial F}{\partial \phi(x)} \right\} \end{aligned}$$

for $F = F(\phi) \in C^2(\mathbb{R}^\Lambda)$ is symmetric under the measure $\mu_\Lambda^{\psi,\rho}$ and the associated Dirichlet form is given by

$$\begin{aligned} \mathcal{E}(F, G) \equiv& \mathcal{E}_\Lambda^{\psi,\rho}(F, G) := -E^{\mu_\Lambda^{\psi,\rho}} \left[F \cdot L_\Lambda^{\psi,\rho} G \right] \\ =& E^{\mu_\Lambda^{\psi,\rho}} \left[(\partial F, \partial G)_\Lambda \right] . \end{aligned} \tag{4.5}$$

Recall that ∂F is defined by

$$\partial_x F(\phi) \equiv \partial F(x, \phi) := \frac{\partial F}{\partial \phi(x)}$$

and

$$(\partial F, \partial G)_\Lambda \equiv (\partial F, \partial G)_\Lambda(\phi) := \sum_{x \in \Lambda} \partial F(x, \phi) \partial G(x, \phi) .$$

For each $\phi \in \mathbb{R}^\Lambda$, the operator $Q_\Lambda^\phi \equiv Q_{\Lambda,0}^{\phi,\psi}$ is introduced by

$$Q_\Lambda^\phi f(x) := \sum_{b \in \overline{\Lambda}^* : y_b = x} V''(\nabla(\phi \vee \psi)(b))\nabla(f \vee 0)(b)$$

$$= \sum_{y \in \overline{\Lambda} : |x-y|=1} V''(\phi(x) - (\phi \vee \psi)(y))\{(f \vee 0)(y) - f(x)\},$$

for $x \in \Lambda$ and $f = \{f(x); x \in \Lambda\} \in \mathbb{R}^\Lambda$ under the boundary conditions $\phi(x) = \psi(x)$ and $f(x) = 0$ for $x \in \partial^+\Lambda$. In particular, when $V(\eta) = \frac{1}{2}c\eta^2, c > 0$, $Q_{\Lambda,0}^{\phi,\psi} = c\Delta_{\Lambda,0}$, which is independent of ϕ and ψ. We further consider the operator

$$\mathcal{L} \equiv \mathcal{L}_\Lambda^{\psi,\rho} := L_\Lambda^{\psi,\rho} + Q_{\Lambda,0}^\psi$$

acting on the functions $F = F(x, \phi)$ on $\Lambda \times \mathbb{R}^\Lambda$, where $Q_{\Lambda,0}^\psi F(x, \phi) := Q_{\Lambda,0}^{\phi,\psi} F(x, \phi)$ is the operator acting on functions with two variables. The next lemma is simple, but explains the reason why the operator $Q_{\Lambda,0}^{\phi,\psi}$ is useful.

Lemma 4.1. *For every $x \in \Lambda$ and $F = F(\phi)$, we have*

$$[\partial_x, L_\Lambda^{\psi,\rho}] \equiv \partial_x L_\Lambda^{\psi,\rho} - L_\Lambda^{\psi,\rho}\partial_x = -\sum_{y \in \Lambda} \frac{\partial^2 H_\Lambda^{\psi,\rho}}{\partial\phi(x)\partial\phi(y)}\partial_y, \tag{4.6}$$

$$\partial L_\Lambda^{\psi,\rho} F(x, \phi) = \mathcal{L}\partial F(x, \phi). \tag{4.7}$$

Proof. (4.6) is obvious from the definition of $L_\Lambda^{\psi,\rho}$. (4.7) follows from (4.6) by noting the symmetry of V'' and

$$\frac{\partial^2 H_\Lambda^{\psi,\rho}}{\partial\phi(x)\partial\phi(y)} = \begin{cases} \sum_{z \in \overline{\Lambda} : |x-z|=1} V''(\phi(x) - (\phi \vee \psi)(z)), & x = y, \\ -V''(\phi(x) - \phi(y)) & , \quad |x - y| = 1, \\ 0 & , \quad \text{otherwise}, \end{cases}$$

for $x, y \in \Lambda$. □

Let $\phi_t \equiv \phi_t^\rho = \{\phi_t(x); x \in \Lambda\}$ be the process on $\mathbb{R}^\Lambda$ generated by $L_\Lambda^{\psi,\rho}$, i.e., the solution of the SDEs (2.9) with $\Gamma = \Lambda$ and $H = H_\Lambda^{\psi,\rho}$:

$$\begin{cases} d\phi_t(x) = -\sum_{y \in \overline{\Lambda} : |x-y|=1} V'(\phi_t(x) - \phi_t(y))dt \\ \qquad\qquad + \rho(x)dt + \sqrt{2}dw_t(x), \quad x \in \Lambda, \\ \phi_t(y) = \psi(y), \quad y \in \partial^+\Lambda. \end{cases} \tag{4.8}$$

Let $X_t, t \geq 0$ be the random walk on Λ (or, more precisely, on $\Lambda \cup \{\Delta\}$) with temporally inhomogeneous generator $Q_\Lambda^{\phi_t}$ (and with killing rate

$\sum_{y \in \partial^+ \Lambda : |x-y|=1} V''(\phi_t(x) - \psi(y))$ at $x \in \partial^- \Lambda$). Then, $\mathcal{L}$ is the generator of (X_t, ϕ_t). Note that the random walk X_t exists since its jump rate $V''(\nabla(\phi_t \vee \psi)(b))$ is positive from our assumption (V3).

Theorem 4.2. (**Helffer-Sjöstrand representation**) *The correlation function of $F = F(\phi)$ and $G = G(\phi)$ under $\mu_\Lambda^{\psi,\rho}$ has the representation*

$$E^{\mu_\Lambda^{\psi,\rho}}[F;G] = \sum_{x \in \Lambda} \int_0^\infty E^{\delta_x \otimes \mu_\Lambda^{\psi,\rho}}[\partial F(x,\phi_0)\partial G(X_t,\phi_t), t < \tau_\Lambda]\, dt. \quad (4.9)$$

In the right hand side, $\delta_x \otimes \mu_\Lambda^{\psi,\rho}$ indicates the initial distribution of (X_t, ϕ_t) and $\delta_x \in \mathcal{P}(\Lambda)$ is defined by $\delta_x(z) = \delta(z-x)$. In particular, the distribution of ϕ_0 is $\mu_\Lambda^{\psi,\rho}$ and the random walk X_t starts at x. $\tau_\Lambda = \inf\{t > 0; X_t \in \Lambda^c\}$ is the exit time of X_t from Λ.

Theorem 4.2 with a special choice of $F(\phi) = \phi(x)$ and $G(\phi) = \phi(y)$ gives the following extension of the formula (3.7) combined with (3.4) for quadratic potentials to general ones; note that $\partial F(z,\phi) = \delta(x-z)$ in this case.

Corollary 4.3. *For every $x, y \in \Lambda$,*

$$E^{\mu_\Lambda^{\psi,\rho}}[\phi(x);\phi(y)] = E^{\delta_x \otimes \mu_\Lambda^{\psi,\rho}}\left[\int_0^{\tau_\Lambda} 1_{\{y\}}(X_t)\, dt\right].$$

The function $F = F(\phi)$ on $\mathbb{R}^\Lambda$ is called **increasing** if it satisfies $\partial F = \partial F(x,\phi) \geq 0$ so that it is nondecreasing under the semi-order on $\mathbb{R}^\Lambda$ determined by "$\phi_1 \geq \phi_2, \phi_1, \phi_2 \in \mathbb{R}^\Lambda \iff \phi_1(x) \geq \phi_2(x)$ for every $x \in \Lambda$". Theorem 4.2 immediately implies the following inequality.

Corollary 4.4. (**FKG inequality**) *If F and G are both (L^2-integrable) increasing functions, then we have*

$$E^{\mu_\Lambda^{\psi,\rho}}[F;G] \geq 0,$$

namely,

$$E^{\mu_\Lambda^{\psi,\rho}}[FG] \geq E^{\mu_\Lambda^{\psi,\rho}}[F]\, E^{\mu_\Lambda^{\psi,\rho}}[G].$$

So far, we are concerned with the representation of correlation functions. The next proposition gives the formula for the expectation of $\phi(x)$, which is an extension of (3.3) with (3.6) for quadratic potentials. See [77] for the proof.

Proposition 4.5. *For $x \in \Lambda$, we have*

$$E^{\mu_\Lambda^\psi}[\phi(x)] = \int_0^1 E^{\delta_x \otimes \mu_\Lambda^{s\psi}}[\psi(X_{\tau_\Lambda})]\, ds.$$

Introducing the external field ρ has an advantage in the next lemma, which is indeed one of the tricks commonly used in statistical mechanics. We shall denote the variance of the random variable X under μ by

$$\text{var}(X;\mu) = E^\mu\left[(X - E^\mu[X])^2\right].$$

Lemma 4.6. *Assume* $\rho, \nu \in \mathbb{R}^\Lambda$. *Then we have*

$$\frac{d}{ds} E^{\mu_\Lambda^{\psi,s\rho}}[\phi(x)] = E^{\mu_\Lambda^{\psi,s\rho}}[\phi(x); (\rho, \phi)_\Lambda], \qquad (4.10)$$

$$E^{\mu_\Lambda^{\psi,\rho}}\left[\exp\left\{(\nu, \phi)_\Lambda - E^{\mu_\Lambda^{\psi,\rho}}[(\nu, \phi)_\Lambda]\right\}\right]$$

$$= \exp\left\{\int_0^1 ds \int_0^s \text{var}\left((\nu, \phi)_\Lambda; \mu_\Lambda^{\psi,\rho+s_1\nu}\right) ds_1\right\}. \qquad (4.11)$$

The left hand side of (4.11) is the generating function of $(\nu, \phi)_\Lambda$ subtracted its mean and sometimes called the free energy in physics.

4.2 Brascamp-Lieb Inequality

A bound on the variances under non-Gaussian φ-Gibbs measures by those under Gaussian φ-Gibbs measures is called the Brascamp-Lieb inequality. More precisely, for every $\nu = \{\nu(x); x \in \Lambda\} \in \mathbb{R}^\Lambda$, we have

$$\text{var}\,(\nu, \phi)_\Lambda \leq \text{var}^*(\nu, \phi)_\Lambda,$$

where ϕ in the left hand side is the field distributed under μ_Λ^ψ determined from the general convex potential V, while it is distributed in the right hand side under the Gaussian φ-Gibbs measures $\mu_\Lambda^{\psi,G}$ determined from the quadratic potential $V^*(\eta) = \frac{1}{2}c_-\eta^2$. Note that the difference of these two potentials $V - V^*$ is still convex. Brascamp-Lieb inequality claims that the stronger convexity of the potential has larger effect on the φ-field to localize it around its mean. This looks plausible from physical point of view. Brascamp-Lieb inequality does not give any information on the mean of ϕ itself.

The original proof due to [33], [34] of this inequality is rather complicated, but now a simpler proof based on Helffer-Sjöstrand representation is available. We prepare a lemma, in which we denote $\mathbb{R}^\Lambda$ by $L^2(\Lambda, dx)$ with counting measure dx on Λ.

Lemma 4.7. (1) *For every* $f, g \in \mathbb{R}^\Lambda$, *we have*

$$(g, -Q_{\Lambda,0}^{\phi,\psi}f)_\Lambda = \frac{1}{4}\sum_{b \in \Lambda^*} V''(\nabla\phi(b))\nabla f(b)\nabla g(b)$$

$$+ \sum_{\substack{x \in \Lambda, y \notin \Lambda \\ |x-y|=1}} V''(\phi(x) - \psi(y))f(x)g(x).$$

In particular, $Q_{\Lambda,0}^{\phi,\psi}$ *is symmetric on* $L^2(\Lambda, dx)$ *for every* $\phi \in \mathbb{R}^\Lambda$.
(2) *As symmetric operators on* $L^2(\Lambda \times \mathbb{R}^\Lambda, dx \times \mu_\Lambda^{\psi,\rho})$, *we have*

$$-\mathcal{L}_\Lambda^{\psi,\rho} \geq -Q_{\Lambda,0}^\psi.$$

(3) *For every $\phi \in \mathbb{R}^\Lambda$, as symmetric operators on $L^2(\Lambda, dx)$, we have*

$$-Q_{\Lambda,0}^{\phi,\psi} \geq -c_- \Delta_{\Lambda,0} \ ,$$

where c_- is the positive constant in (V3).

Proof. (1) is a simple calculation. From (1), we see that $Q_{\Lambda,0}^\psi$ is symmetric on $L^2(\Lambda \times \mathbb{R}^\Lambda, dx \times \mu_\Lambda^{\psi,\rho})$. On the other hand, the identity (4.5) for the Dirichlet form implies that $-L_\Lambda^{\psi,\rho}$ is nonnegative and symmetric on this space. Thus (2) is shown. (3) is obvious from (1) by taking $f = g$ and noting $V'' \geq c_-$. $\square$

Now we state the Brascamp-Lieb inequality between the variances under $\mu_\Lambda^{\psi,\rho}$ and $\mu_\Lambda^{\psi,G}$, which is the Gaussian finite volume φ-Gibbs measure determined from the quadratic potential V^* without external field.

Theorem 4.8. (Brascamp-Lieb inequality) *For every $\nu \in \mathbb{R}^\Lambda$, we have*

$$\mathrm{var}\left((\nu,\phi)_\Lambda; \mu_\Lambda^{\psi,\rho}\right) \leq \mathrm{var}\left((\nu,\phi)_\Lambda; \mu_\Lambda^{\psi,G}\right) . \tag{4.12}$$

Proof. Taking $F = G = (\nu,\phi)_\Lambda - E^{\mu_\Lambda^{\psi,\rho}}[(\nu,\phi)_\Lambda]$ in (4.2) considered on Λ instead of $\mathbb{Z}^d$, since $\partial H = \left(-\mathcal{L}_\Lambda^{\psi,\rho}\right)^{-1} \partial G$, the left hand side of (4.12) is rewritten as

$$\sum_{x \in \Lambda} E^{\mu_\Lambda^{\psi,\rho}}\left[\nu(x)\left(\left(-\mathcal{L}_\Lambda^{\psi,\rho}\right)^{-1}\nu\right)(x,\phi)\right] \ ,$$

which is bounded above by

$$\leq \sum_{x \in \Lambda} E^{\mu_\Lambda^{\psi,\rho}}\left[\nu(x)\left(-Q_{\Lambda,0}^{\phi,\psi}\right)^{-1}\nu(x)\right]$$

$$\leq \sum_{x \in \Lambda} \nu(x)\left(-c_-\Delta_{\Lambda,0}\right)^{-1}\nu(x) \ ,$$

from Lemma 4.7-(2), (3). However, the last term coincides with the right hand side of (4.12), which is indeed independent of the boundary condition ψ, recall (3.4). $\square$

Theorem 4.9. (Brascamp-Lieb inequality for exponential moments)
We have
$$E^{\mu_\Lambda^{\psi,\rho}}\left[e^{(\nu,\phi-\langle\phi\rangle)_\Lambda}\right] \leq e^{\frac{1}{2}\mathrm{var}\left((\nu,\phi)_\Lambda; \mu_\Lambda^{\psi,G}\right)} . \tag{4.13}$$

In particular,
$$E^{\mu_\Lambda^{\psi,\rho}}\left[e^{|(\nu,\phi-\langle\phi\rangle)_\Lambda|}\right] \leq 2e^{\frac{1}{2}\mathrm{var}\left((\nu,\phi)_\Lambda; \mu_\Lambda^{\psi,G}\right)} , \tag{4.14}$$

where $\langle\phi\rangle = E^{\mu_\Lambda^{\psi,\rho}}[\phi]$.

Proof. (4.13) follows from (4.11) and Theorem 4.8. For (4.14), use $e^{|x|} \le e^x + e^{-x}$.   □

Since the right hand side of (4.13) is $E^{\mu_\Lambda^{\psi,G}}[e^{(\nu,\phi-\langle\phi\rangle)_\Lambda}]$, the exponential moments subtracted their means under φ-Gibbs measures are bounded above by those under the Gaussian measures.

Remark 4.1. (1) *An estimate on*

$$E^{\mu_\Lambda^{\psi,\rho}}\left[e^{\epsilon((\nu,\phi-\langle\phi\rangle))_\Lambda^2}\right]$$

for some $\epsilon > 0$ is known. This is a stronger estimate than (4.13), *cf.* [102], *Proposition 1.1.6. See Remark 3.1 or* [198] *for Gaussian case.*
(2) *See* [131] *for some extension of the Brascamp-Lieb inequalities.*
(3) *For FKG or other basic inequalities used in statistical mechanics, see* [138], [235]. *The relation to Witten's Laplacian is discussed in* [143].

We have the Brascamp-Lieb inequality for the $\nabla\varphi$-field[*]:

$$\text{var}\,(\nabla\varphi(b); \mu_{\Lambda,\xi}^\nabla) \le c_-^{-1} \tag{4.15}$$

for every $b \in \overline{\Lambda^*}$ and $\xi \in \mathcal{X}$. Indeed, set $\tilde{H}_\Lambda(\eta) = \sum_{b\in\overline{\Lambda^*}}(\eta(b) + \eta(-b))^2 + \sum_{\mathfrak{P}\cap\overline{\Lambda^*}\neq\emptyset}(\sum_{b\in\mathfrak{P}} \eta \vee \xi(b))^2$ for $\eta = \{\eta(b); b \in \overline{\Lambda^*}\} \in \mathbb{R}^{\overline{\Lambda^*}}$. Then, $\mu_{\Lambda,\xi}^\nabla$ is the weak limit of $\mu_{\Lambda,\xi}^{\nabla,\epsilon}$ as $\epsilon \downarrow 0$, where

$$\mu_{\Lambda,\xi}^{\nabla,\epsilon}(d\eta) = \frac{1}{Z_{\Lambda,\xi}^\epsilon} \exp\left\{-\frac{1}{2}\sum_{b\in\overline{\Lambda^*}} V(\eta(b)) - \frac{1}{\epsilon}\tilde{H}_\Lambda(\eta)\right\} \prod_{b\in\overline{\Lambda^*}} d\eta(b) \in \mathcal{P}(\mathbb{R}^{\overline{\Lambda^*}}).$$

Since $\tilde{H}_\Lambda$ is convex, we have

$$\text{var}\,(\nabla\varphi(b); \mu_{\Lambda,\xi}^{\nabla,\epsilon}) \le \text{var}\,(\nabla\varphi(b); \prod_{b\in\overline{\Lambda^*}} \mu_b) \le \text{var}\,(\eta(b); N(0, c_-^{-1})) = c_-^{-1},$$

where $\mu_b = z^{-1}e^{-V(\eta(b))/2}d\eta(b) \in \mathcal{P}(\mathbb{R})$. Taking the limit $\epsilon \downarrow 0$, (4.15) is obtained. Note that, even in the Gaussian case, it requires some works to have an estimate on $\text{var}\,(\nabla\varphi(b); \mu_{\Lambda,\xi}^\nabla) = \nabla_{i,x}\nabla_{i,y}G_\Lambda(x,y)|_{x=y}$ for $b = x + e_i$.

4.3 Estimates of Nash-Aronson's Type and Long Correlation

This section establishes the long correlation under the φ- and $\nabla\varphi$-Gibbs measures. The Helffer-Sjöstrand representation is combined with Nash-Aronson's type estimates on the transition probability of the random walk in random environments.

---

[*]communicated by Giacomin.

Let $X = (X_t)_{t \geq 0}$ be the random walk on $\mathbb{Z}^d$ with jump rates $c_{x, \pm e_i}(t)$ from x to its adjacent sites $y = x \pm e_i$ at time t satisfying the symmetry

$$c_{x, \pm e_i}(t) = c_{x \pm e_i, \mp e_i}(t)$$

and the uniformity

$$0 < c_- \leq c_{x, \pm e_i}(t) \leq c_+ . \tag{4.16}$$

Then its transition probability $p(s, x; t, y) = P(X_t = y | X_s = x), t \geq s \geq 0, x, y \in \mathbb{Z}^d$ admits the following three estimates (Propositions 4.10-4.12) of Nash-Aronson's type. The constants C_1, C_2 and $C > 0$ depend only on d and $c_{\pm}$.

Proposition 4.10. (*Giacomin-Olla-Spohn [135], Proposition B3, B4*) *For every $t \geq s \geq 0$ and $x, y \in \mathbb{Z}^d$, we have*

$$p(s, x; t, y) \leq \frac{C_1}{(t-s)_*^{d/2}} \exp \left\{ -\frac{|x-y|}{C_1 \sqrt{(t-s)_*}} \right\} .$$

In addition, if $|x - y| \leq \sqrt{(t-s)_*}$ is satisfied, then we have

$$p(s, x; t, y) \geq \frac{C_2}{(t-s)_*^{d/2}} ,$$

where $(t-s)_* := (t-s) \vee 1$.

Proposition 4.11. (*Delmotte-Deuschel [63]*) *For every $t \geq s \geq 0$ and $x, y \in \mathbb{Z}^d$, we have*

$$p(s, x; t, y) \leq \frac{C}{(t-s)^{d/2}} e^{-\Gamma(|x-y|, t-s)/C} ,$$

where

$$\Gamma(x, t) \geq \begin{cases} \dfrac{|x|^2}{t}, & t \geq |x| \quad (\text{long time regime}) , \\[2mm] \dfrac{|x|}{\sqrt{t}}, & t < |x| \quad (\text{Poisson regime}) . \end{cases}$$

Especially when the jump rates are given by $c_{x, \pm e_i}(t) = V''(\nabla \phi_t(\langle x \pm e_i, x \rangle))$, the transition probability is denoted by $p^{\phi \cdot}(s, x; t, y)$. The above two estimates on p are quenched, i.e., hold uniformly in the environment ϕ_t. For its gradient $\nabla p^{\phi \cdot}$, we have only annealed estimate:

Proposition 4.12. (*[63]*) *For every $t \geq s \geq 0, b \in (\mathbb{Z}^d)^*$ and $y \in \mathbb{Z}^d$, we have*

$$E^{\mu_\Lambda^\psi} \left[\left| \nabla p^{\phi \cdot}(s, \cdot; t, y)(b) \right|^2 \right]^{\frac{1}{2}} \leq \frac{C}{(t-s)^{(d+1)/2}} e^{-\Gamma(|x_b - y|, t-s)/C} ,$$

where $\nabla = \nabla_x$ acts on the variable x.

Nash-Aronson's type estimates are applicable to derive estimates on the correlation functions under the Gibbs measures. The next theorem is an extension of Proposition 3.5 for Gaussian fields to general convex potentials. The lower estimate implies the long correlations.

Theorem 4.13. (*Naddaf and Spencer* [202]) *Assume $d \geq 3$. Then there exist $C, c > 0$, which depend only on d and $c_\pm$, such that*

$$0 \leq E^{\mu_\Lambda^\psi} [\phi(x); \phi(y)] \leq \frac{C}{|x-y|^{d-2}} ,$$

for every $x, y \in \mathbb{Z}^d$. Furthermore, for the φ-Gibbs measure μ on $\mathbb{Z}^d$ which is tempered, shift invariant, mean 0 and ergodic under the spatial shift (cf. Sect. 4.5), we have

$$\frac{c}{|x-y|^{d-2}} \leq E^\mu [\phi(x); \phi(y)] \leq \frac{C}{|x-y|^{d-2}}$$

for every $x, y \in \mathbb{Z}^d$.

Proof. (Giacomin [130] §3.3) From Corollary 4.3 of Helffer-Sjöstrand representation and Proposition 4.10, we have that

$$E^{\mu_\Lambda^\psi} [\phi(x); \phi(y)] = \int_0^\infty E^{\mu_\Lambda^\psi} \left[p_\Lambda^\phi (0, x; t, y) \right] dt$$

$$\leq \int_0^\infty E^{\mu_\Lambda^\psi} \left[p^{\phi^\cdot} (0, x; t, y) \right] dt$$

$$\leq C_1 e^{-\frac{1}{C_1}|x-y|} + \int_1^\infty \frac{C_1}{t^{d/2}} e^{-\frac{|x-y|}{C_1 t^{1/2}}} dt ,$$

where $p_\Lambda^\phi (0, x; t, y) = P(X_t = y, t < \tau_\Lambda | X_0 = x)$ is the transition probability of the random walk killed at the time when it goes outside of Λ. The last integral is, after making a simple change of variables $|x-y|^{-2} t = t'$, bounded by

$$\leq \frac{C_1}{|x-y|^{d-2}} \int_0^\infty \frac{1}{t^{d/2}} e^{-\frac{1}{C_1 t^{1/2}}} dt = \frac{C'}{|x-y|^{d-2}} .$$

Note that this integral converges since $d \geq 3$. The lower bound follows again from Proposition 4.10:

$$E^\mu [\phi(x); \phi(y)] = \int_0^\infty E^\mu \left[p^{\phi^\cdot} (0, x; t, y) \right] dt$$

$$\geq C_2 \int_{|x-y|^2}^\infty \frac{dt}{t^{d/2}} = \frac{2C_2}{(d-2)|x-y|^{d-2}} .$$

$\square$

For the $\nabla\varphi$-fields, the following estimate holds.

Proposition 4.14. *The dimensions* $d \geq 1$ *are arbitrary. We have for every* $x, y \in \mathbb{Z}^d$

$$\left| E^{\mu_\Lambda^\psi} \left[\nabla\phi(b); \nabla\phi(b') \right] \right| \leq \frac{C}{|x_b - x_{b'}|^{d-1}} .$$

Proof. Take $F(\phi) = \nabla\phi(b)$ and $G(\phi) = \nabla\phi(b')$ in Helffer-Sjöstrand representation (Theorem 4.2). Then, since $\partial F(x, \phi) = 1_{\{x=x_b\}} - 1_{\{x=y_b\}}$, one obtains

$$E^{\mu_\Lambda^\psi} \left[\nabla\phi(b); \nabla\phi(b') \right] = \int_0^\infty E^{\mu_\Lambda^\psi} \left[\nabla_x \nabla_y p_\Lambda^{\phi\cdot}(0, \cdot; t, \cdot)(b, b') \right] dt .$$

To estimate this integral, we divide the interval $[0, \infty)$ into the sum of $[0, 1)$ and $[1, \infty)$ as in the proof of Theorem 4.13, and for the latter integral we rudely estimate $|\nabla_x \nabla_y p_\Lambda^{\phi\cdot}(b')| \leq |\nabla_x p_\Lambda^{\phi\cdot}(x_{b'})| + |\nabla_x p_\Lambda^{\phi\cdot}(y_{b'})|$ and use Proposition 4.12. $\square$

Delmotte and Deuschel [64] have elaborated the estimate in Proposition 4.14 as

$$\left| E^{\mu_\Lambda^\psi} \left[\nabla\phi(b); \nabla\phi(b') \right] \right| \leq \frac{C}{|x_b - x_{b'}|^d} ,$$

see the final remark of Giacomin [130] §3.

As we have seen, the correlation functions of Gibbs measures decay slowly and this makes the proof of CLT or LDP difficult. One can say that the loop condition (see (L) in Sect. 2.3) for $\nabla\varphi$-field yields the long dependence.

Remark 4.2. *For the potential* $V(\eta) = \eta^2 + \lambda\eta^4, \lambda > 0$, *the decay of correlations is discussed in* [39, 40, 42, 127, 196]; *see also* [102] *and* (3.5) *of* [124].

4.4 Thermodynamic Limit and Construction of $\nabla\varphi$-Gibbs Measures

The next theorem is shown by taking thermodynamic limit for a sequence of finite volume $\nabla\varphi$-Gibbs measures with periodic boundary conditions, see Sect. 9.2 (and Definition 2.2) for the class (ext $\mathcal{G}^\nabla$)$_u$ of measures and recall Definition 2.3 for shift invariance and ergodicity. The tightness of the sequence of measures is a consequence of Brascamp-Lieb inequality. This method is applicable only to strictly convex potentials; see Remark 4.4 for nonconvex potentials.

Theorem 4.15. [124] *(existence of* $\nabla\varphi$-*Gibbs measures) For every* $u \in \mathbb{R}^d$, *there exists* $\mu^\nabla =: \mu_u^\nabla \in$ (ext $\mathcal{G}^\nabla$)$_u$, *i.e., a tempered, shift invariant, mean* u *and ergodic* $\nabla\varphi$-*Gibbs measure* μ_u^∇ *exists. Furthermore, it satisfies* $E^{\mu_u^\nabla} [e^{\beta(\eta(b) - u_b)^2}] < \infty$ *for some* $\beta > 0$, *where* $u_b = u_i$ *if the bond b is i-directed.*

Proof. Let $\mathbb{T}_N^d = (\mathbb{Z}/N\mathbb{Z})^d$ be the lattice torus of size N and let $\mathbb{T}_N^{d,*}$ be the set of all directed bonds in $\mathbb{T}_N^d$. Let $\mathcal{X}_{\mathbb{T}_N^d}$ be the family of all $\eta \in \mathbb{R}^{\mathbb{T}_N^{d,*}}$ satisfying the loop condition on the torus ($\sum_{b \in \mathfrak{C}} \eta(b) = 0$ for all closed loops $\mathfrak{C}$ in $\mathbb{T}_N^d$, see Sect. 2.3), and define $\tilde{\mu}_{N,u}^\nabla \in \mathcal{P}(\mathcal{X}_{\mathbb{T}_N^d})$ by

$$\tilde{\mu}_{N,u}^\nabla(d\tilde{\eta}) := \frac{1}{Z_{N,u}} \exp\left\{ -\frac{1}{2} \sum_{b \in \mathbb{T}_N^{d,*}} V(\tilde{\eta}(b) + u_b) \right\} d\tilde{\eta}_N , \qquad (4.17)$$

where $Z_{N,u}$ is the normalization constant and $d\tilde{\eta}_N$ is the uniform measure on the affine space $\mathcal{X}_{\mathbb{T}_N^d}$. Let $\mu_{N,u}^\nabla$ be the distribution of $\{\eta(b) := \tilde{\eta}(b) + u_b\}$ under $\tilde{\mu}_{N,u}^\nabla$. One can regard $\mu_{N,u}^\nabla \in \mathcal{P}(\mathcal{X})$ by extending it periodically. Then, from the Brascamp-Lieb inequality (on the torus), we have for every $\lambda > 0$

$$\sup_{N, u \in \mathbb{R}^d} E^{\mu_{N,u}^\nabla}[e^{\lambda|\eta(b) - u_b|}] < \infty . \qquad (4.18)$$

This implies the tightness of the measures $\{\mu_{N,u}^\nabla\}_N$. Accordingly, along a proper subsequence $N' \to \infty$, $\mu_{N',u}^\nabla$ weakly converges to a certain measure μ_u^∇. One can easily show that $\mu_u^\nabla \in \mathcal{G}^\nabla$, $E^{\mu_u^\nabla}[\eta(e_i)] = u_i$, $E^{\mu_u^\nabla}[e^{\lambda|\eta(b) - u_b|}] < \infty$ and from Proposition 4.14

$$\left| E^{\mu_u^\nabla}[\eta(b); \eta(b')] \right| \leq \frac{C}{|x_b - x_{b'}|^{d-1}} .$$

We may now suppose $d \geq 2$, since μ_u^∇ is a linear combination of Bernoulli measures when $d = 1$ (see Remark 4.5 below), and in this case this bound implies the ergodicity of μ_u^∇. □

Remark 4.3. *The periodic boundary conditions are taken for the limit measure μ_u^∇ to be automatically shift invariant. Instead, one may consider the sequence $\{\mu_{\Lambda_\ell}^{\psi_u}\}_\ell$ with the Dirichlet boundary conditions $\psi_u(x) \equiv u \cdot x$ and the distributions $\{\mu_{\Lambda_\ell}^{\nabla,\psi_u}\}_\ell$ of $\nabla\varphi$-field under $\mu_{\Lambda_\ell}^{\psi_u}$. Then the tightness of $\{\mu_{\Lambda_\ell}^{\nabla,\psi_u}\}_\ell$ is similar, however, the shift invariance of the limit measures seems nontrivial.*

Remark 4.4. *The general theory in statistical mechanics seems to work for the construction of the $\nabla\varphi$-Gibbs measures even for nonconvex potentials V, which diverge sufficiently rapidly as $|\eta| \to \infty$, cf. [230]. Indeed, the argument due to Giacomin is the following. First define the specific free energy $f_\ell(\beta)$ for $\nabla\varphi$-fields on $\mathbb{T}_N^d$ at inverse temperature $\beta > 0$ in a similar manner to $\sigma_\ell^*(u)$ in (5.1) for φ-fields below. Then, the derivative $f'(\beta)$ (or $f'(\beta\pm)$) exists for the limit $f(\beta) = \lim_{\ell \to \infty} f_\ell(\beta)$, from which one obtains a uniform bound. This implies the tightness of $\{\mu_{N,u}^\nabla\}$.*

Remark 4.5. *In one dimension μ_u^∇ is the Bernoulli measure, i.e., under μ_u^∇, $\{\nabla\phi(x) \equiv \phi(x+1) - \phi(x); x \in \mathbb{Z}\}$ is an i.i.d. sequence and the distribution*

of each $\nabla\phi(x)$ is given by the Cramér transform of $\nu(d\eta) = Z^{-1}e^{-V(\eta)}\,d\eta \in$
$\mathcal{P}(\mathbb{R})$. In other words, if we define $\hat{\nu}_\lambda \in \mathcal{P}(\mathbb{R})$ for $\lambda \in \mathbb{R}$ by (5.25) and deter-
mine the function $\lambda = \lambda(u)$ of $u \in \mathbb{R}$ by the relation $E^{\hat{\nu}_\lambda}[\eta] = u$ (see Sect.
5.5), then we have

$$\mu_u^\nabla(d\eta) = \prod_{x \in \mathbb{Z}} \nu_u(d\eta(x)) \in \mathcal{P}(\mathbb{R}^{\mathbb{Z}}), \quad \eta = \{\eta(x); x \in \mathbb{Z}\},$$

where $\nu_u = \hat{\nu}_{\lambda(u)}$. Indeed, [141] Theorem 3.5 shows that $\mu_{\Lambda_\ell}^{\nabla,\psi_u}$ in Remark 4.3
converges to this μ_u^∇ as $\ell \to \infty$. In particular, in one dimension, the potential
V needs not be convex, but the conditions (1.3)–(1.6) in [141] (with $V(\eta)$ in
place of $\phi(x)$) are sufficient.

4.5 Construction of φ-Gibbs Measures

The Gibbs measures for $\nabla\varphi$-field have been constructed for all dimensions
d, but the Gibbs measures for φ-field are unnormalizable (i.e., finite volume
Gibbs measures for φ-field do not converge as $\Lambda \nearrow \mathbb{Z}^d$) if $d \leq 2$ and normalizable if $d \geq 3$. We have indeed the following theorem.

Theorem 4.16. *If $d \geq 3$, for every $h \in \mathbb{R}$, there exists a φ-Gibbs measure*
$\mu \equiv \mu_h$ on $\mathbb{Z}^d$ with mean h, i.e., $E^\mu[\phi(x)] = h$ for all $x \in \mathbb{Z}^d$.

Proof. Consider the sequence of finite volume φ-Gibbs measures $\{\mu_{\Lambda_\ell}^0 \in$
$\mathcal{P}(\mathbb{R}^{\mathbb{Z}^d})\}_\ell$ with 0-boundary conditions. Then, by the symmetry of V, the mean
is 0: $E^{\mu_{\Lambda_\ell}^0}[\phi(x)] = 0$. When $d \geq 3$, since the variance $G_{\Lambda_\ell}(x,x)$ of the Gaussian
system is uniformly bounded in ℓ, Brascamp-Lieb inequality (Theorem 4.9)
proves

$$\sup_{x \in \mathbb{Z}^d} \sup_{\ell \in \mathbb{N}} E^{\mu_{\Lambda_\ell}^0}\left[e^{\lambda|\phi(x)|}\right] < \infty, \quad \lambda > 0.$$

Therefore, the sequence $\{\mu_{\Lambda_\ell}^0\}_\ell$ is tight and has a limit μ along a proper
subsequence $\ell' \to \infty$. It is obvious that μ is a φ-Gibbs measure with mean 0.
The distribution μ_h of $\phi + h$, where ϕ is μ-distributed, is a φ-Gibbs measure
with mean h. □

Remark 4.6. *It may be possible to construct shift invariant μ_h (cf. Re-*
mark 4.3). In fact, Giacomin suggests to consider $\mu_{\Lambda_\ell;m}^0$ associated with the
massive Hamiltonian instead of $\mu_{\Lambda_\ell}^0$ in the proof of Theorem 4.16. Then,
$\mu_m = \lim_{\ell' \to \infty} \mu_{\Lambda_{\ell'};m}^0$ exists and is shift invariant. Finally, if $d \geq 3$, the
limit $\mu = \lim_{m \downarrow 0} \mu_m$ exists by means of the Brascamp-Lieb inequality and we
see the shift invariance of μ_h.

5 Surface Tension

The surface tension $\sigma = \sigma(u), u \in \mathbb{R}^d$ physically describes the macroscopic energy of a surface with tilt u, i.e., a d dimensional hyperplane located in $\mathbb{R}^{d+1}$ with normal vector $(-u, 1) \in \mathbb{R}^{d+1}$. It is mathematically a fundamental quantity, as it will appear in several limit theorems, e.g., in the rate functional of LDP or diffusion coefficient in the hydrodynamic limit.

5.1 Definition of Surface Tension

The surface tension will be defined thermodynamically from the Hamiltonian $H(\phi)$ in such a way to reflect statistical property of random interfaces with mean tilt u. Let $\psi_u \in \mathbb{R}^{\mathbb{Z}^d}, u = (u_i)_{i=1}^d \in \mathbb{R}^d$, be tilted height variables determined by $\psi_u(x) = u \cdot x, x \in \mathbb{Z}^d$ and set for $\ell \in \mathbb{N}$

$$
\sigma_\ell^*(u) = -\frac{1}{|\Lambda_\ell|} \log Z_{\Lambda_\ell}^{\psi_u}
$$

$$
= -\frac{1}{(2\ell+1)^d} \log \int_{\mathbb{R}^{\Lambda_\ell}} \exp\left\{-H_{\Lambda_\ell}^{\psi_u}(\phi)\right\} d\phi_{\Lambda_\ell}, \tag{5.1}
$$

where $\Lambda_\ell = [-\ell, \ell]^d \cap \mathbb{Z}^d$ is a cube with side length $2\ell + 1$ and $Z_{\Lambda_\ell}^{\psi_u}$ is the normalization constant given by (2.5) with the boundary condition ψ_u. The function $\sigma_\ell^*(u)$ is the specific free energy of interfaces with tilt u insisted through the boundary condition. Note that $|\Lambda_\ell| = (2\ell+1)^d$ is the order of the surface area of the boundary of interfaces settled in $d+1$ dimensional space $\mathbb{R}^{d+1}$.

One can show, based on the subadditivity of σ_ℓ^* in ℓ, that its limit as $\ell \to \infty$ exists.

Theorem 5.1. [124] *The limit*

$$
\sigma^*(u) = \lim_{\ell \to \infty} \sigma_\ell^*(u) \in [-\infty, \infty)
$$

exists.

We shall normalize the limit function $\sigma^*(u)$ as $\sigma(u) = \sigma^*(u) - \sigma^*(0)$ so that $\sigma(0) = 0$; note that $\sigma^*(u) \in (-\infty, \infty)$ is shown comparing with the case of quadratic potentials, see Sect. 5.2. The function $\sigma(u), u \in \mathbb{R}^d$ is called the **(normalized) surface tension**. By Theorem 5.1, we have

$$
\sigma(u) = -\lim_{\ell \to \infty} \frac{1}{|\Lambda_\ell|} \log \frac{Z_{\Lambda_\ell}^{\psi_u}}{Z_{\Lambda_\ell}^{\psi_0}}. \tag{5.2}
$$

The ratio $Z_{\Lambda_\ell}^{\psi_u}/Z_{\Lambda_\ell}^{\psi_0}$ is roughly equal to the probability to find interfaces with mean tilt u under $\mu_{\Lambda_\ell}^{\psi_0}$ and therefore

$$\mu\left(\text{tilt of } h^N \sim u\right) \underset{N \to \infty}{\asymp} \exp\{-N^d \sigma(u)\}, \tag{5.3}$$

for the φ-Gibbs measure μ (with tilt 0) and macroscopically scaled height variables h^N defined by (2.16). This broadly explains the meaning of the surface tension.

Another Definition

Sheffield [230] gives a different but actually an equivalent definition for the surface tension. Let μ_Λ^{free} be the finite volume φ-Gibbs measure on $\Lambda \Subset \mathbb{Z}^d$ given by (2.4) and determined from the free boundary condition, i.e., the sum for the Hamiltonian (2.1) is taken for all $\langle x, y \rangle \subset \Lambda$ only. Denote the distribution of $\{\nabla\phi(b); b \in \Lambda^*\}$ under μ_Λ^{free} by $\mu_{\Lambda^*}^{\nabla, free} \in \mathcal{P}(\mathbb{R}^{\Lambda^*})$. Recall that for two probability measures μ and ν

$$\mathcal{H}(\mu|\nu) = E^\mu \left[\log \frac{d\mu}{d\nu}\right] \tag{5.4}$$

defines the relative entropy of μ with respect to ν. Then, the **specific free energy** of shift invariant measure $\mu^\nabla \in \mathcal{P}(\mathbb{R}^{(\mathbb{Z}^d)^*})$ is defined by the relative entropy with respect to the $\nabla\varphi$-Gibbs measure with free boundary condition per site:

$$F(\mu^\nabla) = \lim_{\ell \to \infty} \frac{1}{|\Lambda_\ell|} \mathcal{H}(\mu_{\Lambda_\ell^*}^\nabla | \mu_{\Lambda_\ell^*}^{\nabla, free}), \tag{5.5}$$

where $\mu_{\Lambda_\ell^*}^\nabla$ stands for the marginal distribution of $\{\nabla\phi(b); b \in \Lambda_\ell^*\}$ under μ^∇. The surface tension has another expression

$$\sigma(u) = \inf_{\mu^\nabla : \text{mean } u} F(\mu^\nabla), \tag{5.6}$$

where the infimum is taken over all shift invariant $\mu^\nabla \in \mathcal{P}(\mathbb{R}^{(\mathbb{Z}^d)^*})$ with mean tilt u: $E^{\mu^\nabla}[\eta(e_i)] = u_i$ for every $1 \le i \le d$.

Sheffield establishes the variational characterization for the $\nabla\varphi$-Gibbs measures for general tilt u: the minimizer $\mu^\nabla \equiv \mu^{\nabla,(u)}$ of (5.6) is in fact the $\nabla\varphi$-Gibbs measure for each u. This tells that $\mu_{\Lambda_\ell^*}^{\nabla,(u)} = \mu_{\Lambda_\ell^*,u}^\nabla$ (= the distribution of $\{\nabla\phi(b); b \in \Lambda_\ell^*\}$ under $\mu_{\Lambda_\ell}^{\psi_u}$) and for such μ^∇, since one can expect that the free boundary condition may be replaced with the 0-boundary condition, the specific free energy $F(\mu^\nabla)$ is the limit of

$$\frac{1}{|\Lambda_\ell|} \mathcal{H}(\mu_{\Lambda_\ell^*,u}^\nabla | \mu_{\Lambda_\ell^*,0}^\nabla) \sim -\frac{1}{|\Lambda_\ell|} \log \frac{Z_{\Lambda_\ell}^{\psi_u}}{Z_{\Lambda_\ell}^{\psi_0}}$$

as $\ell \to \infty$. This coincides with the definition (5.2) of the normalized surface tension.

5.2 Quadratic Potentials

The surface tension is explicitly computable for quadratic potentials. Recall that $\Delta_\Lambda \equiv \Delta_{\Lambda,0}$ denotes the discrete Laplacian on Λ with the boundary condition 0, see Sect. 3.1.

Proposition 5.2. *Assume* $V(\eta) = \frac{c}{2}\eta^2, c > 0$. *Then, the corresponding unnormalized surface tension* $\sigma^*(u) \equiv \sigma_c^*(u)$ *is given by*

$$\sigma_c^*(u) = \frac{1}{2}c|u|^2 - \frac{1}{2}\log 2\pi c^{-1} + \frac{1}{2}\lim_{\ell\to\infty}\frac{1}{|\Lambda_\ell|}\log\det(-\Delta_{\Lambda_\ell}),\qquad(5.7)$$

where $\det(-\Delta_{\Lambda_\ell})$ *denotes the determinant of* $-\Delta_{\Lambda_\ell}$ *regarding it as a* $|\Lambda_\ell| \times |\Lambda_\ell|$ *matrix and* $|u| = \sqrt{\sum_{i=1}^d u_i^2}$ *stands for the Euclidean norm of u. The eigenvalues of* Δ_{Λ_ℓ} *are known (see, e.g., [138] p.185, (9.5.12) or [108]) and therefore* $\det(-\Delta_{\Lambda_\ell})$ *is specifically computable.*

Proof. Since ψ_u is harmonic, we have $\overline{\phi}_{\Lambda_\ell,\psi_u} = \psi_u$ and therefore, from the proof of Proposition 3.1

$$Z_{\Lambda_\ell}^{\psi_u} = e^{-\frac{c}{2}B_{\ell,u}}\int_{\mathbb{R}^{\Lambda_\ell}} e^{\frac{c}{2}((\phi-\psi_u),\Delta_{\Lambda_\ell}(\phi-\psi_u))_{\Lambda_\ell}}\,d\phi_{\Lambda_\ell},\qquad(5.8)$$

where $B_{\ell,u}$ is the boundary term defined by

$$B_{\ell,u} = \sum_{\substack{x\in\Lambda_\ell,y\notin\Lambda_\ell\\|x-y|=1}} \psi_u(y)\nabla\psi_u(\langle y,x\rangle).$$

However, the integral in the right hand side of (5.8) can be rewritten as

$$= \int_{\mathbb{R}^{\Lambda_\ell}} e^{-\frac{c}{2}(\phi,-\Delta_{\Lambda_\ell}\phi)_{\Lambda_\ell}}\,d\phi_{\Lambda_\ell} = (2\pi c^{-1})^{|\Lambda_\ell|/2}\big(\det(-\Delta_{\Lambda_\ell})\big)^{-1/2},$$

while the boundary term is equal to

$$B_{\ell,u} = 2\sum_{i=1}^d \sum_{|y_j|\le\ell,j\neq i} (u_jy_j + u_i(\ell+1))u_i$$
$$= 2(\ell+1)(2\ell+1)^{d-1}|u|^2.$$

In this way $\sigma_\ell^*(u)$ in (5.1) is explicitly calculated and the conclusion follows by taking the limit $\ell\to\infty$. $\square$

By the basic conditions (2.2) on V, the potential V is in general estimated by quadratic functions both from above and below:

$$V(0) + \frac{1}{2}c_-\eta^2 \le V(\eta) \le V(0) + \frac{1}{2}c_+\eta^2.$$

This proves the following bounds on the unnormalized surface tension $\sigma^*(u)$ corresponding to V

$$\sigma_{c_-}^*(u) + d \cdot V(0) \leq \sigma^*(u) \leq \sigma_{c_+}^*(u) + d \cdot V(0) ,$$

which, in particular, implies $\sigma^*(u) \in (-\infty, \infty)$.

5.3 Fundamental Properties of Surface Tension

Here we summarize several properties of the surface tension. In the case of quadratic potentials, the (normalized) surface tension is given by $\sigma_c(u) = \frac{c}{2}|u|^2$ as we have seen in Proposition 5.2 and all properties listed below are obvious.

The following theorem and its corollary indicate that the function σ is strictly convex, symmetric, $\sigma \in C^1$ and its derivative $\nabla\sigma$ is Lipschitz continuous; in particular, a surface with gentle slope has low energy.

Theorem 5.3. ([124]; [77], [135] *for* (3)) *The function* $\sigma = \sigma(u) \in [0, \infty)$ *enjoys the following properties.*
(1) *(regularity)* $\sigma \in C^1(\mathbb{R}^d)$ *and* $\nabla\sigma = (\partial\sigma/\partial u_i)_{i=1}^d$ *is Lipschitz continuous, i.e., for some* $C > 0$,

$$|\nabla\sigma(u) - \nabla\sigma(v)| \leq C|u - v|, \quad u, v \in \mathbb{R}^d . \tag{5.9}$$

(2) *(symmetry)* $\sigma(-u) = \sigma(u)$.
(3) *(strict convexity)* *With constants* $c_-, c_+ > 0$ *in* (V3) *of* (2.2), *we have for every* $u, v \in \mathbb{R}^d$ *that*

$$\frac{1}{2}c_-|u - v|^2 \leq \sigma(v) - \sigma(u) - (v - u) \cdot \nabla\sigma(u) \leq \frac{1}{2}c_+|u - v|^2 . \tag{5.10}$$

The Lipschitz continuity of $\nabla\sigma$ is shown based on the coupling used for the proof of Theorem 9.3 with the help of (5.14) below. The strict convexity of σ is a consequence of uniform strict convexity of σ_ℓ^* in ℓ, cf. [87] for the pyramid inequality and also [199]. See the next subsection for the proof. Theorem 5.3-(3) implies the following.

Corollary 5.4. *For every* $u, v \in \mathbb{R}^d$, *we have that*

$$c_-|u - v|^2 \leq (u - v) \cdot (\nabla\sigma(u) - \nabla\sigma(v)) \leq c_+|u - v|^2 . \tag{5.11}$$

In particular,
$$c_-|u|^2 \leq u \cdot \nabla\sigma(u) \leq c_+|u|^2 . \tag{5.12}$$

Proof. The first estimate (5.11) is immediate by taking the sum of each side of (5.10) with itself, but with u and v replaced by each other. The second (5.12) is from (5.11) with $v = 0$ noting that $\nabla\sigma(0) = 0$. $\square$

Problem 5.1. *Theorem 5.3-(1) nearly establishes "$\sigma \in C^2(\mathbb{R}^d)$", but this is actually not yet proved and, indeed, remains to be one of the important open problems. Such problem on the regularity of σ is related to the CLT (Sects. 8, 11). See also a recent approach by Caputo and Ioffe [48].*

Remark 5.1. *(Physical argument on σ) For the SOS model or the sine-Gordon model, which is its continuous-spin version in a sense, the surface tension may have a cusp at $u = 0$ and therefore $\sigma \notin C^1$ in general, see Fröhlich and Spencer [107], [240]. This reflects the phenomena called roughening transition, cf. Remark 6.7. Moreover, if the potential V is nonconvex, it is conjectured that σ need not be in C^1; i.e., the singularity of σ is related to the phase transition of first order exhibited by the system. See Fernández et al. [100].*

Theorem 5.5. [124] *(Thermodynamic identities)*

$$u_i = E^{\mu_u^\nabla} [\eta(e_i)] , \tag{5.13}$$

$$\frac{\partial \sigma}{\partial u_i}(u) = E^{\mu_u^\nabla} [V'(\eta(e_i))] , \tag{5.14}$$

$$u \cdot \nabla \sigma(u) + 1 = E^{\mu_u^\nabla} \left[\sum_{i=1}^{d} \eta(e_i) V'(\eta(e_i)) \right] , \tag{5.15}$$

where $\mu_u^\nabla \in \mathcal{P}(\mathcal{X})$ is the unique probability measure in $(\text{ext } \mathcal{G}^\nabla)_u$, in other words, $\nabla\varphi$-pure phase, see Sect. 9.2.

The identity (5.13) is just by definition, while (5.15) is shown by the integration by parts for similar integrals appearing in σ_ℓ^*, see Lemma 5.7 below.

5.4 Proof of Theorems 5.3 and 5.5

The first lemma is to replace the boundary conditions ψ_u in (5.2) with the periodic boundary conditions as we have done in the proof of Theorem 4.15. The proof of the lemma can be found at Appendix II of [124].

Lemma 5.6. *Let $Z_{N,u}$ be the normalization constant in (4.17) and set*

$$\sigma_N(u) = -\frac{1}{|\mathbb{T}_N^d|} \log \frac{Z_{N,u}}{Z_{N,0}} .$$

Then we have

$$\sigma(u) = \lim_{N \to \infty} \sigma_N(u) .$$

This lemma is more convenient for us than (5.2), since the $\nabla\varphi$-Gibbs measure μ_u^∇ was constructed under the periodic boundary conditions.

Lemma 5.7. *Let* $\mu_{N,u}^{\nabla} \in \mathcal{P}(\mathcal{X}_{\mathbb{T}_N^d})$ *be the measure introduced in the proof of Theorem 4.15. Then we have the following three identities:*

$$u_b = E^{\mu_{N,u}^{\nabla}}[\eta(b)], \tag{5.16}$$

$$\frac{\partial \sigma_N}{\partial u_i}(u) = E^{\mu_{N,u}^{\nabla}}[V'(\eta(e_i))], \tag{5.17}$$

$$u \cdot \nabla \sigma_N(u) + 1 = E^{\mu_{N,u}^{\nabla}}\left[\sum_{i=1}^d \eta(e_i)V'(\eta(e_i))\right]. \tag{5.18}$$

Proof. Since $\sum_{b \in \mathfrak{C}_i} \tilde{\eta}(b) = 0$ holds for every $\tilde{\eta} \in \mathcal{X}_{\mathbb{T}_N^d}$ along with the closed loop $\mathfrak{C}_i$ parallel to the ith axis, the shift invariance of $\tilde{\mu}_{N,u}^{\nabla}$ (on the torus) implies $E^{\tilde{\mu}_{N,u}^{\nabla}}[\tilde{\eta}(b)] = 0$. This shows (5.16). To see (5.17), noting that $V'(\eta) = -V'(-\eta)$, we rewrite its left hand side as

$$-\frac{1}{|\mathbb{T}_N^d|Z_{N,u}}\frac{\partial}{\partial u_i}\int e^{-\frac{1}{2}\sum_{b \in \mathbb{T}_N^{d,*}} V(\tilde{\eta}(b)+u_b)} \, d\tilde{\eta}_N$$

$$= \frac{1}{|\mathbb{T}_N^d|Z_{N,u}}\int \sum_{b' \in \mathbb{T}_N^{d,*}:b'\|e_i} V'(\tilde{\eta}(b')+u_{b'})e^{-\frac{1}{2}\sum_{b \in \mathbb{T}_N^{d,*}} V(\tilde{\eta}(b)+u_b)} \, d\tilde{\eta}_N$$

and this coincides with the right hand side, where $b' \parallel e_i$ means that the bond b' is i-directed. Finally, the third identity (5.18) follows from

$$\sum_{b \in \mathbb{T}_N^{d,*}} E^{\tilde{\mu}_{N,u}^{\nabla}}[\tilde{\eta}(b)V'(\tilde{\eta}(b)+u_b)] = 2|\mathbb{T}_N^d|, \tag{5.19}$$

since the right hand side of (5.18) is equal to

$$\frac{1}{2|\mathbb{T}_N^d|}\sum_{b \in \mathbb{T}_N^{d,*}} E^{\tilde{\mu}_{N,u}^{\nabla}}[(\tilde{\eta}(b)+u_b)V'(\tilde{\eta}(b)+u_b)]$$

and if we note (5.17). However, the left hand side of (5.19) can be rewritten as

$$\sum_{b \in \mathbb{T}_N^{d,*}}\int_{\mathbb{R}^{\mathbb{T}_N^d}\setminus\{O\}} \nabla\tilde{\phi}(b)V'(\nabla\tilde{\phi}(b)+u_b)F(\tilde{\phi}) \prod_{x \in \mathbb{T}_N^d\setminus\{O\}} d\tilde{\phi}(x)$$

$$= -2\sum_{y \in \mathbb{T}_N^d}\int_{\mathbb{R}^{\mathbb{T}_N^d}\setminus\{O\}} \tilde{\phi}(y)\frac{\partial F}{\partial\tilde{\phi}(y)}(\tilde{\phi}) \prod_{x \in \mathbb{T}_N^d\setminus\{O\}} d\tilde{\phi}(x)$$

$$= 2\sum_{y \in \mathbb{T}_N^d}\int_{\mathbb{R}^{\mathbb{T}_N^d}\setminus\{O\}} F(\tilde{\phi}) \prod_{x \in \mathbb{T}_N^d\setminus\{O\}} d\tilde{\phi}(x) = 2|\mathbb{T}_N^d|,$$

by the integration by parts, where $\tilde{\phi}(O) = h$ is arbitrarily taken and we set

$$F(\tilde{\phi}) = \frac{1}{Z_{N,u}} e^{-\frac{1}{2}\sum_{b\in\mathbb{T}_N^{d,*}} V(\nabla\tilde{\phi}(b)+u_b)} ,$$

for $\tilde{\phi} = \{\tilde{\phi}(x); x \in \mathbb{T}_N^d \setminus \{O\}\}$. This concludes (5.19). $\square$

Proof (Theorems 5.3 and 5.5). As we have seen in the proof of Theorem 4.15, $\mu_{N,u}^\nabla$ weakly converges to μ_u^∇ as $N \to \infty$ (we actually need not to take the subsequence). Therefore, noting the uniform estimate (4.18), we obtain

$$\lim_{N\to\infty} E^{\mu_{N,u}^\nabla}[V'(\eta(e_i))] = E^{\mu_u^\nabla}[V'(\eta(e_i))] . \tag{5.20}$$

This convergence is uniform in u in any bounded set of $\mathbb{R}^d$. Let us take the limit $N \to \infty$ in the following trivial identity

$$\sigma_N(u) - \sigma_N(v) = \int_0^1 (u - v) \cdot \nabla\sigma_N(tu + (1-t)v)\, dt . \tag{5.21}$$

From Lemma 5.6 the left hand side converges to $\sigma(u) - \sigma(v)$, while the limit of the right hand side is computable by (5.17) and (5.20), and we have

$$\sigma(u) - \sigma(v) = \int_0^1 \sum_{i=1}^d (u_i - v_i) E^{\mu_{tu+(1-t)v}^\nabla}[V'(\eta(e_i))]\, dt .$$

However, the uniqueness of the (tempered, shift invariant) ergodic $\nabla\varphi$-Gibbs measure for each tilt u (see Theorem 9.5 and Corollary 9.6 below) implies that, as $u_n \to u$, $\mu_{u_n}^\nabla$ weakly converges to μ_u^∇. This, in particular, proves that $E^{\mu_u^\nabla}[V'(\eta(e_i))]$ is continuous in u. Thus we have shown that $\sigma \in C^1(\mathbb{R}^d)$ and the identity (5.14). The symmetry of $\sigma(u)$ follows from that of $\sigma_\ell^*(u)$, which is readily seen from the symmetry of the potential V. The identity (5.15) is a consequence of (5.18) by letting $N \to \infty$.

To prove the Lipschitz continuity (5.9) of $\nabla\sigma$, we need to apply the dynamic coupling (see Sect. 9). In fact, by noting (5.14) and

$$|V'(\eta(e_i)) - V'(\bar{\eta}(e_i))| \leq c_+ |\eta(e_i) - \bar{\eta}(e_i)| ,$$

we obtain (5.9) from Proposition 9.8.

The proof of the strict convexity (5.10) is only left. To this end, it suffices to show (5.10) for σ_N in place of σ. Consider the Hessian of σ_N:

$$D^2\sigma_N(u) = \left(\frac{\partial^2\sigma_N}{\partial u_i \partial u_j}\right)_{1\leq i,j\leq d} .$$

Then, for every $\lambda = (\lambda_i)_{i=1}^d \in \mathbb{R}^d$, we have that

$$(D^2\sigma_N(u)\lambda, \lambda) = \sum_{i=1}^d \lambda_i^2 E^{\mu_{N,u}^\nabla}[V''(\eta_i(O))]$$

$$- \frac{1}{|\mathbb{T}_N^d|} \text{var}\left(\sum_{i=1}^d \lambda_i \sum_{x\in\mathbb{T}_N^d} V'(\eta_i(x)); \mu_{N,u}^\nabla\right) . \tag{5.22}$$

Indeed, this identity follows by computing $\partial^2 \log Z_{N,u}/\partial u_i \partial u_j$. In the left hand side $(\, , \,)$ denotes the inner product of $\mathbb{R}^d$ (which is usually denoted by $\cdot$), while $\eta_i(x) := \eta(\langle x + e_i, x \rangle)$ in the right hand side. Since the second term in the right hand side of (5.22) is nonpositive, we obtain from $V'' \le c_+$ the upper bound

$$\left(D^2 \sigma_N(u) \lambda, \lambda \right) \le c_+ |\lambda|^2 . \tag{5.23}$$

In order to get the lower bound, we use the Helffer-Sjöstrand representation for the φ-field on $\mathbb{T}_N^d \setminus \{O\}$ defined by

$$\phi(x) = \sum_{b \in \mathfrak{C}_{O,x}} \eta(b), \quad x \ne O ,$$

where $\mathfrak{C}_{O,x}$ is a chain connecting O and x. We may think of $\phi(O) = 0$. Set

$$F(\phi) = \sum_{i=1}^d \lambda_i \sum_{x \in \mathbb{T}_N^d} V'(\phi(x + e_i) - \phi(x)) .$$

Then, since

$$\partial F(x, \phi) = \sum_{i=1}^d \lambda_i \left\{ V''(\phi(x) - \phi(x - e_i)) - V''(\phi(x + e_i) - \phi(x)) \right\} ,$$

from Theorem 4.2 (and its proof), we have the representation

$$\mathrm{var}\, (F; \mu_{N,u}^\nabla) = \sum_{x \in \mathbb{T}_N^d \setminus \{O\}} \langle \partial F(x, \phi)(-\mathcal{L})^{-1} \partial F(x, \phi) \rangle .$$

Here $\langle \cdot \rangle$ denotes the expectation under $\mu_{N,u}^\nabla$ and $\mathcal{L} = L^{0,0}_{\mathbb{T}_N^d \setminus \{O\}} + Q^{\phi,0}_{\mathbb{T}_N^d \setminus \{O\},0}$. The right hand side can be further rewritten into

$$\sup_{f=f(x,\phi)} \left\{ 2 \sum_{x \in \mathbb{T}_N^d \setminus \{O\}} \langle f(x, \phi) \partial F(x, \phi) \rangle - \sum_{x \in \mathbb{T}_N^d \setminus \{O\}} \langle f(x, \phi)(-\mathcal{L})f(x, \phi) \rangle \right\} ,$$

where the functions f satisfy the condition $f(O, \phi) = 0$. However, each term in the above supremum can be rewritten as

$$\sum_{x \in \mathbb{T}_N^d \setminus \{O\}} \langle f(x, \phi) \partial F(x, \phi) \rangle = \sum_{x \in \mathbb{T}_N^d} \sum_{i=1}^d \lambda_i \langle \nabla_i f(x, \phi) V''(\eta_i(x)) \rangle ,$$

$$\sum_{x \in \mathbb{T}_N^d \setminus \{O\}} \langle f(x, \phi)(-\mathcal{L})f(x, \phi) \rangle = \sum_{x \in \mathbb{T}_N^d} \sum_{i=1}^d \langle V''(\eta_i(x))(\nabla_i f(x, \phi))^2 \rangle$$

$$+ \sum_{x \in \mathbb{T}_N^d \setminus \{O\}} \left\langle \left(\frac{\partial f}{\partial \phi(x)} \right)^2 \right\rangle ,$$

and, therefore, we have from (5.22) that

$$
\left(D^2\sigma_N(u)\lambda, \lambda\right) = \frac{1}{|\mathbb{T}_N^d|} \inf_f \left\{ \sum_{x \in \mathbb{T}_N^d} \sum_{i=1}^d \langle V''(\eta_i(x))(\lambda_i - \nabla_i f(x, \phi))^2 \rangle \right.
$$

$$
\left. + \sum_{x \in \mathbb{T}_N^d \setminus \{O\}} \left\langle \left(\frac{\partial f}{\partial \phi(x)} \right)^2 \right\rangle \right\}.
$$

Since $V'' \geq c_-$, this identity implies

$$
\left(D^2\sigma_N(u)\lambda, \lambda\right) \geq \frac{c_-}{|\mathbb{T}_N^d|} \inf_f \sum_{x \in \mathbb{T}_N^d} \sum_{i=1}^d \langle (\lambda_i - \nabla_i f)^2 \rangle.
$$

However, by estimating

$$
\sum_{x \in \mathbb{T}_N^d} \sum_{i=1}^d \langle (\lambda_i - \nabla_i f)^2 \rangle = \sum_{x \in \mathbb{T}_N^d} \sum_{i=1}^d (\lambda_i^2 + \langle (\nabla_i f)^2 \rangle) \geq |\mathbb{T}_N^d| \cdot |\lambda|^2,
$$

we finally get the lower bound

$$
\left(D^2\sigma_N(u)\lambda, \lambda\right) \geq c_-|\lambda|^2. \tag{5.24}
$$

Now (5.23) and (5.24) establish (5.10) for σ_N, since we have

$$
\sigma_N(v) - \sigma_N(u) - (v - u) \cdot \nabla\sigma_N(u)
$$

$$
= \int_0^1 dt \int_0^t \left(D^2\sigma_N(u + s(v - u))(v - u), v - u\right) ds
$$

for every $u, v \in \mathbb{R}^d$. Letting $N \to \infty$ shows (5.10) for σ. $\square$

5.5 Surface Tension in one Dimensional Systems

In one dimension (i.e., for interfaces in $1+1$ dimensional space), the $\nabla\varphi$-Gibbs measures are simple Bernoulli measures (cf. Remark 4.5) and the features exhibited by them are completely different from the higher dimensional systems. In some cases, however, the surface tension $\sigma = \sigma(u), u \in \mathbb{R}$ is explicitly computable and this might be useful to see.

Define $\hat{\nu}_\lambda \in \mathcal{P}(\mathbb{R}), \lambda \in \mathbb{R}$ and the normalization constant $\hat{Z}_\lambda$ by

$$
\hat{\nu}_\lambda(d\eta) = \frac{1}{\hat{Z}_\lambda} e^{-V(\eta)+\lambda\eta} d\eta, \tag{5.25}
$$

$$
\hat{Z}_\lambda = \int_\mathbb{R} e^{-V(\eta)+\lambda\eta} d\eta,
$$

respectively, and introduce a function $u = u(\lambda)$ as

$$u = E^{\hat{\nu}_\lambda}[\eta] \equiv \frac{d}{d\lambda} \log \hat{Z}_\lambda, \quad \lambda \in \mathbb{R} .$$

Then, since

$$u'(\lambda) = E^{\hat{\nu}_\lambda}\left[\left(\eta - E^{\hat{\nu}_\lambda}[\eta]\right)^2\right] > 0 ,$$

u is strictly increasing in λ and therefore it admits an inverse function $\lambda = \lambda(u)$. The function λ actually coincides with the differential of the surface tension σ:

$$\sigma'(u) = \lambda(u) .$$

Indeed, from (5.14), we have an expression $\sigma'(u) = E^{\hat{\nu}_\lambda}[V'(\eta)]$ in one dimension and, by integration by parts, one can easily find that the expectation in the right hand side is equal to $\lambda(u)$. The normalized surface tension is thus given by

$$\sigma(u) = \int_0^u \lambda(v) \, dv, \quad u \in \mathbb{R} . \tag{5.26}$$

Since $\sigma'' = \lambda' > 0$, one can see that σ is strictly convex and smooth.

Except for the normalization, the surface tension can be expressed as the Legendre transform of $\log \hat{Z}_\lambda$:

$$\sigma(u) = \sup_\lambda \left\{ \lambda u - \log \hat{Z}_\lambda \right\} . \tag{5.27}$$

Indeed, denoting the right hand side of (5.27) by $\hat{\sigma}(u)$, $\sup_\lambda$ is attained at $\lambda = \lambda(u)$ and

$$\lambda = \hat{\sigma}'(u) \quad \Longleftrightarrow \quad u = u(\lambda)$$

holds, see [141], (1.12) or [200], (1.2).

We give three examples of V for which $\sigma = \sigma(u)$ is explicitly computable based on the formula (5.26).

Example 5.1. $V(\eta) = \frac{c}{2}\eta^2$ with $c > 0$. A simple computation shows $u(\lambda) = \frac{1}{c}\lambda$ whose inverse function is $\lambda(u) = cu$. Therefore, we have $\sigma(u) = \frac{1}{2}cu^2$ and this coincides with Proposition 5.2.

Example 5.2. $V(\eta) = c|\eta|$ with $c > 0$. This potential does not satisfy the conditions (V1) and (V3) in (2.2), but $\sigma(u)$ are computable. The measure $\hat{\nu}_\lambda$ is defined only for $|\lambda| < c$ and, by an explicit computation, we have

$$u(\lambda) = \frac{2\lambda}{c^2 - \lambda^2}, \quad |\lambda| < c .$$

Its inverse function is

$$\lambda(u) = \frac{\sqrt{1 + c^2 u^2} - 1}{u}, \quad u \in \mathbb{R} ,$$

so that the surface tension is given by

$$\sigma(u) = \int_0^{|u|} \frac{\sqrt{1 + c^2 v^2} - 1}{v} \, dv, \quad u \in \mathbb{R}.$$

Incidentally, since $\lambda(u) \to \pm c$ as $u \to \pm\infty$, the function σ is linearly growing as $|u| \to \infty$.

Example 5.3. $V(\eta) = c_1\eta \, (\eta \geq 0)$ *and* $V(\eta) = -c_2\eta \, (\eta \leq 0)$ *with* $c_1, c_2 > 0$. *This potential is even asymmetric and as the Hamiltonian we adopt the sum of positively directed bonds only:* $H(\phi) = \sum_x V(\nabla\phi(x))$, *where* $\nabla\phi(x) = \phi(x + 1) - \phi(x)$, *cf. Remark 2.1. Then, similarly to Example 5.2, we have*

$$u(\lambda) = \frac{2\lambda + c_2 - c_1}{(c_2 + \lambda)(c_1 - \lambda)}, \quad -c_2 < \lambda < c_1.$$

Especially if $c_2 = +\infty$, we have

$$u(\lambda) = \frac{1}{c_1 - \lambda}, \quad \lambda(u) = c_1 - \frac{1}{u},$$

and therefore, except normalization,

$$\sigma(u) = c_1 u - \log u, \quad u > 0.$$

The condition $c_2 = +\infty$ means that $\eta = \nabla\phi(x) \geq 0$ is only realizable under the Gibbs measures μ_u^∇. In other words, the graph of interfaces is always increasing.

6 Large Deviation and Concentration Properties

This section starts the analysis on the limit procedure under the scaling (2.16) which connects microscopic interface height variables $\phi = \{\phi(x)\}$ with macroscopic ones $h^N = \{h^N(\theta)\}$. We shall establish the LDP for h^N as $N \to \infty$ and, as its application, obtain two types of LLNs under the (canonical) φ-Gibbs measures.

The LDP for the $\nabla\varphi$ interface model was first studied by Ben Arous and Deuschel [12] in Gaussian case. They considered the field $\{\phi(x); x \in D_N\}$, $D_N = (0, N)^d \cap \mathbb{Z}^d$, which is distributed under the finite volume φ-Gibbs measure $\mu_N^0 \equiv \mu_{D_N}^0$ with $V(\eta) = \frac{1}{2}\eta^2$ having 0-boundary condition $\psi = 0$. The field is then conditioned in such a manner that $\phi(x) \geq 0$ and macroscopic total volume $= v$ (i.e. $N^{-d-1} \sum_{x \in D_N} \phi(x) = v$). They proved for the conditioned field that on several kind of scalings the macroscopic height variables $h^N = \{h^N(\theta); \theta \in D = (0, 1)^d\}$ converge as $N \to \infty$ to $\overline{h}_{D,v} = \{\overline{h}_{D,v}(\theta)\}$ which minimizes the total surface tension $\frac{1}{2} \int_D |\nabla h|^2 \, d\theta$ under the three conditions: $h = 0$ at ∂D, $h \geq 0$ and $\int_D h \, d\theta = v$. The function $\overline{h}_{D,v}$ describes the **Wulff shape**.

Deuschel, Giacomin and Ioffe [77] generalized the results to the non-Gaussian setting. They considered the finite volume φ-Gibbs measure μ_N^0 with 0-boundary conditions for general macroscopic domain D and general potential V satisfying (2.2), and proved the LD estimates, that is, the probability that h^N is close to a given macroscopic surface $h \in H_0^1(D)$ behaves as

$$\mu_N^0\left(h^N \sim h\right) \underset{N \to \infty}{\asymp} \exp\{-N^d \Sigma_D(h)\}, \qquad (6.1)$$

where $\Sigma_D(h)$ is the (integrated) **total surface tension** (or sometimes called surface free energy) of h defined by

$$\Sigma_D(h) = \int_D \sigma(\nabla h(\theta)) \, d\theta, \qquad (6.2)$$

and $\sigma = \sigma(u)$ is the (normalized) surface tension with tilt $u \in \mathbb{R}^d$ introduced in Sect. 5. Roughly saying, the asymptotic behavior (6.1) is obtained by patching the relation (5.3) for localized systems. Mathematically precise formulation for (6.1) is the usual LD upper and lower bounds, see Theorem 6.1 below. This LDP result is an analogue of that by Dobrushin, Kotecký and Shlosman [86] for the Ising model and $\Sigma_D(h)$ corresponds to the Wulff functional (1.1).

These results can be further generalized for the system with weak self potential (one-body potential) under general Dirichlet boundary conditions ψ. We therefore state the LDP result in such settings; Sect. 6.1 for higher dimensions and Sect. 6.3 for one dimension. As an application, the LLN is proved for the finite volume φ-Gibbs measures (without conditioning) and the limit profile is characterized by a variational problem which was studied by Alt and Caffarelli [5] and others. The minimizers generate free boundaries inside the domain; Sects. 6.2 and 6.3. We also discuss the $\nabla\varphi$ interface model for δ-pinning with quadratic potential in one dimension; Sect. 6.4. Sect. 6.5 outlines the proof of Theorem 6.1. Sect. 6.6 is devoted to the LDP for empirical measures of φ-field distributed under the Gaussian φ-Gibbs measures on $\mathbb{Z}^d$ when $d \geq 3$. This is sometimes called the third level LDP.

6.1 LDP with Weak Self Potentials

Setting and Assumptions

A bounded domain D in $\mathbb{R}^d$ with piecewise Lipschitz boundary is given and microscopic regions $D_N, \overline{D_N}$ and $\partial^+ D_N$, $N \in \mathbb{N}$ in $\mathbb{Z}^d$ are defined from D, recall Sect. 1.4. The regularity assumption on ∂D is needed to employ some results in the theory of partial differential equations (PDEs), see Lemma 6.17. The boundary condition $\psi = \{\psi(x); x \in \partial^+ D_N\}$ for microscopic height variables is given.

We assume the space, which is $d + 1$ dimensional, is filled by a media changing in the distances from the hyperplane D_N. Such situation can be realized by adding self potentials (one-body potentials) $U : D \times \mathbb{R} \to \mathbb{R}$ to the original Hamiltonian $H_N^\psi(\phi) \equiv H_{D_N}^\psi(\phi)$ introduced in (2.1) with the boundary condition ψ in the following manner

$$H_N^{\psi,U}(\phi) = \sum_{\langle x,y \rangle \subset \overline{D_N}} V(\phi(x) - \phi(y)) + \sum_{x \in D_N} U\left(\frac{x}{N}, \phi(x)\right) . \qquad (6.3)$$

The first term in the right hand side is $H_N^\psi(\phi)$ and the interaction potential V is always assumed to satisfy the conditions (2.2). The statistical ensemble for the height variables ϕ is then defined by the finite volume φ-Gibbs measure on D_N

$$\mu_N^{\psi,U}(d\phi) = \frac{1}{Z_N^{\psi,U}} \exp\left\{-H_N^{\psi,U}(\phi)\right\} d\phi_{D_N} , \qquad (6.4)$$

where $Z_N^{\psi,U} \equiv Z_{D_N}^{\psi,U}$ is the normalization constant so that $\mu_N^{\psi,U} \in \mathcal{P}(\mathbb{R}^{D_N})$. We shall regard $\mu_N^{\psi,U} \in \mathcal{P}(\mathbb{R}^{\overline{D_N}})$ by considering $\phi(x) = \psi(x)$ for $x \in \partial^+ D_N$ under $\mu_N^{\psi,U}$ as before. When $U \equiv 0$, $\mu_N^{\psi,0}$ coincides with $\mu_N^\psi \equiv \mu_{D_N}^\psi$ defined by (2.4).

We consider the case that the self potential U is represented as a product $U(\theta, r) = Q(\theta)W(r)$ of two functions $Q : D \to [0, \infty)$ and $W : \mathbb{R} \to \mathbb{R}$ and assume the following conditions on Q and W, respectively:

(Q) Q is bounded and piecewise continuous,

(W) W is measurable and there exists $A \geq 0$ such   (6.5)

that $\lim_{r \to +\infty} W(r) = 0$, $\lim_{r \to -\infty} W(r) = -A$ and

$W(r) \in [-A, 0]$ for every $r \in \mathbb{R}$.

The self potential U is called weak since it is bounded. A typical example of W we have in mind is a function of the form

$$W(r) = -A \cdot 1_{\{r<0\}}, \quad r \in \mathbb{R} . \qquad (6.6)$$

This potential describes the situation that the space is filled by two different media above and below the hyperplane D_N. Since we assume $A \geq 0$, the negative values are more favorable than the positive ones for the interface height variables ϕ under the Gibbs measures. In other words the interface is weakly attracted to the negative side, namely by the media below the hyperplane D_N. The opposite case $A \leq 0$ can be easily reduced to our case $A \geq 0$ by turning the interfaces upside down by the map $\phi \mapsto -\phi$ and $\psi \mapsto -\psi$. Moreover, adding a constant to W does not make any change in the Gibbs measure $\mu_N^{\psi,U}$ so that, without loss of generality, we have assumed $\lim_{r \to +\infty} W(r) = 0$ in (6.5)-(W).

The microscopic boundary condition ψ should be scaled to have macroscopic limits. We therefore assume that the following conditions hold for $\psi \in \mathbb{R}^{\partial^+ D_N}$ with some $C > 0$, $g \in C^\infty(\mathbb{R}^d)$ and $p_0 > 2$:

$(\psi 1)$ $\qquad \max_{x \in \partial^+ D_N} |\psi(x)| \leq CN$,

$(\psi 2)$ $\qquad \sum_{x \in \partial^+ D_N} \left| \psi(x) - Ng\left(\dfrac{x}{N}\right) \right|^{p_0} \leq CN^d$.

$$\text{(6.7)}$$

These conditions roughly mean that $\psi(x)/N \sim g(x/N)$ at $x \in \partial^+ D_N$.

Scaling and Polilinear Interpolation

The aim is to study the macroscopic behavior of the microscopic height variables $\phi = \{\phi(x); x \in \overline{D_N}\}$ under the Gibbs measures $\mu_N^{\psi, U}$ as $N \to \infty$. The scaling connecting microscopic and macroscopic levels was introduced by (2.16) associating the macroscopic height variables $h^N = \{h^N(\theta); \theta \in \overline{D}\}$ with ϕ as step functions on $\overline{D}$ satisfying

$$h^N\left(\frac{x}{N}\right) = \frac{1}{N}\phi(x), \quad x \in \overline{D_N} . \tag{6.8}$$

However, from certain technical reasons, it turns out to be more tractable to define h^N by polilinear interpolation of the macroscopic variables on $\frac{1}{N}\overline{D_N}$ determined by (6.8), i.e., for general $\theta \in \overline{D}$, we set

$$h^N(\theta) = \sum_{\lambda \in \{0,1\}^d} \left[\prod_{i=1}^d (\lambda_i \{N\theta_i\} + (1 - \lambda_i)(1 - \{N\theta_i\})) \right] h^N\left(\frac{[N\theta] + \lambda}{N}\right),$$

$$\text{(6.9)}$$

where $\{\cdot\}$ denotes the fractional part, see (1.17) of [77]. In one dimension, (6.9) is just the usual polygonal approximation of $\{h^N(x/N)\}$, see (6.21).

LDP Result

Now we are in the position to state the LDP result. Define $H_g^1(D) = \{h \in H^1(D); h - g|_D \in H_0^1(D)\}$ for $g \in C^\infty(\mathbb{R}^d)$, where $H_0^1(D)$ stands for the Sobolev space on D determined from the 0-boundary condition. The function $g|_{\partial D}$ serves for the macroscopic boundary condition as in (6.7).

Theorem 6.1. [77, 123] *The family of random surfaces $\{h^N(\theta); \theta \in D\}$ distributed under $\mu_N^{\psi, U}$ satisfies the LDP on the space $L^2(D)$ with speed N^d and the rate functional $I_D^U(h)$, that is, for every closed set $\mathcal{C}$ and open set $\mathcal{O}$ of $L^2(D)$ we have that*

$$\limsup_{N \to \infty} \frac{1}{N^d} \log \mu_N^{\psi, U}(h^N \in \mathcal{C}) \leq -\inf_{h \in \mathcal{C}} I_D^U(h) , \tag{6.10}$$

$$\liminf_{N \to \infty} \frac{1}{N^d} \log \mu_N^{\psi, U}(h^N \in \mathcal{O}) \geq -\inf_{h \in \mathcal{O}} I_D^U(h) . \tag{6.11}$$

The functional $I_D^U(h)$ is given by

$$I_D^U(h) = \begin{cases} \Sigma_D^U(h) - \inf\limits_{H_g^1(D)} \Sigma_D^U & \text{if } h \in H_g^1(D) \,, \\ +\infty & \text{otherwise} \,, \end{cases}$$

where $\inf\limits_{H_g^1(D)} \Sigma_D^U = \inf\{\Sigma_D^U(h); h \in H_g^1(D)\}$ and

$$\Sigma_D^U(h) = \int_D \sigma(\nabla h(\theta))\, d\theta - A \int_D Q(\theta) 1(h(\theta) \le 0)\, d\theta \,. \tag{6.12}$$

The first term in the right hand side is $\Sigma_D(h)$ defined by (6.2) for $h \in H^1(D)$.

The unnormalized rate functional $\Sigma_D^U(h)$ is lower semicontinuous on $L^2(D)$, which can be shown by (5.10).

Remark 6.1. *Consider the case where $\psi \equiv 0$ (i.e., $g \equiv 0$) and $U \equiv 0$. Then the LDP rate functional is given by $\Sigma_D(h)$. Since the surface tension $\sigma(u)$ attains its minimal value 0 at $u = 0$, the minimizer is $h \equiv 0$. In fact, since Poincaré's inequality for $h \in H_0^1(D)$ and then (5.10) with $u = 0$ imply*

$$\|h\|_{L^2(D)}^2 \le C\|\nabla h\|_{L^2(D)}^2 \le \frac{2C}{c_-}\Sigma_D(h) \,,$$

taking $C = \{h \in L^2(D); \|h\|_{L^2(D)} \ge a\}, a > 0$ in (6.10), we obtain

$$\mu_N^{0,0}(\|h^N\|_{L^2(D)} \ge a) \le e^{-\left(\frac{a^2 c_-}{2C} - \epsilon\right)N^d}$$

for every $\epsilon > 0$ and sufficiently large N. This means that the macroscopic interface is flat with high probability and tilted surface appears with extremely small probability.

Remark 6.2. *Since ∂D is piecewise Lipschitz and $g|_{\bar{D}} \in C^\infty(\bar{D})$, by Theorems 8.7 and 8.9 of [253], there exists a continuous linear trace operator $T_0 : H^1(D) \to H^{\frac{1}{2}}(\partial D)$ such that $T_0 u = u|_{\partial D}$ for every $u \in C^\infty(\bar{D})$ and it holds that $H_g^1(D) = \{h \in H^1(D); T_0 h = g|_{\partial D}\}$.*

Remark 6.3. *The Gaussian case with 0-boundary condition was studied by [12] regarding $h^N \in L^p(D), 2 \le p < 2d/(d-2)$ when $D = (0,1)^d$. For the general Dirichlet boundary condition, the mean of the Gaussian field is shifted by a harmonic function and therefore one can easily establish the LDP applying the contraction principle, see the proof of Lemma 6.6 below.*

Remark 6.4. *Sheffield [230] improved the topology for the LDP using Orlicz-Sobolev spaces. In addition, he discussed the LDP jointly for macroscopic height variables and empirical measures (cf. Sect. 6.6).*

Remark 6.5. *If $Q \equiv 1$ and U is given by $U(\theta, r) = W(r)$, then it holds that*

$$-A = -\lim_{\ell \to \infty} \frac{1}{|\Lambda_\ell|} \log \frac{Z_{\Lambda_\ell}^{0,U}}{Z_{\Lambda_\ell}^0} \,. \tag{6.13}$$

The right hand side represents the difference of the free energies of the interface in two cases with and without self potential, see (6.33) with $a = b = 0$ in one dimension. In this sense, $\Sigma_D^U(h)$ above represents macroscopic total surface energy of the profile h.

The LDP result of Deuschel et al. [77] is a special case of Theorem 6.1: $\psi \equiv 0$ and $U \equiv 0$ so that $A = 0$. However, the actual proof of Theorem 6.1 is given in a converse way. We reduce it to the case of $U \equiv 0$, since the potential U is weak and can be treated as a rather simple perturbation. The main effort in [123] was therefore made for the treatment of the general boundary conditions. By a simple shift the problem can be reduced to the 0-boundary case, however with bond-depending interaction potentials. The proof of Theorem 6.1 will be outlined in Sect. 6.5. Instead, a complete proof will be given for one dimensional system with quadratic potentials in Sect. 6.3. The author takes this way, since it may be acceptable for a wide variety of readers including nonexperts. In one dimension, one can prove the LDP under uniform topology rather than the L^2-topology.

6.2 Concentration Properties

We can deduce from Theorem 6.1 the LLNs for h^N distributed under $\mu_N^{\psi,U}$ (i.e., φ-Gibbs measure) or under its conditional probability (i.e., canonical φ-Gibbs measure) as $N \to \infty$ and the limits $h = \{h(\theta); \theta \in D\}$ are characterized by certain variational principles.

Wulff Shape

The macroscopic shape of the droplet put on a hard wall and having a definite volume $v(> 0)$ can be determined by the LLN for a conditioned field of $h^N = \{h^N(\theta); \theta \in D\}$. The conditions are introduced in such a manner that $h^N \geq 0$ (wall condition; wall is put at the height level $h \equiv 0$) and $\int_D h^N(\theta) \, d\theta \geq v$ (or $= v$, constant volume condition).

Corollary 6.2. (*Wall and constant volume conditions*) *For every $v \geq 0$, under the conditional probability $\mu_{N,v}^+ = \mu_N^{0,0} \left(\cdot \, \big| h^N \geq 0, \int_D h^N(\theta) \, d\theta \geq v \right)$ (note that we take $\psi \equiv 0, U \equiv 0$), the LLN*

$$\lim_{N \to \infty} \mu_{N,v}^+ (\|h^N - \overline{h}_{D,v}\|_{L^2(D)} > \delta) = 0 \, ,$$

holds for every $\delta > 0$, where $\overline{h}_{D,v}$ is the unique minimizer called **Wulff shape** *of the variational problem*

$$\min \left\{ \Sigma_D(h); h \in H_0^1(D), h \geq 0, \int_D h(\theta) \, d\theta = v \right\} \, . \tag{6.14}$$

Proof. We first notice that, if $d \geq 2$, denoting $\mu_N \equiv \mu_N^{0,0}$

$$\mu_N(\Omega^+(D_N)) \geq e^{-CN^{d-1}} \tag{6.15}$$

for some $C > 0$, where $\Omega^+(D_N) = \{\phi \in \mathbb{R}^{D_N}; \phi(x) \geq 0 \text{ for every } x \in D_N\}$, see [76] and Theorem 7.2 (entropic repulsion). This bound claims that the probability $\mu_N(\Omega^+(D_N))$ is large enough compared with the LDP estimate (at the order of e^{-CN^d}). Setting the volume condition

$$A_v = \left\{ \phi \in \mathbb{R}^{D_N}; \int_D h^N(\theta) \, d\theta \geq v \right\},$$

we have that

$$\begin{aligned}
&\mu_{N,v}^+(\|h^N - \overline{h}_{D,v}\|_{L^2(D)} > \delta) \\
&= \frac{\mu_N(\|h^N - \overline{h}_{D,v}\|_{L^2(D)} > \delta, \Omega^+(D_N) \cap A_v)}{\mu_N(\Omega^+(D_N) \cap A_v)} \\
&\leq \frac{\mu_N(\|h^N - \overline{h}_{D,v}\|_{L^2(D)} > \delta, A_v)}{\mu_N(\Omega^+(D_N))\mu_N(A_v)}.
\end{aligned}$$

The last inequality is a consequence of the FKG inequality for the denominator. However, Theorem 6.1 implies

$$\limsup_{N\to\infty} \frac{1}{N^d} \log \mu_N(\|h^N - \overline{h}_{D,v}\|_{L^2(D)} > \delta, A_v) < -\Sigma_v^*,$$

$$\liminf_{N\to\infty} \frac{1}{N^d} \log \left\{ \mu_N(\Omega^+(D_N))\mu_N(A_v) \right\} \geq -\Sigma_v^*,$$

where

$$\Sigma_v^* := \inf \left\{ \Sigma_D(h); h \in H_0^1(D), h \geq 0, \int_D h(\theta) \, d\theta = v \right\}.$$

We have applied (6.15) for the second, and these two estimates prove the conclusion. Note that the value of Σ_v^* is unchanged if the conditions "$h \geq 0, \int_D h(\theta) \, d\theta = v$" are replaced with "$\int_D h(\theta) \, d\theta \geq v$". $\qquad\square$

Remark 6.6. *Bolthausen and Ioffe [31] proved the LLN for the Gibbs measure on the wall with δ-pinning and quadratic potential under the constant volume condition in two dimension (i.e., for interfaces in $2 + 1$ dimensional space). The limit called* **Winterbottom shape** *is uniquely (except translation) characterized by a certain variational problem, see Sect. 7.3.*

The Euler equation for the minimizer $\overline{h}_{D,v}$ of the variational problem (6.14) has the following form of the elliptic PDE:

$$\begin{cases} \text{div}\left\{ (\nabla\sigma)(\nabla \overline{h}_{D,v}(\theta)) \right\} = -c_{D,v}, & \theta \in D, \\ \overline{h}_{D,v}(\theta) = 0, & \theta \in \partial D, \end{cases} \tag{6.16}$$

where $c_{D,v}$ is an appropriate constant. Indeed, the minimizer h satisfies

$$\frac{d}{d\epsilon} \Sigma_D(h + \epsilon g) \bigg|_{\epsilon=0} = 0$$

for all g such that $\int_D g(\theta) \, d\theta = 0$. This implies that

$$\int_D \nabla\sigma(\nabla h(\theta)) \cdot \nabla g(\theta) \, d\theta = -\int_D \operatorname{div}\left[\nabla\sigma(\nabla h)\right] g \, d\theta = 0 \, ,$$

and leads us to (6.16).

Dobrushin and Hryniv [85] studied the **fluctuation of the Wulff shape** when $d = 1$. They adopted the random walk model, i.e. the SOS type model $\phi :$ $\{0, 1, \ldots, N\} \to \mathbb{Z}$ in one dimension under the condition that the macroscopic volume of ϕ is always constant:

$$\frac{1}{N} \sum_{x=1}^{N} \frac{1}{N}\phi(x) = v, \quad v \in \mathbb{R} \, .$$

They proved, under the one-sided Dirichlet boundary condition (i.e., $\phi(0) = 0$) or under the two-sided conditions (i.e., $\phi(0) = 0, \phi(N) = Nb$), the LLN and the CLT for the macroscopic height variables $h^N = \{h^N(\theta); 0 \le \theta \le 1\}$ defined by the polygonal approximation of $\{\phi(x)/N\}$:

(1) LLN: $h^N(\theta) \to \overline{h}(\theta)$ $(N \to \infty)$, where $\overline{h}$ is the Wulff shape.
(2) CLT (Fluctuation of h^N around $\overline{h}$):

$$\sqrt{N}(h^N(\theta) - \overline{h}(\theta)) \Longrightarrow \text{Gaussian process} \, .$$

They did not impose the wall condition. See Higuchi et al. [146] for the extension to the two dimensional lattice Widom-Rowlinson model.

Remark 6.7. (*Wulff shape from the Ising model, cf. Sect. 1.1*) *For the two dimensional Ising model with nearest neighbor and ferromagnetic (attractive) interactions, it is well-known that there exists the critical temperature $T_c > 0$ such that if $T < T_c$ the system has the positive spontaneous magnetization $m^* = m^*(T) > 0$. Let $\mu_{N,m}$ be the canonical Gibbs measure for such Ising model on $[-N, N]^2 \cap \mathbb{Z}^2$ with $+$ boundary condition, which is obtained by conditioning the finite volume Gibbs measure in a way that the sample average of the spins is m for some $|m| < m^*$. Dobrushin, Kotecký and Shlosman [86] proved under $\mu_{N,m}$ the macroscopic region occupied by $-$ spins converges to the Wulff shape, except translations, as $N \to \infty$. Afterward, Ioffe [150, 151], Ioffe and Schonmann [152] extended this result for all $T < T_c$ applying the method of percolation.*

Pisztora [213] invented the so-called L^1-theory on the local sample averages of the spins. This method is applicable to three dimension and higher, and uses the idea based on the renormalization group called Pisztora's coarse graining, see also Cerf and Pisztora [52]. It is believed that the Wulff shape has facets if $T < T_R$ (roughening transition, $\exists T_R < T_c$), see [233]. The review paper by Bodineau, Ioffe and Velenik [22] is recommended to catch the whole picture of the results on the Ising model, including the derivation of the Winterbottom shape, see Sect. 7.3. See [1] for results by the middle of 1980s.

Cerf and Pisztora [53] studied the LDP under phase coexistence for Ising, Potts and random cluster models in dimensions $d \geq 3$ for $T < T_c$. See also [3, 51].

Remark 6.8. Cohn et al. [58] considered the SOS type model on $\mathbb{Z}^2$ (i.e., $\phi : \mathbb{Z}^2 \to \mathbb{Z}$) induced from the domino tiling with equal probabilities for all possible tilings, and proved the LLN and the LDP for the corresponding macroscopic height variables. See also Kenyon [168].

Remark 6.9. The Wulff construction at zero temperature, but for a wider class of Gibbs models, was studied by Descombes and Pechersky [73].

Alt-Caffarelli's Variational Problems

The upper bound (6.10) in Theorem 6.1 implies the LLN for h^N distributed under $\mu_N^{\psi,U}$.

Corollary 6.3. If Σ_D^U has a unique minimizer $\bar{h}$ in $H_g^1(D)$, then the LLN holds under $\mu_N^{\psi,U}$, namely,

$$\lim_{N \to \infty} \mu_N^{\psi,U}(\|h^N - \bar{h}\|_{L^2(D)} > \delta) = 0 \,,$$

for every $\delta > 0$.

The variational problems for minimizing Σ_D^U were thoroughly studied by Alt and Caffarelli [5] for nonnegative macroscopic boundary data g (one phase problem) with $A > 0$ and by Alt, Caffarelli and Friedman [6] for general g (two phases problem) especially when σ is quadratic: $\sigma(u) = |u|^2$, and by Weiss [252] for more general σ. The minimizer $h = \bar{h}$ of Σ_D^U generates the **free boundaries** inside D. If the surface tension $\sigma = \sigma(u)$ is smooth enough (i.e., $\sigma \in C^{2,\gamma}(\mathbb{R}^d), \gamma > 0$) and if the free boundary $\partial\{h > 0\}$ of the minimizer h is locally C^2, then h satisfies the Euler equation

$$\text{div}\,\{\nabla\sigma(\nabla h)\} = 0$$

in $D \setminus \partial\{h > 0\}$ and the condition

$$\Psi(\nabla h^+) - \Psi(\nabla h^-) = AQ \tag{6.17}$$

on the free boundary $D \cap \partial\{h > 0\}$, where $\Psi(u) = u \cdot \nabla\sigma(u) - \sigma(u)$. The Lipschitz continuity of the minimizer h and the regularity of its free boundary were studied by the papers listed above and others. In our case, for the regularity of the surface tension, $\sigma \in C^{1,1}(\mathbb{R}^d)$ is only known in general, recall Theorem 5.3-(1).

6.3 LDP with Weak Self Potentials in one Dimension

In this section we reformulate Theorem 6.1 in one dimension (i.e., we consider interfaces in $1+1$ dimensional space) and give a complete proof of the theorem. As we have already noticed, in one dimension, one can argue under a stronger topology determined by the uniform norm.

Reformulation of the Results

Let us take $D = (0,1) \subset \mathbb{R}$ so that $D_N = \{1, 2, \ldots, N-1\}$ and $\partial^+ D_N = \{0, N\}$. For simplicity, we consider the case of the quadratic potential $V(\eta) = \frac{1}{2}\eta^2$ with $Q \equiv 1$ in the self potential $U(\theta, r)$. The corresponding Gibbs measure for the height variables $\phi = \{\phi(x); x \in D_N\}$ is then defined by

$$\mu_N^{a,b,W}(d\phi) = \frac{1}{Z_N^{a,b,W}} \exp\left\{-H_N^{a,b,W}(\phi)\right\} d\phi_{D_N}, \tag{6.18}$$

under the boundary conditions

$$\psi(0) = aN, \quad \psi(N) = bN \tag{6.19}$$

for some $a, b \in \mathbb{R}$. The corresponding macroscopic boundary conditions are $g(0) = a$ and $g(1) = b$ at $\partial D = \{0, 1\}$, recall (6.7)-($\psi2$). The Hamiltonian $H_N^{a,b,W}$ is given by

$$H_N^{a,b,W}(\phi) = \frac{1}{2} \sum_{x=0}^{N-1} (\phi(x+1) - \phi(x))^2 + \sum_{x=1}^{N-1} W(\phi(x)), \tag{6.20}$$

and $Z_N^{a,b,W}$ is the normalization constant. The formulas (6.18) and (6.20) correspond to (6.4) and (6.3), respectively. The function W satisfies the condition (W) in (6.5).

The macroscopic height variable $h^N = \{h^N(\theta); \theta \in [0,1]\}$ is defined from ϕ by the interpolation (6.9) which is, in one dimension, the polygonal approximation of $\{h^N(x/N) = \phi(x)/N; x \in \overline{D_N}\}$:

$$h^N(\theta) = \left(\theta - \frac{x}{N}\right)\phi(x+1) + \left(\frac{x+1}{N} - \theta\right)\phi(x), \quad \frac{x}{N} \le \theta \le \frac{x+1}{N}. \tag{6.21}$$

Introduce two function spaces

$$C_{a,b} = \{h \in C([0,1]); h(0) = a, h(1) = b\},$$
$$H_{a,b}^1 = \{h \in C_{a,b}; h \text{ is absolutely continuous}$$
$$\text{and its derivative } h' \in L^2([0,1])\}.$$

The space $C_{a,b}$ is endowed with the topology determined by the uniform-norm $\|\cdot\|_\infty$. The function h^N belongs to $C_{a,b}$. Since the normalized surface tension is $\sigma(u) = \frac{1}{2}u^2$ for $V(\eta) = \frac{1}{2}\eta^2$ (recall Proposition 5.2 and Example 5.1), the total surface tension of $h \in C_{a,b}$ defined by the formula (6.2) has the form

$$\Sigma(h) \equiv \Sigma_{(0,1)}(h) = \frac{1}{2}\int_0^1 (h')^2(\theta)\, d\theta$$

for $h \in H^1_{a,b}$ and $\Sigma_{(0,1)}(h) = +\infty$, otherwise. We set

$$\Sigma^W(h) \equiv \Sigma^W_{(0,1)}(h) = \Sigma(h) - A|\{h \leq 0\}|, \qquad (6.22)$$

where $|\cdot|$ stands for the Lebesgue measure and $\{h \leq 0\} = \{\theta \in [0,1]; h(\theta) \leq 0\}$. Note that Σ^W corresponds to Σ^U_D defined by (6.12). It is lower semicontinuous in $h \in C_{a,b}$ and good in the sense that $\{h \in C_{a,b}; \Sigma^W(h) \leq \ell\}$ is compact in $C_{a,b}$ for each $\ell \in \mathbb{R}$.

Theorem 6.4. *Under $\mu_N^{a,b,W}$, the family of the macroscopic height variables h^N defined by (6.21) satisfies the LDP on the space $C_{a,b}$ with speed N and the unnormalized rate functional Σ^W, that is, for every closed set C and open set $\mathcal{O}$ of $C_{a,b}$ we have that*

$$\limsup_{N\to\infty} \frac{1}{N}\log\mu_N^{a,b,W}(h^N \in C) \leq -\inf_{h\in C} I^W(h), \qquad (6.23)$$

$$\liminf_{N\to\infty} \frac{1}{N}\log\mu_N^{a,b,W}(h^N \in \mathcal{O}) \geq -\inf_{h\in\mathcal{O}} I^W(h), \qquad (6.24)$$

where $I^W(h) = \Sigma^W(h) - \inf_{H^1_{a,b}} \Sigma^W$ is the normalized functional of Σ^W.

Concentrations

Before giving the proof of Theorem 6.4, we reformulate Corollary 6.3 in one dimensional setting and study the minimizers of the functional Σ^W, which exhibit different aspects depending on the boundary conditions a and b.

Corollary 6.5. *If Σ^W has a unique minimizer $\bar{h}$ in $H^1_{a,b}$, then the LLN holds under $\mu_N^{a,b,W}$, namely,*

$$\lim_{N\to\infty} \mu_N^{a,b,W}\left(\|h^N - \bar{h}\|_\infty > \delta\right) = 0,$$

for every $\delta > 0$.

This result is related to those obtained by Pfister and Velenik [212]. They considered the two dimensional Ising model at low temperature on a large box with attractive wall set at the bottom line. This line segment corresponds to our hyperplane D_N, although it has an effect of hard wall at the same time, since the interfaces separating $\pm$-phases can not go down beyond the bottom line in their setting. One of the motivations of [212] was to understand the so-called wetting or pinning/depinning transition, which will be discussed in Sect. 7.3 for the $\nabla\varphi$ interface model.

Minimizers of Σ^W

The functional Σ^W is essentially the same as $W(C)$ defined by (4.1) in [212], which is derived from the two dimensional Ising model; note that $h(\theta) \geq 0$ in their case. The minimizer of $W(C)$ was studied in Proposition 4.1 of [212].

Case 1. $a, b > 0$: The straight line $h^{(1)}$ connecting $(0, a)$ and $(1, b)$, i.e.,

$$h^{(1)}(\theta) = (1 - \theta)a + b\theta, \quad \theta \in [0, 1]$$

is a critical point of Σ^W; see Fig. 6.1. In fact, $\Sigma^W(h) = \Sigma(h)$ for h always staying in the positive side $\{h > 0\}$ and, for such h, the Euler equation $\delta\Sigma^W/\delta h(\theta) = -h''(\theta) = 0$ means that h is a linear function.

Since the second term of Σ^W in (6.22) makes it smaller if h stays longer in the nonpositive side $\{h \leq 0\}$, there is another candidate for minimizers. Let $h^{(2)}$ be the curve composed of three straight line segments connecting four points $(0, a), P_1(\theta_1, 0), P_2(1 - \theta_2, 0)$ and $(1, b)$ in this order; see Fig. 6.2. The angles at two corners P_1 and P_2 are both equal to $\alpha \in [0, \pi/2]$, which is

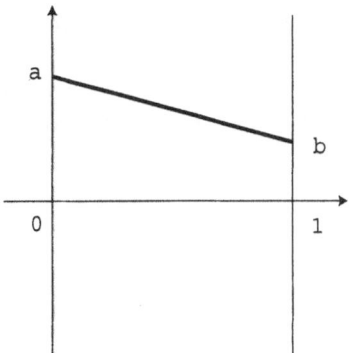

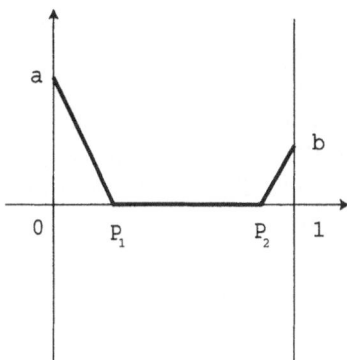

Fig. 6.1. The function $h^{(1)}$ **Fig. 6.2.** The function $h^{(2)}$

determined by the **Young's relation**:

$$\tan \alpha = \sqrt{2A} . \tag{6.25}$$

Then, $h^{(2)}$ is a critical point of Σ^W. Indeed, as we have explained above, if the curve is straight in the positive side $\{h > 0\}$, the energy is smaller. Once h reaches the side $\{h \leq 0\}$, the energy is minimal if it stays on $\{h = 0\}$, since it is a straight line and gives no contribution to $\Sigma(h)$. To derive the angle relation (6.25), set

$$\Sigma^W(h^{(2)}) \equiv F(\theta_1, \theta_2) = \frac{a^2}{2\theta_1} - A(1 - \theta_1 - \theta_2) + \frac{b^2}{2\theta_2} .$$

Then, $\partial F/\partial\theta_1 = \partial F/\partial\theta_2 = 0$ shows

$$\frac{a}{\theta_1} = \frac{b}{\theta_2} = \sqrt{2A} , \qquad (6.26)$$

which implies (6.25). One can derive (6.25) from the free boundary condition (6.17) as well. Indeed, by $\Psi(u) = |u|^2 - \frac{1}{2}|u|^2 = \frac{1}{2}|u|^2$ and $Q = 1$, (6.17) reads $|\nabla h^+|^2 - |\nabla h^-|^2 = 2A$ and this implies $\nabla h^+ = \sqrt{2A}$ since $\nabla h^- = 0$. The condition (6.25) is the same as [49] discussed.

Since $h^{(2)}(\theta)$ is described as

$$h^{(2)}(\theta) = (a - \sqrt{2A}\theta)1_{\{\theta \leq \theta_1\}} + (b - \sqrt{2A}(1 - \theta))1_{\{\theta \geq 1-\theta_2\}}$$

with θ_1 and θ_2 defined by (6.26), $h \equiv h^{(2)}$ satisfies an equation

$$h'' = \nu, \quad \nu = \sqrt{2A} \sum_{\bar{\theta} \in \partial\{\theta; h(\theta)=0\}} \delta_{\bar{\theta}} , \qquad (6.27)$$

in the sense of generalized functions, namely,

$$\langle h, J'' \rangle = \sqrt{2A}(J(\theta_1) + J(1 - \theta_2))$$

for every $J \in C_0^\infty(0,1)$. The equation (6.27) may be regarded as the Euler equation for the functional Σ^W.

Case 2. $a > 0, b < 0$: Let $h^{(3)}$ be the curve composed of two straight line segments connecting three points $(0, a), P(\theta_1, 0)$ and $(1, b)$ in this order; see Fig. 6.3. The angles at the corner P of the first and second segments to the horizontal line are denoted by α and $\beta \in [0, \pi/2]$, respectively, and obey the relation

$$\tan^2 \alpha - \tan^2 \beta = 2A . \qquad (6.28)$$

These two angles depend on the boundary conditions in such a way that

$$\frac{a}{\tan \alpha} - \frac{b}{\tan \beta} = 1 .$$

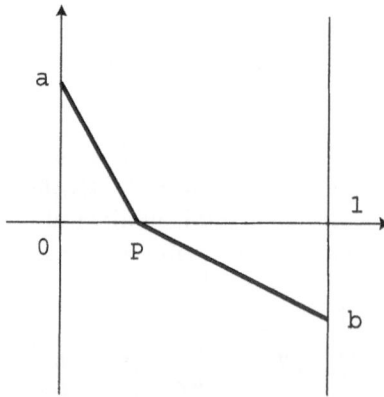

Fig. 6.3. The function $h^{(3)}$

Then, $h^{(3)}$ is a critical point of Σ^W. Indeed, the critical curve has to be straight both in the positive and nonpositive sides. The relation (6.28) is derived similarly to Case 1: Set

$$\Sigma^W(h^{(3)}) \equiv F(\theta_1) = \frac{a^2}{2\theta_1} - A(1 - \theta_1) + \frac{b^2}{2(1 - \theta_1)}$$

and $F'(\theta_1) = 0$ implies (6.28). The function $h \equiv h^{(3)}$ satisfies an equation

$$h'' = \nu, \quad \nu = (\tan \alpha - \tan \beta)\delta_{\theta_1} \ ,$$

in the sense of generalized functions.

Case 3. $a, b < 0$: The minimizer of Σ^W is the straight line connecting $(0, a)$ and $(1, b)$.

Proof of Theorem 6.4

Now let us give the proof of Theorem 6.4. We prepare three lemmas, first of which discusses the LDP for the finite volume Gibbs measure $\mu_N^{a,b,0}$ defined by (6.18) taking $W \equiv 0$ in the Hamiltonian.

Lemma 6.6. *Under* $\mu_N^{a,b,0}$, *the family of surfaces* h^N *satisfies the LDP on* $C_{a,b}$ *with speed* N *and the rate functional* $\Sigma^{a,b,0}(h) = \Sigma(h) - (b - a)^2/2$.

Proof. Let $w = \{w(x); x \in [0, N]\}$ be the one dimensional standard Brownian motion starting at 0 and set $\bar{h}^N(\theta) = w(N\theta)/N, \theta \in [0, 1]$. Then, by Schilder's theorem (see, e.g., Theorem 5.1 of [246]; Mogul'skii [200] discusses the random walk with general transition probabilities), the LDP holds for $\bar{h}^N$ on $C_0 = C([0, 1]) \cap \{h(0) = 0\}$ with the rate functional $\Sigma(h)$. Define $\phi = \{\phi(x); x \in [0, N]\}$ from w as $\phi(x) = w(x) - xw(N)/N + (N - x)a + xb$. Then, ϕ is the pinned Brownian motion satisfying (6.19) and $\{\phi(x); x \in D_N\}$ is $\mu_N^{a,b,0}$-distributed; see Proposition 2.2. Set $\tilde{h}^N(\theta) = \phi(N\theta)/N, \theta \in [0, 1]$, and consider the mapping $\Phi : \bar{h} \in C_0 \mapsto \tilde{h} \in C_{a,b}$ defined by

$$\Phi(\bar{h})(\theta) = \bar{h}(\theta) - \theta\bar{h}(1) + (1 - \theta)a + \theta b \ .$$

Then, Φ is continuous and $\tilde{h}^N = \Phi(\bar{h}^N)$ holds. Therefore, by the contraction principle (cf. [246], [79] and [72, Theorem 4.2.1]), the LDP holds for $\tilde{h}^N$ with the rate functional $\tilde{\Sigma}(h) = \inf_{\bar{h} \in C_0 : \Phi(\bar{h}) = h} \Sigma(\bar{h})$, which coincides with $\Sigma^{a,b,0}(h)$.

The proof of lemma is completed by showing a super exponential estimate for the difference between h^N and $\tilde{h}^N$ as in p.17 of [246]: For every $\delta > 0$,

$$P\left(\|h^N - \tilde{h}^N\|_\infty \geq \delta\right) \leq \sum_{x=0}^{N-1} P\left(\sup_{\theta \in [x/N, (x+1)/N]} \left|\left(\theta - \frac{x}{N}\right) w(x+1)\right.\right.$$

$$\left.\left. + \left(\frac{x+1}{N} - \theta\right) w(x) - \frac{1}{N} w(N\theta)\right| \geq \delta\right)$$

$$= NP\left(\sup_{\theta \in [0,1]} |w(\theta) - \theta w(1)| \geq N\delta\right)$$

$$\leq 4NP\left(|w(1)| \geq N\delta/2\right) = \exp\left\{-\frac{N^2\delta^2}{8} + o(N^2)\right\},$$

as $N \to \infty$. $\qquad\qquad\qquad\qquad\qquad\qquad\qquad\qquad\qquad\qquad\qquad\qquad\square$

For $g \in C_{a,b}$ and $\delta > 0$, set $B_\infty(g, \delta) = \{h \in C_{a,b}; \|h - g\|_\infty < \delta\}$. The next lemma estimates the second term in the Hamiltonian $H_N^{a,b,W}$ in (6.20).

Lemma 6.7. *Let $g \in C_{a,b}$ and $0 < \delta < 1$ be fixed. If $h^N \in B_\infty(g, \delta)$ for N large enough, then there exists some constant $C > 0$ such that*

$$-AN|\{g \leq 3\delta\}| - CN\delta \leq \sum_{x \in D_N} W(\phi(x)) \leq -AN|\{g \leq -3\delta\}| + CN\delta,$$

for every N large enough.

Proof. The condition (6.5)-(W) on W is needed. In fact, the upper bound is shown as

$$\sum_{x \in D_N} W(\phi(x)) \leq \sum_{x \in D_N : g(x/N) \leq -2\delta} W(\phi(x))$$

$$\leq (-A + \delta) \cdot \#\{x \in D_N : g(x/N) \leq -2\delta\}$$

$$\leq -AN|\{g \leq -3\delta\}| + CN\delta,$$

for every sufficiently large N. Here, the first inequality follows from $W \leq 0$, the second one is because, if $h^N \in B_\infty(g, \delta)$, $\phi(x) = Nh^N(x/N) \leq -N\delta$ so that $W(\phi(x)) \to -A$ as $N \to \infty$ uniformly in $x \in D_N$ in the sum, and the third one is by the uniform continuity of g. The lower bound is similar:

$$\sum_{x \in D_N} W(\phi(x)) \geq \sum_{x \in D_N : g(x/N) > 2\delta} W(\phi(x)) - A \cdot \#\{x \in D_N : g(x/N) \leq 2\delta\}$$

$$\geq -CN\delta - AN|\{g \leq 3\delta\}|,$$

where the first inequality is from $W \geq -A$ and the second one is because, if $h^N \in B_\infty(g, \delta)$, $\phi(x) \geq N\delta$ so that $W(\phi(x)) \to 0$ as $N \to \infty$ uniformly in $x \in D_N : g(x/N) > 2\delta$. $\qquad\qquad\qquad\qquad\qquad\qquad\qquad\square$

Let Σ_-^W be the functional Σ^W with $|\{h \leq 0\}|$ replaced by $|\{h < 0\}|$, i.e.,

$$\Sigma_-^W(h) = \Sigma(h) - A|\{h < 0\}|. \qquad\qquad\qquad\qquad (6.29)$$

Lemma 6.8. *For every open set $\mathcal{O}$ of $C_{a,b}$, we have that*

$$\inf_{h \in \mathcal{O}} \Sigma^W(h) = \inf_{h \in \mathcal{O}} \Sigma_-^W(h) .$$

Proof. Since $\Sigma^W(h) \le \Sigma_-^W(h)$ is obvious for every $h \in C_{a,b}$, the conclusion follows once we can show that

$$\inf_{h \in \mathcal{O}} \Sigma^W(h) \ge \inf_{h \in \mathcal{O}} \Sigma_-^W(h) . \tag{6.30}$$

To this end, for every $\epsilon > 0$, take $h \in \mathcal{O}(\cap H_{a,b}^1)$ such that $\Sigma^W(h) \le \inf_{\mathcal{O}} \Sigma^W + \epsilon$. We approximate such h by a sequence $\{h^n\}_{n \ge 1}$ defined by $h^n(\theta) = h(\theta) - f^n(\theta)$, where $f^n \in C_0^\infty((0,1))$ are functions such that $f^n(\theta) \equiv \frac{1}{n}$ for $\theta \in [\frac{1}{n}, 1 - \frac{1}{n}]$ and $|(f^n)'(\theta)| \le 2$ for every $\theta \in (0, \frac{1}{n}) \cup (1 - \frac{1}{n}, 1)$; note that $h^n \in H_{a,b}^1$. Then, since $\lim_{n \to \infty} \Sigma(h^n) = \Sigma(h)$ and

$$-|\{h^n < 0\}| \le -|\{h \le 0\}| + \frac{2}{n} ,$$

we obtain that $\limsup_{n \to \infty} \Sigma_-^W(h^n) \le \Sigma^W(h)$. However, $\mathcal{O}$ is an open set of $C_{a,b}$, so that $h^n \in \mathcal{O}$ for n large enough and thus (6.30) is shown. $\qquad\square$

Proof (Theorem 6.4). Let $\Sigma^{a,b,W}$ and $\Sigma_-^{a,b,W}$ be the functionals Σ^W and Σ_-^W defined by (6.22) and (6.29) with Σ replaced by $\Sigma^{a,b,0}$, respectively.

Step 1 (Lower bound). Let $g \in C_{a,b}$ and $0 < \delta < 1$ be given. Then, by the upper bound in Lemma 6.7 and the LD lower bound for $\mu_N^{a,b,0}$ (Lemma 6.6), we have

$$\liminf_{N \to \infty} \frac{1}{N} \log \left[\frac{Z_N^{a,b,W}}{Z_N^{a,b,0}} \mu_N^{a,b,W} \left(h^N \in B_\infty(g, \delta) \right) \right]$$

$$\ge \liminf_{N \to \infty} \frac{1}{N} \log \left[\exp \left(AN|\{g \le -3\delta\}| - CN\delta \right) \cdot \mu_N^{a,b,0} \left(h^N \in B_\infty(g, \delta) \right) \right]$$

$$\ge A|\{g \le -3\delta\}| - C\delta - \inf_{h \in B_\infty(g,\delta)} \Sigma^{a,b,0}(h) .$$

Now, suppose that an open set $\mathcal{O}$ of $C_{a,b}$ is given. Then, for every $h \in \mathcal{O}$ and $\delta > 0$ such that $B_\infty(h, \delta) \subset \mathcal{O}$, we have that

$$\liminf_{N \to \infty} \frac{1}{N} \log \left[\frac{Z_N^{a,b,W}}{Z_N^{a,b,0}} \mu_N^{a,b,W} \left(h^N \in \mathcal{O} \right) \right] \ge -\Sigma^{a,b,0}(h) + A|\{h \le -3\delta\}| - C\delta .$$

Letting $\delta \downarrow 0$, since $h \in \mathcal{O}$ is arbitrary, we have

$$\liminf_{N \to \infty} \frac{1}{N} \log \left[\frac{Z_N^{a,b,W}}{Z_N^{a,b,0}} \mu_N^{a,b,W} \left(h^N \in \mathcal{O} \right) \right] \ge - \inf_{h \in \mathcal{O}} \Sigma_-^{a,b,W}(h) . \tag{6.31}$$

However, by Lemma 6.8, $\Sigma_-^{a,b,W}(h)$ can be replaced with $\Sigma^{a,b,W}(h)$ in the right hand side of (6.31).

Step 2 (Upper bound). Similarly, by the lower bound in Lemma 6.7 and the LD upper bound for $\mu_N^{a,b,0}$ (Lemma 6.6), we have

$$\limsup_{N\to\infty} \frac{1}{N} \log \left[\frac{Z_N^{a,b,W}}{Z_N^{a,b,0}} \mu_N^{a,b,W} \left(h^N \in B_\infty(g,\delta) \right) \right]$$

$$\leq \limsup_{N\to\infty} \frac{1}{N} \log \left[\exp\left(AN|\{g \leq 3\delta\}| + CN\delta \right) \cdot \mu_N^{a,b,0} \left(h^N \in B_\infty(g,\delta) \right) \right]$$

$$\leq A|\{g \leq 3\delta\}| + C\delta - \inf_{h\in\overline{B}_\infty(g,\delta)} \Sigma^{a,b,0}(h) ,$$

where $\overline{B}_\infty(g,\delta) = \{h \in C_{a,b}; \|h-g\|_\infty \leq \delta\}$ is the closure of $B_\infty(g,\delta)$ in $C_{a,b}$. Then, by using the lower semicontinuity of $\Sigma^{a,b,0}(h)$ and the right-continuity of $|\{g \leq 3\delta\}|$ in δ, we see that for every $g \in C_{a,b}$ and $\epsilon > 0$, there exists $\delta > 0$ such that

$$\limsup_{N\to\infty} \frac{1}{N} \log \left[\frac{Z_N^{a,b,W}}{Z_N^{a,b,0}} \mu_N^{a,b,W} \left(h^N \in B_\infty(g,\delta) \right) \right] \leq -\Sigma^{a,b,W}(g) + \epsilon .$$

Therefore, the standard argument in the theory of LDP [72, 79, 246] yields the upper bound

$$\limsup_{N\to\infty} \frac{1}{N} \log \left[\frac{Z_N^{a,b,W}}{Z_N^{a,b,0}} \mu_N^{a,b,W} \left(h^N \in \mathcal{C} \right) \right] \leq - \inf_{h\in\mathcal{C}} \Sigma^{a,b,W}(h) \qquad (6.32)$$

for every compact set $\mathcal{C}$ of $C_{a,b}$. However, since W is bounded, the exponential tightness for $\mu_N^{a,b,0}$ implies that for $\mu_N^{a,b,W}$: For every $M > 0$, there exists a compact set $\mathcal{K} \subset C_{a,b}$ such that

$$\limsup_{N\to\infty} \frac{1}{N} \log \mu_N^{a,b,W} \left(\mathcal{K}^c \right) \leq -M .$$

Thus, (6.32) holds for every closed set $\mathcal{C}$ of $C_{a,b}$.

Taking $\mathcal{O} = \mathcal{C} = C_{a,b}$ in (6.31) and (6.32), we see that

$$\lim_{N\to\infty} \frac{1}{N} \log \frac{Z_N^{a,b,W}}{Z_N^{a,b,0}} = - \inf_{h\in C_{a,b}} \Sigma^{a,b,W}(h) \qquad (6.33)$$

and this concludes the proof of the theorem.   □

6.4 LDP for δ-Pinning in one Dimension

Gibbs Measures with Pinning Potentials

Let us go back to the d dimensional setting in Sect. 6.1. The **pinning** is an effect of weak force which attracts interfaces ϕ toward the level of height 0, i.e., to a neighborhood of the hyperplane D_N. Such effect is again realized by adding self potentials U to the original Hamiltonian. We assume $Q \equiv 1$ so that $U(\theta, r) = W(r)$ and denote the finite volume φ-Gibbs measure $\mu_N^{\psi, U}$ introduced in (6.4) by $\mu_N^{\psi, W}$. Specifically, we consider the following two types of pinning potentials.

Square-well pinning: The potential W has a form

$$W(r) = -b \mathbf{1}_{\{|r| \le a\}}, \quad r \in \mathbb{R} \tag{6.34}$$

with $a, b > 0$. The constant $s = 2a(e^b - 1)$ is called the strength of pinning. As s increases, the effect of pinning becomes stronger.

δ-pinning: Under the limit $a \downarrow 0, b \to \infty$ keeping $s = e^J$ constant for $J \in \mathbb{R}$, we have that

$$e^{b \mathbf{1}_{\{|\phi(x)| \le a\}}} d\phi(x) \implies e^J \delta_0(d\phi(x)) + d\phi(x)$$

for each $x \in D_N$. In this way, the finite volume Gibbs measure with δ-pinning is introduced as a weak limit of the Gibbs measure $\mu_N^{\psi, W}$ with square-well pinning and has the following form

$$\mu_N^{\psi, J}(d\phi) = \frac{1}{Z_N^{\psi, J}} \exp\left\{ -H_N^\psi(\phi) \right\} \prod_{x \in D_N} \left(e^J \delta_0(d\phi(x)) + d\phi(x) \right). \tag{6.35}$$

We regard $\mu_N^{\psi, J} \in \mathcal{P}(\mathbb{R}^{\overline{D_N}})$ by considering $\phi(x) = \psi(x)$ for $x \in \partial^+ D_N$ as before. The larger J gives stronger pinning. When $J = -\infty$, there is no pinning and $\mu_N^{\psi, -\infty}$ coincides with $\mu_N^\psi = \mu_{D_N}^\psi$ defined by (2.4).

The square-well pinning potential W of (6.34) obviously does not satisfy the condition (W) in (6.5) and the LDP is not established yet. Several properties of the Gibbs measures $\mu_N^{\psi, W}, \mu_N^{\psi, J}$ with square-well or δ-pinnings will be discussed in the subsequent section, Sect. 7.2.

LDP Result

The LDP is established for δ-pinning in one dimension with quadratic potential $V(\eta) = \frac{1}{2}\eta^2$. Let us take $D = (0, 1) \subset \mathbb{R}$ so that $D_N = \{1, 2, \ldots, N - 1\}$, and consider the Gibbs measure $\mu_N^{\psi, J}$ under the boundary conditions (6.19): $\psi(0) = aN, \psi(N) = bN$ for $a, b \in \mathbb{R}$. We denote $\mu_N^{\psi, J}, Z_N^{\psi, J}$ and $Z_N^\psi(= Z_N^{\psi, -\infty})$ as $\mu_N^{a, b, J}, Z_N^{a, b, J}$ and $Z_N^{a, b}$, respectively.

Theorem 6.9. (*Funaki and Sakagawa* [123]) *Under* $\mu_N^{a, b, J}$, *the family of macroscopic random surfaces* $\{h^N(\theta); \theta \in [0, 1]\}$ *defined by (6.21) satisfies the LDP on* $C_{a, b}$ *with speed N and the rate functional* $I^J(h) \equiv I_{(0, 1)}^J(h) = \Sigma^J(h) - \inf_{H_{a, b}^1} \Sigma^J$ (*if* $h \in H_{a, b}^1$ *and* $= +\infty$ *otherwise*), *where*

$$\Sigma^J(h) \equiv \Sigma^J_{(0,1)}(h) = \Sigma_{(0,1)}(h) + \tau^{pin}(J)|\{h = 0\}|\,,$$

and

$$\tau^{pin}(J) = -\lim_{N\to\infty}\frac{1}{N}\log\frac{Z_N^{0,0,J}}{Z_N^{0,0}}\,. \tag{6.36}$$

The function $\tau^{pin}(J)$ is called the **pinning free energy**. It is known that the limit exists and $\tau^{pin}(J) < 0$ for every $J \in \mathbb{R}$, see [123].

Minimizers of Σ^J

The minimizers of Σ^J are computable in a similar manner to what we did in Sect. 6.3 for Σ^W. For instance, in the case where $a > 0$ and $b < 0$, the minimizer is either the straight line $h^{(4)}$ connecting $(0, a)$ and $(1, b)$ or the curve $h^{(5)}$ composed of three straight line segments connecting four points $(0, a), P_1(\theta_1, 0), P_2(1 - \theta_2, 0)$ and $(1, b)$ in this order; see Figs. 6.4 and 6.5, respectively. If $h^{(5)}$ is realized, the angles at two corners P_1 and P_2 are both equal to $\alpha \in [0, \pi/2]$ and obey the Young's relation:

$$\tan\alpha = \sqrt{-2\tau^{pin}(J)}\,.$$

6.5 Outline of the Proof of Theorem 6.1

LDP without Self Potentials

Similarly to the one dimensional case, the proof of Theorem 6.1 can be reduced to the LDP for the finite volume φ-Gibbs measure $\mu_N^{\psi}(= \mu_N^{\psi,0})$ without self potentials. Indeed, the following proposition substitutes for Lemma 6.6. The topology under which the LDP holds becomes weaker when $d \geq 2$.

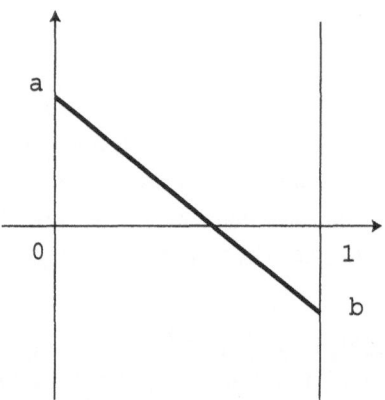

Fig. 6.4. The function $h^{(4)}$

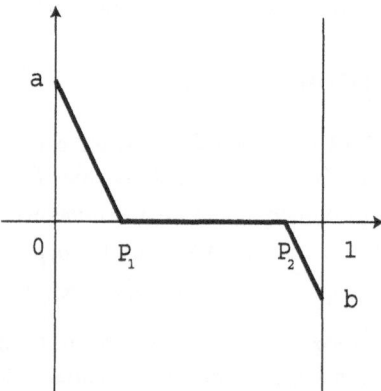

Fig. 6.5. The function $h^{(5)}$

Proposition 6.10. *The family of random surfaces* $\{h^N(\theta); \theta \in D\}$ *distributed under* μ_N^ψ *satisfies the LDP on* $L^2(D)$ *with speed* N^d *and the rate functional*

$$
I_D(h) = \begin{cases} \Sigma_D(h) - \displaystyle\inf_{H_g^1(D)} \Sigma_D & \text{if } h \in H_g^1(D), \\ +\infty & \text{otherwise}. \end{cases}
$$

Once Proposition 6.10 is established, the lower bound in Theorem 6.1 is shown essentially in the same way as Theorem 6.4 with small modifications. For instance, the second estimate in Lemma 6.7 is changed as follows: If $h^N \in B_2(g, \delta) = \{h \in L^2(D); \|h - g\|_{L^2(D)} < \delta\}$ for some $g \in L^2(D)$, $0 < \delta < 1$ and for N large enough, then there exists some constant $C > 0$ such that

$$
\sum_{x \in D_N} U(\frac{x}{N}, \phi(x)) \leq -AN^d \int_D Q(\theta) 1(g(\theta) \leq -\sqrt{\delta}) \, d\theta + CN^d \delta \, ,
$$

for every N large enough. Furthermore, Lemma 6.8 holds for the functional $\Sigma_{D,-}^U(h)$ defined by (6.12) with $1(h(\theta) \leq 0)$ replaced by $1(h(\theta) < 0)$, i.e., $\inf_{h \in \mathcal{O}} \Sigma_D^U(h) = \inf_{h \in \mathcal{O}} \Sigma_{D,-}^U(h)$ for every open set $\mathcal{O}$ of $L^2(D)$.

The proof of the upper bound, to treat the self potential term as a perturbation, requires slightly more careful consideration than for the lower bound, since the LDP upper bound for μ_N^ψ (Proposition 6.10) is shown only in the space $L^2(D)$ and not under the uniform topology. Since U is bounded, the exponential tightness for $\mu_N^{\psi,U}$ can be proved from that for μ_N^ψ which follows from Lemma 6.14 below.

Treatment of Boundary Conditions

Proposition 6.10 is first established by [77] when the boundary condition $\psi \equiv 0$. By shifting the field, the problem with the general boundary condition ψ can be reduced to the 0-boundary case. Indeed, define $\bar\phi$ as $\bar\phi(x) = Ng(\frac{x}{N})$ for $x \in \overline{D_N}$ (recall $g \in C^\infty(\mathbb{R}^d)$) and, by shifting $\phi \mapsto \phi - \bar\phi$, introduce

$$
\widetilde{H}_N^\psi(\phi) = \frac{1}{2} \sum_{b \in \overline{D_N}^*} V(\nabla(\phi \vee 0)(b) + \nabla(\bar\phi \vee \psi)(b)) \, ,
$$

and consider the associated finite volume φ-Gibbs measure having 0-boundary condition:

$$
\tilde\mu_N^\psi(d\phi) = \frac{1}{\widetilde{Z}_N^\psi} \exp\{-\widetilde{H}_N^\psi(\phi)\} \prod_{x \in D_N} d\phi(x) \, .
$$

Then the following LDP holds for $\tilde\mu_N^\psi$.

Proposition 6.11. *The family of random surfaces* $\{h^N(\theta); \theta \in D\}$ *distributed under* $\tilde\mu_N^\psi$ *satisfies the LDP on* $L^2(D)$ *with speed* N^d *and the rate functional*

$$\tilde{I}_D(h) = \begin{cases} \widetilde{\Sigma}_D(h) - \inf_{H_0^1(D)} \widetilde{\Sigma}_D & \text{if } h \in H_0^1(D), \\ +\infty & \text{otherwise}, \end{cases}$$

where

$$\widetilde{\Sigma}_D(h) = \int_D \sigma(\nabla h(\theta) + \nabla g(\theta)) \, d\theta \, .$$

Under the continuous map $\Phi_g : L^2(D) \to L^2(D)$ given by $\Phi_g(h) = h + g$, we have $I_D(h) = \inf\{\tilde{I}_D(\tilde{h}); \tilde{h} \in L^2(D), \Phi_g(\tilde{h}) = h\}$. Hence, by the contraction principle, Proposition 6.10 follows from Proposition 6.11.

Proof of Proposition 6.11

(a) *Convergence of Average Profiles*

The proof of Proposition 6.11 is further reduced to showing the convergence of average profiles, Lemma 6.12. We shall follow the strategy of [77]. For $f \in C_0^\infty(D)$, set $\rho_N = \{\rho_N(x) := f(x/N)/N; x \in D_N\}$ and define the Hamiltonian $H_{N,f}^\psi(\phi)$ by (4.3) with $\Lambda = D_N$ and $\rho = \rho_N$. Another Hamiltonian $\widetilde{H}_{N,f}^\psi(\phi)$ is defined in a similar manner from $\widetilde{H}_N^\psi(\phi)$. Then, consider the following two finite volume φ-Gibbs measures:

$$\mu_{N,f}^\psi(d\phi) = \frac{1}{Z_{N,f}^\psi} \exp\{-H_{N,f}^\psi(\phi)\} \prod_{x \in D_N} d\phi(x),$$

$$\tilde{\mu}_{N,f}^\psi(d\phi) = \frac{1}{\widetilde{Z}_{N,f}^\psi} \exp\{-\widetilde{H}_{N,f}^\psi(\phi)\} \prod_{x \in D_N} d\phi(x) \, ,$$

having the different boundary conditions $\phi(x) = \psi(x)$ and $\phi(x) = 0$ for $x \in \partial^+ D_N$, respectively; recall that ψ and g satisfy the conditions (6.7). Two probability measures $\mu_{N,f}^\psi$ and $\tilde{\mu}_{N,f}^\psi$ are called Cramér transforms of μ_N^ψ and $\tilde{\mu}_N^\psi$. We write the averages of the profile h^N defined by (6.9) under $\mu_{N,f}^\psi$ and $\tilde{\mu}_{N,f}^\psi$ as $\bar{h}_{N,f}^\psi(\theta) = E^{\mu_{N,f}^\psi}[h^N(\theta)]$ and $\tilde{h}_{N,f}^\psi(\theta) = E^{\tilde{\mu}_{N,f}^\psi}[h^N(\theta)]$, respectively. For $f \in L^2(D)$, h_f denotes the unique weak solution $h = h(\theta)$ in $H_0^1(D)$ of the following elliptic PDE:

$$\text{div}\{(\nabla\sigma)(\nabla h(\theta) + \nabla g(\theta))\} = -f(\theta), \quad \theta \in D \, .$$

The crucial step in the proof of Proposition 6.11 is the following lemma.

Lemma 6.12. (1) $\tilde{h}_{N,f}^\psi \to h_f$ *in* $H_0^1(D)$ *as* $N \to \infty$.

(2) *(LLN)* $\displaystyle\lim_{N\to\infty} E^{\tilde{\mu}_{N,f}^\psi}[\|h^N - h_f\|_{L^2(D)}^2] = 0.$

Postponing the proof of this lemma later, we set

$$\Xi_{N,f}^{\psi} \equiv \frac{\widetilde{Z}_{N,f}^{\psi}}{\widetilde{Z}_{N}^{\psi}} = E^{\tilde{\mu}_{N}^{\psi}}\left[\exp\left\{\frac{1}{N}\sum_{x \in D_N} f\left(\frac{x}{N}\right)\phi(x)\right\}\right].$$

Then, by calculating the functional derivative of $\widetilde{\Sigma}_D(h)$ and using the trick to compute $\frac{d}{dt}\log\widetilde{Z}_{N,tf}^{\psi}$ and integrate it in $t \in [0,1]$, Lemma 6.12-(1) yields the following lemma.

Lemma 6.13. *The limit $\Lambda_D(f) \equiv \lim_{N\to\infty}\frac{1}{N^d}\log\Xi_{N,f}^{\psi}$ exists and it holds that*

$$\Lambda_D(f) = \int_D d\theta \int_0^1 h_{tf}(\theta)f(\theta)\,dt$$
$$= \sup_{h \in H_0^1(D)} \{\langle h, f\rangle - \widetilde{\Sigma}_D(h)\} + \inf_{H_0^1(D)} \widetilde{\Sigma}_D,$$

where $\langle h, f\rangle = \int_D h(\theta)f(\theta)\,d\theta$. The supremum is attained at $h = h_f$.

(b) *Exponential Tightness*

For the proof of the LDP upper bound in Proposition 6.11, we need the following uniform exponential estimate.

Lemma 6.14. *There exists $\varepsilon > 0$ such that*

$$\sup_{N \geq 1}\frac{1}{N^d}\log E^{\tilde{\mu}_{N,f}^{\psi}}\left[\exp\left\{\varepsilon\sum_{x \in \overline{D}_N}\left(\left|h^N\left(\frac{x}{N}\right)\right|^2 + \left|\nabla^N h^N\left(\frac{x}{N}\right)\right|^2\right)\right\}\right] < \infty,$$

where $\nabla^N u(\frac{x}{N}) = \{\nabla_i^N u(\frac{x}{N}) := N(u(\frac{x+e_i}{N}) - u(\frac{x}{N}))\}_{1 \leq i \leq d}$ denotes the discrete gradient of a scalar lattice field $u = \{u(\frac{x}{N}); x \in \overline{D}_N\}$.

(c) *Proof of Proposition 6.11*

The upper bound is shown based on the exponential Chebyshev's inequality combined with Lemmas 6.13 and 6.14. For the lower bound, we rely on the usual Cramér's trick: By Lemmas 6.12-(1) and 6.13, it is easy to see that

$$\lim_{N\to\infty}\frac{1}{N^d}\mathcal{H}(\tilde{\mu}_{N,f}^{\psi}|\tilde{\mu}_N^{\psi}) = \widetilde{I}_D(h_f),$$

where $\mathcal{H}(\tilde{\mu}_{N,f}^{\psi}|\tilde{\mu}_N^{\psi})$ is the relative entropy of $\tilde{\mu}_{N,f}^{\psi}$ with respect to $\tilde{\mu}_N^{\psi}$; recall (5.4) and also see (5.4) in [77]. Combining this with Lemma 6.12-(2) and applying the entropy inequality (cf. [79, Lemma 5.4.21]), we obtain

$$\liminf_{N\to\infty}\frac{1}{N^d}\log\tilde{\mu}_N^{\psi}(h^N \in \mathcal{O}) \geq -\inf_{\substack{f \in C_0^\infty(D) \\ \text{s.t. } h_f \in \mathcal{O}}}\widetilde{I}_D(h_f) = \inf_{h \in \mathcal{O}}\widetilde{I}_D(h),$$

for every open set $\mathcal{O} \subset L^2(D)$.

Proof of Lemma 6.12

(a) *Reduction to Lemma 6.16*

We shall prove Lemma 6.12-(1). Lemma 6.12-(2) is shown from it by applying Brascamp-Lieb inequality. The following lemma follows from (5.10).

Lemma 6.15. *Let* $\{h_n\}_{n \geq 1}$ *be a sequence of* $H_0^1(D)$ *and define* $\widetilde{\Sigma}_f(h)$ $= \widetilde{\Sigma}_D(h) - \langle h, f \rangle$. *If* $\lim\limits_{n \to \infty} \widetilde{\Sigma}_f(h_n) = \inf\limits_{H_0^1(D)} \widetilde{\Sigma}_f$, *then* $h_n \to h_f$ *in* $H_0^1(D)$ *as* $n \to \infty$.

Also by (5.10), we have

$$\widetilde{\Sigma}_f(q) - \widetilde{\Sigma}_f(\tilde{h}_{N,f}^{\psi}) \geq \int_D (\nabla q(\theta) - \nabla \tilde{h}_{N,f}^{\psi}(\theta)) \cdot (\nabla \sigma)(\nabla \tilde{h}_{N,f}^{\psi}(\theta) + \nabla g(\theta)) \, d\theta$$

$$- \int_D (q(\theta) - \tilde{h}_{N,f}^{\psi}(\theta)) f(\theta) \, d\theta \ ,$$

for every $q \in C_0^\infty(D)$. Once we can prove that the right hand side goes to 0 as $N \to \infty$ for every $q \in C_0^\infty(D)$, we have $\lim\limits_{N \to \infty} \widetilde{\Sigma}_f(\tilde{h}_{N,f}^{\psi}) = \inf\limits_{H_0^1(D)} \widetilde{\Sigma}_f$. This combined with Lemma 6.15 completes the proof of Lemma 6.12. Hence, all we have to prove are summarized in the following lemma.

Lemma 6.16. (1) *For every* $q \in C_0^\infty(D)$,

$$\lim_{N \to \infty} \int_D \nabla q(\theta) \cdot (\nabla \sigma)(\nabla \tilde{h}_{N,f}^{\psi}(\theta) + \nabla g(\theta)) \, d\theta = \int_D q(\theta) f(\theta) \, d\theta \ .$$

(2) *Moreover, we have that*

$$\lim_{N \to \infty} \int_D \nabla \tilde{h}_{N,f}^{\psi}(\theta) \cdot (\nabla \sigma)(\nabla \tilde{h}_{N,f}^{\psi}(\theta) + \nabla g(\theta)) \, d\theta = \lim_{N \to \infty} \int_D \tilde{h}_{N,f}^{\psi}(\theta) f(\theta) \, d\theta \ .$$

For the proof of Lemma 6.16, we need three lemmas.

(b) *A Priori Bounds*

The assumption that the domain D has a piecewise Lipschitz boundary is necessary to show the following lemma by employing the PDE techniques like Caccioppoli and inverse Hölder inequalities.

Lemma 6.17. (1) $(L^p$-*estimates*) *There exists some* $p \in (2, p_0)$ *such that*

$$\sup_{N \geq 1} \|\nabla \tilde{h}_{N,f}^{\psi}\|_{L^p(D)} < \infty \quad and \quad \sup_{N \geq 1} \|\nabla \bar{h}_{N,f}^{\psi}\|_{L^p(D)} < \infty \ ,$$

where $p_0 > 2$ *is the constant appearing in the condition* $(\psi 2)$.
(2) (*Fluctuation inequalities*) *For every* $e \in \mathbb{Z}^d$ *with* $|e| = 1$, *we have*

$$\lim_{N\to\infty} \frac{1}{N^d} \sum_{x\in D_N} \left| \nabla^N \tilde{h}^{\psi}_{N,f}\left(\frac{x+e}{N}\right) - \nabla^N \tilde{h}^{\psi}_{N,f}\left(\frac{x}{N}\right) \right|^2 = 0 \,,$$

$$\lim_{N\to\infty} \frac{1}{N^d} \sum_{x\in D_N} \left| \nabla^N \bar{h}^{\psi}_{N,f}\left(\frac{x+e}{N}\right) - \nabla^N \bar{h}^{\psi}_{N,f}\left(\frac{x}{N}\right) \right|^2 = 0 \,.$$

(c) *Local Equilibria*

Define $Q_N \in \mathcal{M}_+(D \times \mathcal{X})$ and $V_N \in \mathcal{M}_+(\mathbb{R}^d \times \mathcal{X})$ by

$$Q_N(d\theta d\eta) = \frac{1}{N^d} \sum_{x\in D_N} \delta_{\frac{x}{N}}(d\theta)\mu^{\psi,\nabla}_{N,f} \circ \tau_x^{-1}(d\eta),$$

$$V_N(dv d\eta) = \frac{1}{N^d} \sum_{x\in D_N} \delta_{\nabla^N \bar{h}^{\psi}_{N,f}(\frac{x}{N})}(dv)\mu^{\psi,\nabla}_{N,f} \circ \tau_x^{-1}(d\eta) \,,$$

where $\mathcal{X}$ is the state space for the $\nabla\varphi$-field introduced in Sect. 2.3, $\mathcal{M}_+(\mathcal{E})$ stands for the class of all nonnegative measures on $\mathcal{E}$, $\mu^{\psi,\nabla}_{N,f}(d\eta)$ is the distribution of $\eta = \nabla\phi$ on $\mathcal{X}$ under $\mu^{\psi}_{N,f}$ and $\tau_x : \mathcal{X} \to \mathcal{X}$ denotes the shift on $\mathbb{Z}^d$, cf. Definition 2.3. We regard $\mu^{\psi}_{N,f} \in \mathcal{P}(\mathbb{R}^{\mathbb{Z}^d})$ by considering $\phi(x) = \psi(x)(= g(\frac{x}{N}))$ for $x \in \mathbb{Z}^d \setminus D_N$ as before.

Then, one can prove the following lemma, cf. Sect. 10.3-(c) for dynamics.

Lemma 6.18. *For each $r > 0$ both the families of measures $\{Q_N\}$ on $D \times \mathcal{X}_r$ and $\{V_N\}$ on $\mathbb{R}^d \times \mathcal{X}_r$ are tight, see Sect. 9.1 for the precise definition of the space $\mathcal{X}_r$. Moreover, for every limit point Q of $\{Q_N\}$, there exists $\nu_Q \in \mathcal{M}_+(D \times \mathbb{R}^d)$ such that Q is represented as*

$$Q(d\theta d\eta) = \int_{\mathbb{R}^d} \nu_Q(d\theta dv)\, \mu^{\nabla}_v(d\eta) \,,$$

where μ^{∇}_v is the unique probability measure in $(\text{ext } \mathcal{G}^{\nabla})_v$, i.e., $\nabla\varphi$-pure phase with mean $v \in \mathbb{R}^d$. Similarly, for each limit point V of $\{V_N\}$, there exists $\nu_V \in \mathcal{M}_+(\mathbb{R}^d \times \mathbb{R}^d)$ such that V is represented as

$$V(dv d\eta) = \int_{\mathbb{R}^d} \nu_V(dv du)\, \mu^{\nabla}_u(d\eta) \,.$$

Now by Lemma 6.17-(1), along some subsequence, $\{\nabla\tilde{h}^{\psi}_{N,f}(\theta)\}_N$ generates the family of Young measures $\tilde{\nu}(\theta, dv) \in \mathcal{P}(\mathbb{R}^d)$, i.e., it holds that

$$\lim_{N\to\infty} \int_D q(\theta)G(\nabla\tilde{h}^{\psi}_{N,f}(\theta))\, d\theta = \int_{D\times\mathbb{R}^d} q(\theta)G(v)\, \tilde{\nu}(\theta, dv)d\theta$$

for every $q \in L^{\infty}(D)$ and $G \in C_0(\mathbb{R}^d)$ (cf. [77, Sect. 4.3], [9]). Then, the following lemma holds.

Lemma 6.19. *If the subsequence $\{N\}$ is commonly taken, the limits ν_Q and ν_V which have appeared in Lemma 6.18 can be represented as*

$$\nu_Q(d\theta dv) = \tilde{\nu}(\theta, dv - \nabla g(\theta)) \, d\theta \, ,$$

and

$$\nu_V(dvdu) = \delta_v(du) \int_D \tilde{\nu}(\theta, dv - \nabla g(\theta)) \, d\theta \, .$$

(d) *Proof of Lemma 6.16*

We are now in the position to prove Lemma 6.16. For every $q \in C_0^\infty(D)$, since $(\nabla^N)^* E^{\mu_{N,f}^\psi}[V'(\nabla\phi(x))] = -f(x/N)$, we have by summation by parts

$$\int_D q(\theta)f(\theta) \, d\theta = \lim_{N\to\infty} \frac{1}{N^d} \sum_{x\in D_N} \nabla^N q\left(\frac{x}{N}\right) \cdot E^{\mu_{N,f}^\psi}[V'(\nabla\phi(x))] \, ,$$

where $(\nabla^N)^* \equiv \sum_{i=1}^d \nabla_i^{N*}$ denotes the dual operator of ∇^N (cf. Sect. 10.2-(b)) and $V'(\nabla\phi(x)) = \{V'(\nabla_i\phi(x))\}_{1\le i\le d} \in \mathbb{R}^d$. Now by the definition of Q_N, Lemmas 6.18, 6.19 and the property (5.14) of the surface tension, we obtain

$$\int_D q(\theta)f(\theta) \, d\theta = \int_{D\times\mathcal{X}} \nabla q(\theta) \cdot E^{\mu_v^\nabla}[V'(\nabla\phi(0))] \, \nu_Q(d\theta dv)$$

$$= \int_{D\times\mathbb{R}^d} \nabla q(\theta) \cdot (\nabla\sigma)(v + \nabla g(\theta)) \, \tilde{\nu}(\theta, dv) d\theta$$

$$= \lim_{N\to\infty} \int_D \nabla q(\theta) \cdot (\nabla\sigma)(\nabla\tilde{h}_{N,f}^\psi(\theta) + \nabla g(\theta)) \, d\theta \, .$$

This shows (1). We similarly have

$$\lim_{N\to\infty} \int_D \tilde{h}_{N,f}^\psi(\theta)f(\theta) \, d\theta = \lim_{N\to\infty} \frac{1}{N^d} \sum_{x\in D_N} \nabla^N \tilde{h}_{N,f}^\psi\left(\frac{x}{N}\right) \cdot E^{\mu_{N,f}^\psi}[V'(\nabla\phi(x))]$$

$$= \lim_{N\to\infty} \frac{1}{N^d} \sum_{x\in D_N} \nabla^N \tilde{h}_{N,f}^\psi\left(\frac{x}{N}\right) \cdot E^{\mu_{N,f}^\psi}[V'(\nabla\phi(x))]$$

$$- \lim_{N\to\infty} \frac{1}{N^d} \sum_{x\in D_N} \nabla(\bar{\phi} \vee \psi)(x) \cdot E^{\mu_{N,f}^\psi}[V'(\nabla\phi(x))]$$

$$\equiv S_1 - S_2 \, .$$

Now, since $\bar{\phi}(x) = Ng(\frac{x}{N})$, we see by the assumptions on V and ψ that

$$S_2 = \lim_{N\to\infty} \frac{1}{N^d} \sum_{x\in D_N} \nabla^N g\left(\frac{x}{N}\right) \cdot E^{\mu_{N,f}^\psi}[V'(\nabla\phi(x))]$$

$$= \lim_{N\to\infty} \int_D \nabla g(\theta) \cdot (\nabla\sigma)(\nabla\tilde{h}_{N,f}^\psi(\theta) + \nabla g(\theta)) \, d\theta \, .$$

Also, by Lemmas 6.18, 6.19 and the property of the surface tension σ, in a similar way to the proof of (1) we can prove that

$$S_1 = \lim_{N \to \infty} \int_D (\nabla \tilde{h}_{N,f}^{\psi}(\theta) + \nabla g(\theta)) \cdot (\nabla \sigma)(\nabla \tilde{h}_{N,f}^{\psi}(\theta) + \nabla g(\theta)) \, d\theta \, .$$

Therefore, the proof of (2) is also completed.

6.6 Critical LDP

Bolthausen and Deuschel [27] discussed the LDP for the empirical measures of the field distributed under the Gaussian φ-Gibbs measure on $\mathbb{Z}^d$, when $d \geq 3$. Here we summarize their results.

Let $\mu \in \mathcal{P}(\mathbb{R}^{\mathbb{Z}^d})$ be the Gaussian measure on $\mathbb{R}^{\mathbb{Z}^d}$ with mean 0 and covariance $(-\Delta)^{-1}(x,y), x, y \in \mathbb{Z}^d$, recall Sect. 3.2. Define the **empirical distribution functional** of φ-field by

$$R_N(\phi) := \frac{1}{N^d} \sum_{x \in D_N} \delta_{\tau_x \phi_N} \in \mathcal{P}(\mathbb{R}^{\mathbb{Z}^d}) \, , \tag{6.37}$$

for $\phi = \{\phi(x); x \in \mathbb{Z}^d\} \in \mathbb{R}^{\mathbb{Z}^d}$, where $D_N \equiv ND \cap \mathbb{Z}^d = [0, N-1]^d \cap \mathbb{Z}^d$ with the choice of $D = [0,1)^d$. Note that $N^d = |D_N|$. In (6.37), ϕ_N is the periodic extension to $\mathbb{Z}^d$ of $\phi|_{D_N} = \{\phi(x); x \in D_N\}$, the restriction of ϕ on D_N, and $\tau_x : \mathbb{R}^{\mathbb{Z}^d} \to \mathbb{R}^{\mathbb{Z}^d}$ denotes the shift. Note that $R_N(\phi)$ is a shift invariant probability measure. Since μ is ergodic under shifts, the LLN:

$$R_N \Longrightarrow \mu \quad (N \to \infty)$$

holds for μ-a.s. ϕ, where $\Longrightarrow$ denotes the weak convergence of measures on $\mathbb{R}^{\mathbb{Z}^d}$. The aim of [27] is to study the corresponding LDP.

First Result

The first result is at the order of N^d, i.e., the weak LDP of **volume order**:

$$\mu \left(R_N \in \text{"neighborhood of } \nu \text{"} \right) \underset{N \to \infty}{\asymp} e^{-N^d \mathbf{h}(\nu|\mu)} \quad (N \to \infty) \, . \tag{6.38}$$

Here the rate functional is the specific entropy (specific free energy)

$$\mathbf{h}(\nu|\mu) := \lim_{N \to \infty} \frac{1}{N^d} \mathcal{H}_N(\nu|\mu) \, ,$$

for shift invariant $\nu \in \mathcal{P}(\mathbb{R}^{\mathbb{Z}^d})$ and $\mathcal{H}_N(\nu|\mu) = \int \log \frac{d\nu}{d\mu}|_{\mathcal{F}_{D_N}} d\nu$ is the relative entropy defined by restricting ν and μ to the σ-field $\mathcal{F}_{D_N} = \sigma\{\phi(x); x \in D_N\}$. The asymptotic property (6.38) is rudely stated, but it can be precisely formulated as the LDP upper and lower bounds as usual. The "weak" LDP means that the upper bound is available only for compact sets. Similar results are known for lattice systems with bounded spins (e.g., Ising model) and for Markov processes (Donsker-Varadhan theory).

Let us denote the class of all tempered (i.e., square integrable) and shift invariant φ-Gibbs measures by $\mathcal{G}$ and that of all ergodic $\mu \in \mathcal{G}$ by ext $\mathcal{G}$, respectively; recall that we are assuming $V(\eta) = \frac{1}{2}\eta^2$. From Theorem 9.10 below, we have

$$\text{ext } \mathcal{G} = \{\mu_h; h \in \mathbb{R}\},$$

where $\mu_h \in \mathcal{P}(\mathbb{R}^{\mathbb{Z}^d})$ is the Gaussian measure with mean h and covariance $(-\Delta)^{-1}(x, y)$. It is however known that φ-Gibbs measure has an entropic characterization (cf. Sheffield for general potentials):

$$\mathbf{h}(\nu|\mu) = 0 \iff \nu \in \mathcal{G}.$$

In particular, the LDP estimate (6.38) gives no useful information when ν is the φ-Gibbs measure, since the rate functional is 0 for such ν.

Second Result

The order (speed) of the LDP for $\nu \in \mathcal{G}$ is N^{d-2}, i.e., the LDP of **capacity order** holds:

$$\mu\left(R_N \in \text{``neighborhood of } \nu\text{''}\right) \underset{N \to \infty}{\asymp} e^{-N^{d-2}\mathcal{C}(\nu|\mu)}. \tag{6.39}$$

The rate functional is given by

$$\mathcal{C}(\nu|\mu) = \inf\left\{\frac{1}{2}\mathcal{E}_D(h); \, h \in L^2(D) \text{ s.t. } \nu = \int_D \mu_{h(\theta)} \, d\theta\right\},$$

where

$$\mathcal{E}_D(h) = \inf\left\{\frac{1}{2}\|\nabla\tilde{h}\|^2_{L^2(\mathbb{R}^d)}; \tilde{h} \in H^1(\mathbb{R}^d), \tilde{h}(\theta) = h(\theta) \text{ a.e. } \theta \in D\right\}.$$

In particular, if $\nu = \mu_h \in \text{ext } \mathcal{G}$, since $h(\theta) \equiv h$, we have

$$\mathcal{C}(\mu_h|\mu) = h^2 \text{Cap}_{\mathbb{R}^d}(D),$$

see (7.1) below for the definition of the capacity.

Third Result

If ν is the mixture of finitely many $\{\mu_h\}$'s, then $h(\theta)$ in $\mathcal{C}(\nu|\mu)$ is a step function so that the condition $\tilde{h}(\theta) \in H^1(\mathbb{R}^d)$ is never fulfilled and therefore $\mathcal{C}(\nu|\mu) = +\infty$. This means that (6.39) is not at the correct order for such ν. Indeed, for such ν, the LDP with speed N^{d-1}, i.e., the LDP of **surface order** holds and we have that

$$\lim_{N \to \infty} \frac{1}{N^{d-1}} \mathcal{H}_N(\mu_{h_N}|\mu) < \infty \, ,$$

for the sequence of Gaussian fields $\mu_{h_N} \in \mathcal{P}(\mathbb{R}^{D_N})$ on D_N with mean $h(x/N)$ and covariance $(-\Delta)^{-1}(x,y)$.

Remark 6.10. *The LDP for the Gaussian φ-field on $\mathbb{Z}^d$, whose covariance is given by the Green function of a long-range random walk, is studied by* [46], [47].

7 Entropic Repulsion, Pinning and Wetting Transition

In this section we discuss the subjects of entropic repulsion and pinning. The **entropic repulsion** is the problem to study, when a hard wall is settled at the height level 0, how high the interfaces are pushed up by the random fluctuations naturally caused by the Lebesgue measure $d\phi_\Lambda$ in the Gibbs measure (2.4), in other words, by the entropic effects of the measure. The **pinning** is, on the other hand, the problem to study, under the effect of weak force attracting interfaces to the height level 0, whether the field is really localized or not. This is the energy effect. These two effects conflict with each other and therefore, it is an interesting question to know which effect is dominant in the system. This leads us to the problem of the **wetting transition**.

We first briefly summarize the results. Let $\mu_N = \mu_{D_N}^0$ be the finite volume φ-Gibbs measure on D_N with 0-boundary conditions. The goal is to investigate the asymptotic behavior as $N \to \infty$ of height variables ϕ under μ_N and under μ_N with wall or/and pinning effects.

(a) No wall nor pinning (cf. (3.5), Theorem 4.13):
$\qquad d = 2 \Longrightarrow$ delocalized $(|\phi(x)| \approx \sqrt{\log N})$
$\qquad d \geq 3 \Longrightarrow$ localized $(\phi(x) = O(1))$
$\qquad\qquad$ massless (algebraic decay of two-point correlations)
(b) Wall effect only (Entropic repulsion, Theorem 7.3):
$\qquad d = 2 \Longrightarrow$ delocalized $(\phi(x) \approx \log N)$
$\qquad d \geq 3 \Longrightarrow$ delocalized $(\phi(x) \approx \sqrt{\log N})$
(c) Pinning effect only (Theorems 7.4, 7.5):
$\qquad d = 2 \Longrightarrow$ localized and mass generation
$\qquad\qquad$ (exponential decay of two-point correlations)
$\qquad d \geq 3 \Longrightarrow$ localized, mass generation
(d) Both wall and pinning effects (Theorem 7.7):
$\qquad d = 1, 2 \Longrightarrow$ wetting transition occurs, i.e., if the strength of
$\qquad\qquad$ pinning is strong, the field is localized; while, if
$\qquad\qquad$ it is weak, the field is delocalized.
$\qquad d \geq 3 \Longrightarrow$ no wetting transition and the field is always localized.

7.1 Entropic Repulsion

First let us remind some notation. For $\Lambda \Subset \mathbb{Z}^d$, the finite volume φ-Gibbs measure $\mu_\Lambda^0 \in \mathcal{P}(\mathbb{R}^{\mathbb{Z}^d})$ is defined by (2.4) with 0-boundary condition. We shall denote by $\mathcal{D}$ the class of all connected and bounded domains D in $\mathbb{R}^d$ with piecewise smooth boundaries ∂D. We take $D \in \mathcal{D}$ and fix it, and simply write μ_N for $\mu_{D_N}^0$ with the choice of $\Lambda = D_N$. When $V(\eta) = \frac{1}{2}c_-\eta^2$, μ_Λ^0 and μ_N are denoted by $\mu_\Lambda^{0,G}$ and μ_N^G, respectively. For $\tilde{\Lambda} \Subset \mathbb{Z}^d$, consider the **entropic repulsion event** defined by

$$\Omega^+(\tilde{\Lambda}) = \{\phi; \phi(x) \geq 0 \text{ for every } x \in \tilde{\Lambda}\} .$$

This event is realized by putting a wall at height level 0 on the region $\tilde{\Lambda}$.

We explain the results on the entropic repulsion following Deuschel and Giacomin [76]. The conditions (V1)-(V3) in (2.2) are always assumed on the potential V. The **first result** is for the probability estimate on the entropic repulsion event when the wall is put at strictly inside of D: $A \Subset D$ means that $A \subset D$ and $\text{dist}(A, D^c) > 0$. We shall write

$$\log_d(N) = \begin{cases} \log N, & d \geq 3 , \\ (\log N)^2, & d = 2 . \end{cases}$$

See Remark 7.2 below for $d = 1$.

Theorem 7.1. [76] *Assume* $A, D \in \mathcal{D}$ *and* $A \Subset D$. *Then there exist* $0 < C_1 \leq C_2 < \infty$ *such that*

$$-C_2 \leq \liminf_{N \to \infty} \frac{1}{N^{d-2} \log_d(N)} \log \mu_N(\Omega^+(A_N))$$

$$\leq \limsup_{N \to \infty} \frac{1}{N^{d-2} \log_d(N)} \log \mu_N(\Omega^+(A_N)) \leq -C_1 .$$

Proof (Partially). We only outline the proof of the lower bound for the Gaussian case: $\mu_N = \mu_N^G$ (with $c_- = 1$) when $d \geq 3$. The bound can be shown with $C_2 = 2dG(O, O) \text{Cap}_D(A)$, where G denotes the Green function of Δ on $\mathbb{Z}^d$ and

$$\text{Cap}_D(A) = \inf \left\{ \frac{1}{2} \|\nabla h\|_{L^2(D)}^2; h \in C_0^1(D), h \geq 1_A \right\} \tag{7.1}$$

is the capacity; note that $\frac{1}{2}\|\nabla h\|_{L^2(D)}^2 = \Sigma_D(h)$ for the Gaussian case (the proof of the lower bound goes essentially in a similar way also for $d = 2$ if $G(O, O)$ is suitably modified). In fact, transforming the measure μ_N^G into $\mu_N^{G,\tilde{\phi}}$ under the map $\phi \mapsto \phi + \tilde{\phi}$ where $\tilde{\phi}(x) = \sqrt{a \log N} h(x/N)$ with $h(\theta) \geq 1_A(\theta)$, one can show the LLN for the transformed measure $\mu_N^{G,\tilde{\phi}}$:

$$\lim_{N \to \infty} \mu_N^{G,\tilde{\phi}}(\Omega^+(A_N)) = 1 , \tag{7.2}$$

if $a > 2dG(O, O)$; i.e., the probability that the interfaces touch the wall becomes negligible if they are pushed up to the sufficiently high level by adding $\tilde{\phi}$. The price to adding $\tilde{\phi}$ should be paid by the relative entropy which behaves as

$$\lim_{N \to \infty} \frac{1}{N^{d-2} \log N} \mathcal{H}(\mu_N^{G,\tilde{\phi}} | \mu_N^G) = \frac{a}{2} \|\nabla h\|_{L^2(D)}^2 . \tag{7.3}$$

Applying the entropy inequality, (7.2) combined with (7.3) shows

$$\frac{1}{N^{d-2} \log N} \log \frac{\mu_N^G(\Omega^+(A_N))}{\mu_N^{G,\tilde{\phi}}(\Omega^+(A_N))}$$

$$\geq -\frac{1}{N^{d-2} \log N} \frac{1}{\mu_N^{G,\tilde{\phi}}(\Omega^+(A_N))} \left\{ \mathcal{H}(\mu_N^{G,\tilde{\phi}} | \mu_N^G) + \frac{1}{e} \right\}$$

$$\xrightarrow[N \to \infty]{} -\frac{a}{2} \|\nabla h\|_{L^2(D)}^2 .$$

Thus the lower bound is obtained with $C_2 = 2dG(O, O) \operatorname{Cap}_D(A)$. $\qquad \square$

Remark 7.1. *The problem of the entropic repulsion was posed by Lebowitz and Maes* [186]. *They notify that the probability of* $\Omega^{+,\delta}(A_N) = \{\phi; \phi(x) \geq \delta$ *for every* $x \in A_N\}, \delta > 0$ *instead of* $\Omega^+(A_N)$ *may be estimated as follows:*

$$\mu_N(\Omega^{+,\delta}(A_N)) \leq \mu_N(X \geq \delta)$$

$$\approx \exp\left\{-\delta^2/2\mathrm{var}\,(X; \mu_N)\right\}$$

$$\leq \exp\left\{-\delta^2/2\mathrm{var}\,(X; \mu_N^G)\right\} \leq \exp\left\{-C\delta^2 N^{d-2}\right\} ,$$

where $X = \frac{1}{|A_N|} \sum_{x \in A_N} \phi(x)$. *The second line is true at least if the φ-field has Gaussian tail and the Brascamp-Lieb inequality proves the third line. The above calculation roughly explains the capacity order appearing in (7.4), though the logarithmic correction does not come out because of the difference of* $\Omega^{+,\delta}(A_N)$ *from* $\Omega^+(A_N)$.

The **second result** is for the case where the wall is put over the whole domain D, i.e., we take $A = D$, cf. (6.15).

Theorem 7.2. [76] *Assume* $D \in \mathcal{D}$. *Then there exist* $0 < C_1 \leq C_2 < \infty$ *such that*

$$-C_2 \leq \liminf_{N \to \infty} \frac{1}{N^{d-1}} \log \mu_N(\Omega^+(D_N))$$

$$\leq \limsup_{N \to \infty} \frac{1}{N^{d-1}} \log \mu_N(\Omega^+(D_N)) \leq -C_1 .$$

Moreover, we have that

$$\lim_{L \to \infty} \limsup_{N \to \infty} \frac{1}{N^{d-1}} \left| \log \mu_N(\Omega^+(D_N)) - \log \mu_N(\Omega^+(\partial_L D_N)) \right| = 0 ,$$

where $\partial_L D_N = \{x \in D_N; \operatorname{dist}(x, D_N^c) \leq L\}$.

Proof (Partially). In the proof of Theorem 7.1, one can take $h \equiv 1$ so that $\tilde{\phi}(x) \equiv \sqrt{a \log N}$ and

$$\lim_{N \to \infty} \mu_N^{G,\tilde{\phi}}(\Omega^+(D_N)) = 1$$

holds for $a > 2dG(O,O)$. Then, the lower bound is shown with the speed $N^{d-1} \log N$ instead of N^{d-1}. To remove the log factor, further delicate arguments are required. For the non-Gaussian case, the application of the Brascamp-Lieb inequality gives the same result for $a > 2dG(O,O)/c_-$.   □

Theorem 7.1 claims that the probability behaves as

$$\mu_N(\Omega^+(A_N)) \asymp e^{-CN^{d-2} \log_d(N)} , \tag{7.4}$$

i.e., the decay is essentially of **capacity order** except for the logarithmic correction, while Theorem 7.2 indicates that

$$\mu_N(\Omega^+(D_N)) \asymp e^{-CN^{d-1}} , \tag{7.5}$$

i.e., the decay is much faster than (7.4) and it is of **surface order**. The behavior of $\phi(x)$ near the boundary of D_N substantially contributes to the decay of the probability in (7.5). Indeed, since ϕ satisfies the boundary condition $\phi(x) = 0$ at $x \in \partial^+ D_N$, ϕ can be negative near the boundary with high probability and this makes $\mu_N(\Omega^+(D_N))$ smaller. If one can assume that the probability for $\phi(x)$ getting to the positive side at each x near $\partial^+ D_N$ behaves as e^{-c} and is nearly independent when x's are apart from each other, then we get the order in (7.5). Once the interface comes to the positive side, the probability to stay there is governed by (7.4) at the inside of D_N, which is negligible compared with the boundary effect (7.5).

The **third result** is for the estimate giving how high the interfaces are pushed up by the effect of the wall put at $A = D$, i.e., our object is the conditional probability

$$\mu_N^+ = \mu_N(\cdot | \Omega^+(D_N)) .$$

This measure was introduced in Corollary 6.2 imposing the constant volume condition at the same time.

Theorem 7.3. [76] (1) (*Upper bound*) For every $C > \frac{2d}{c_-} G_d^*$,

$$\lim_{N \to \infty} \inf_{x \in D_N} \mu_N^+(\phi(x) < \sqrt{C \log_d(N)}) = 1 ,$$

where $G_d^* = G(O,O)$ (when $d \geq 3$), $G_2^* = \lim_{N \to \infty} G_{D_N}(O,O)/\log N$ (when $d = 2$), and G and G_{D_N} are the Green functions of Δ on $\mathbb{Z}^d$ and D_N, respectively.
(2) (*Lower bound*) There exists $K > 0$ such that

$$\lim_{N \to \infty} \inf_{x \in D_N^{(\delta)}} \mu_N^+(\phi(x) > \sqrt{K \log_d(N)}) = 1 ,$$

for every $\delta \in (0,1)$, where $D_N^{(\delta)} = D_N \setminus \partial_{\delta N} D_N$.

Proof (Outline). We assume $d \geq 3$. Note that the Brascamp-Lieb inequality holds for μ_N^+ since the conditioning under $\Omega^+(D_N)$ is equivalent to adding to the Hamiltonian the self potential term $\sum_x W(\phi(x))$, where $W(r) = 0$ for $r \geq 0$ and $+\infty$ for $r < 0$, and such W can be regarded convex (by approximating it by a sequence of convex potentials). Therefore, we have

$$\mu_N^+(\phi(x) > \sqrt{C \log N})$$

$$\leq \exp\left\{ -\frac{c_- \left((\sqrt{C \log N} - E^{\mu_N^+}[\phi(x)]) \vee 0 \right)^2}{2 G_{D_N}(x,x)} \right\}.$$

This proves the upper bound, if one can show

$$\limsup_{N \to \infty} \sup_{x \in D_N} \frac{E^{\mu_N^+}[\phi(x)]}{\sqrt{C \log N}} < 1. \tag{7.6}$$

To see (7.6), the method of changing the measure (as in the proof of Theorem 7.1), the FKG and Brascamp-Lieb inequalities are applied. The details are omitted. The lower bound is much more delicate. The lattice is divided into even and odd sites, and then the Markov property of the φ-field is effectively used.   □

As Theorem 7.3 suggests, the expectation of the height variables behaves as follows at the inside (in the macroscopic sense) of D_N:

$$E^{\mu_N^+}[\phi(O)] \underset{N \to \infty}{\approx} \sqrt{\log_d(N)}, \tag{7.7}$$

assuming $O \in D^\circ$ (the interior of D) for simplicity. This means that, once the wall is put on D_N, the interfaces are pushed up to the level of order $O(\sqrt{\log_d(N)})$ inside of D_N. The behavior of $\phi(x)$ was given by (3.5) when there was no wall including the non-Gaussian case, see also [21]. Compared with this, we see that the wall pushes up the interfaces further at the order of $\sqrt{\log N}$ for every $d \geq 2$. Bolthausen et al. [28] and Daviaud [60] studied the fine behavior of $\max_x \phi(x)$ for $d = 2$ for Gaussian case.

Remark 7.2. *When $d = 1$, the height variables behave as $|\phi(x)| \approx \sqrt{N}$ under μ_N, recall (3.5). As we shall see in Sect. 14, they behave as $\phi(x) \approx \sqrt{N}$ under μ_N^+ too. Namely, the wall does not change the order of the heights of the φ-field in one dimension.*

Much more precise results are known especially for the Gaussian φ-field in dimensions $d \geq 3$. Deuschel and Giacomin [75] proved that the distribution of $\{\phi(x) - a_N; x \in \mathbb{Z}^d\}$ under μ_N^+ (regarding $\phi(x) = 0$ on D_N^c) weakly converges as $N \to \infty$ to the Gaussian φ-Gibbs measure $\mu \equiv \mu_{\mathbb{Z}^d} = N(0, G(x,y))$ on $\mathbb{Z}^d$, where $a_N := E^{\mu_N^+}[\phi(O)] \sim \sqrt{4G(O,O) \log N}$. In other words, the wall has an effect on the φ-field simply pushing it up by a_N. In addition, the following precise asymptotic estimates on the probabilities $\mu_{\mathbb{Z}^d}(\Omega^+(D_N))$ and $\mu_N(\Omega^+(D_N))$ were established by Bolthausen et al. [29] and Deuschel [74], respectively:

$$\lim_{N\to\infty} \frac{1}{N^{d-2}\log N} \log \mu_{\mathbb{Z}^d}(\Omega^+(D_N)) = -2G(O,O)\,\mathrm{Cap}_{\mathbb{R}^d}(D)\,, \qquad (7.8)$$

$$\lim_{N\to\infty} \frac{1}{N^{d-1}} \log \mu_N(\Omega^+(D_N)) = -c\,, \qquad (7.9)$$

where c is a certain constant determined from the surface tension.

The entropic repulsion for the Gaussian φ-field on a (quenched) random hard wall was discussed by Bertacchi and Giacomin [18], when $d \geq 3$. The entropic repulsion for two interfaces (over the wall) to lie one above the other is discussed by Bertacchi and Giacomin [19] and Sakagawa [227]. The entropic repulsion for interfaces between two walls is studied by Sakagawa [228]. This problem was first discussed by Bricmont et al. [38].

Remark 7.3. (1) *The entropic repulsion for the Gaussian field with covariance $\{P(-\Delta)\}^{-1}(x,y)$ is discussed by Sakagawa [226] for polynomials P in higher dimensions (i.e., at the transient regime: $d \geq 3$ when $P(a) = a$). The physical motivation comes from [145].*
(2) *The probability estimate on the entropic repulsion event is, in general, more delicate in lower dimensions (i.e., at the recurrent regime). Sinai [236] considered the field $\phi = \{\phi(x) \in \mathbb{Z}; x \in \mathbb{Z}_+\}$ with mean 0, covariance $(-\Delta)^{-2}(x,y)$ satisfying $\phi(0) = 0$ in one dimension and proved that*

$$C_1 N^{-1/4} \leq P(\phi(x) \geq 0 \text{ for every } 0 \leq x \leq N) \leq C_2 N^{-1/4}\,, \qquad (7.10)$$

where Δ is the discrete Laplacian on $\mathbb{Z}_+$ determined from the Dirichlet 0-boundary condition at $x = 0$. Theorem 7.1 and (7.8) show the exponential decay of the probability when $d \geq 3$, while (7.10) exhibits the decay in power law which is much delicate. Note that the field ϕ can be constructed as $\phi(x) = \sum_{y=0}^{x} \eta(y)$ from the simple and symmetric random walk $\{\eta(y); y \in \mathbb{Z}_+\}$ on $\mathbb{Z}$ with time parameter y.

7.2 Pinning

Recall that $\mu_N^{\psi,W}$ is the finite volume φ-Gibbs measure (6.4) with $U(\theta,r) = W(r)$. The square-well pinning and δ-pinning were introduced in Sect. 6.4. We first shortly summarize the known results on the pinning problem.

Dunlop et al. [93] first proved the **localization** of the φ-field under the square-well pinning, namely the uniform boundedness in N of the expected height variables $E^{\mu_N^{0,W}}[|\phi(x)|]$ under the φ-Gibbs measures $\mu_N^{0,W}$ with 0-boundary conditions or the existence of infinite volume limit of $\mu_N^{0,W}$ as $N \to \infty$, if the Hamiltonian contains arbitrarily weak pinning potentials W when $d = 2$ for quadratic V. This should be compared with the case without pinning (i.e., $W \equiv 0$) in which the localization occurs only when $d \geq 3$ and also compared with the case of strong pinning (or massive) potentials satisfying $\lim_{|r| \to \infty} W(r) = +\infty$ for which the localization occurs for all dimensions. The result of [93] is extended for general convex potential V by Deuschel and Velenik [81] later. In addition to the localization, the **mass generation**, namely the exponential decay of the correlations of the φ-field is shown by Ioffe and Velenik [153] for $d = 2$ with δ-pinning, see also [92]. Further precise estimates on the asymptotic behaviors of the mass and the degree of localization by means of the variances of the field as the pinning effect becomes smaller were established by Bolthausen and Velenik [32].

Let us state the results in more details. The first result is on the **localization** due to Deuschel and Velenik [81]. The φ-field is localized even without the pinning if $d \geq 3$. Therefore, the interesting case is in the two dimension so that we take $D = [-1, 1]^2$ and consider the φ-field on $D_N = [-N, N]^2 \cap \mathbb{Z}^2$ adding square-well or δ-pinnings. The next theorem gives an estimate on the decay of the tail distribution of the height variables. This implies the localization as we shall see later. We write $\mu_N^{a,b}$ for $\mu_N^{0,W}$ with the square-well pinning potential W given in (6.34) and μ_N^J for $\mu_N^{0,J}$ with δ-pinning, respectively. The 0-boundary conditions are imposed in both measures.

Theorem 7.4. ([81], *Non-Gaussian tail*) (1) *For every $a, b > 0$, there exist constants $C_1 = C_1(s), C_2 = C_2(s) > 0$ such that*

$$e^{-C_1 T^2 / \log T} \leq \mu_N^{a,b}(\phi(x) \geq T) \leq e^{-C_2 T^2 / \log T} , \qquad (7.11)$$

for every sufficiently large $T > 0$ and every $N \in \mathbb{N}$, where the upper bound holds for every $x \in D_N$ while the lower bound holds only for $x \in D_N$ satisfying $\mathrm{dist}\,(x, \partial D_N) \geq d_0 T / \log T$ with some $d_0 > 0$. Recall that $s = 2a(e^b - 1)$ stands for the strength of the pinning.
(2) *Letting $a \downarrow 0$ and $b \to \infty$, a similar estimate holds for μ_N^J with δ-pinning.*

The upper bound implies, in particular, that the exponential moments of $|\phi(x)|$ are uniformly bounded in N under $\mu_N^{a,b}$ or μ_N^J so that these measures are tight and admit the limits $\mu^{a,b}$ and μ^J, respectively, along a proper subsequence $N' \to \infty$. It is obvious that the limits $\mu^{a,b}$ or μ^J are the φ-Gibbs measures with square-well or δ-pinnings on $\mathbb{Z}^2$. The fields are thus localized in two dimension.

Remark 7.4. (1) *The lower bound in (7.11) indicates that the decay of the tail distribution of $\phi(x)$ is slower than that of the Gaussian distribution; i.e., $\phi(x)$ can take rather large values. Without pinning, if $d \geq 3$, the φ-field exists on $\mathbb{Z}^d$ and the Brascamp-Lieb inequality shows its Gaussian tail: $\mu(\phi(x) \geq T) \asymp e^{-CT^2}$. Therefore, with pinning, the tail distributions exhibit completely different behaviors for $d = 2$ and $d \geq 3$.*
(2) *The estimate (7.11) holds for the Gibbs measures $\mu^{a,b}$ and μ^J on $\mathbb{Z}^2$ by taking the limit $N \to \infty$.*

The second result is on the **mass generation** due to Ioffe and Velenik [153] for $d = 2$ and δ-pinning. The case of $d \geq 3$ is discussed by [41].

Theorem 7.5. [153] *For every $J \in \mathbb{R}$, there exist constants $m = m(J), C = C(J) > 0$ such that*

$$E^{\mu_N^J}[\phi(x); \phi(y)] \leq Ce^{-m|x-y|}, \quad x, y \in \mathbb{Z}^2 ,$$

i.e., the covariances under μ_N^J have the exponential decay estimates uniformly in N.

In the proof of Theorems 7.4 and 7.5, the following expansions of the measures μ_N^J and $\mu_N^{a,b}$ play the key role.

Lemma 7.6. (1) *For each $\phi \in \mathbb{R}^{D_N}$ (or $\in \mathbb{R}^{\mathbb{Z}^d}$ regarding $\phi \equiv 0$ on D_N^c) define the **pinned region** (random region) by*

$$\mathcal{A} \equiv \mathcal{A}_N(\phi) = \{x \in D_N; \phi(x) = 0\} . \qquad (7.12)$$

Then the measure μ_N^J admits the expansion

$$\mu_N^J(\cdot) = \sum_{A \subset D_N} \nu_N^J(A)\mu_{D_N \setminus A}^0(\cdot) ,$$

where $\nu_N^J(A) = \mu_N^J(\mathcal{A} = A)$ is the probability that the pinned region is A and $\mu_{D_N \setminus A}^0(\cdot)$ is the φ-Gibbs measure on $D_N \setminus A$ with 0-boundary condition.
(2) *The measure $\mu_N^{a,b}$ admits the similar expansion:*

$$\mu_N^{a,b}(\cdot) = \sum_{A \subset D_N} \nu_N^{a,b}(A)\mu_{N,A^c}^{a,b}(\cdot) ,$$

where $\mathcal{A} = \{x \in D_N; |\phi(x)| \leq a\}$ and $\nu_N^{a,b}(A) = \mu_N^{a,b}(\mathcal{A} = A)$, $\mu_{N,A^c}^{a,b}(\cdot) = \mu_N^{a,b}(\cdot|\mathcal{A} = A)$.

For the proof of two theorems, using Lemma 7.6, Helffer-Sjöstrand representation and Brascamp-Lieb inequality are applied, but the details are omitted.

Remark 7.5. *The constant $m = m(J)$ arising in Theorem 7.5 behaves as $\lim_{J \downarrow -\infty} m(J) = 0$, since the decay of the correlation is algebraic at $J = -\infty$. [32] studied the detailed behavior of $m(J)$ when $V(\eta) = \frac{c}{2}\eta^2$. In particular, if $d = 2$, $\lim_{J \downarrow -\infty} E^{\mu_N^J}[\phi(x)^2] = \infty$ since the field is delocalized at $J = -\infty$. They investigated the fine behavior of the variances as well.*

7.3 Wetting Transition

The problem of the wetting transition, which is studied by Pfister and Velenik [212] for the Ising model, is recently discussed for the $\nabla\varphi$ interface model as well by several authors. The effects of the hard wall and the pinning near 0-level are introduced at the same time. The φ-field can take only nonnegative values. Recall that the field on a hard wall is delocalized for all dimensions d (Theorem 7.3-(2)) while the pinning localizes the field (Theorem 7.4). The former is caused by the entropy effect and the latter is by the energy effect.

Fisher [101] proved the existence of the wetting transition, namely the qualitative change in the localization/delocalization of the field depending on which of these two competitive effects dominate the other, when $d = 1$ for the SOS type discrete model. This result is extended by Caputo and Velenik [49] for $d = 2$. The precise path level behavior is discussed by Isozaki and Yoshida [154] when $d = 1$. Bolthausen et al. [30] showed that, contrarily when $d \geq 3$, no transition occurs and the field is always localized, i.e., only the phase of partial wetting appears. Bolthausen and Ioffe [31] proved the LLN in the partial wetting phase in two dimension (i.e., $d = 2$) under the Gibbs measures with 0-boundary conditions, hard wall, δ-pinning and quadratic V conditioned that the macroscopic total volume of the interfaces is kept constant. They derived the so-called **Winterbottom shape** in the limit and the variational problem characterizing it, cf. Remark 6.6. The one dimensional case with general V was discussed by De Coninck et al. [61].

Let us state the results more precisely. We deal with the δ-pinning only, but the square-well pinning can be treated essentially in the same way. The finite volume φ-Gibbs measure $\mu_N^{J,+} \in \mathcal{P}(\mathbb{R}_+^{\mathbb{Z}^d})$ with hard wall and δ-pinning is defined by

$$\mu_N^{J,+}(\cdot) = \mu_N^J(\cdot|\Omega^+(D_N)),\qquad(7.13)$$

for $J \in [-\infty, \infty)$, where $\mathbb{R}_+ = [0, \infty)$. We take $V(\eta) = \frac{1}{2}\eta^2$ and consider on the region $D_N = [-N, N]^d \cap \mathbb{Z}^d$ determined from $D = [-1, 1]^d$. Recall that $\mu_N^J \in \mathcal{P}(\mathbb{R}^{\mathbb{Z}^d})$ is the finite volume φ-Gibbs measure with δ-pinning and 0-boundary condition, and $\Omega^+(D_N) = \{\phi; \phi(x) \geq 0 \text{ for every } x \in D_N\}$. If $J = -\infty$, there is no pinning effect so that $\mu_N^{-\infty,+} = \mu_N^+$.

The random region $\mathcal{A} \equiv \mathcal{A}_N(\phi)$ defined by (7.12) is regarded as the **dry region**, because the heights are 0 on $\mathcal{A}$ so $\mathcal{A}$ is the region not covered by the matter under our consideration. One can expect that $\mathcal{A}$ is wide if the interface ϕ is localized, while it is narrow if ϕ is delocalized. In this sense, the localized or delocalized states are called **partial wetting** or **complete wetting**, respectively.

For instance, the materials such as gasoline or oil dropped on a flat plane spread over like a film. This is the state of complete wetting. On the other hand, a small droplet of the water does not spread over on the desk. This is the partial wetting. The place not covered by the water is the dry region.

Existence and Nonexistence of the Wetting Transition

Recent researches by Caputo and Velenik [49] and Bolthausen et al. [30] show that when $d = 1$ and 2 the competition between entropy effect and energy effect brings about the phase transition called wetting transition from complete wetting to partial wetting when the parameter J increases, while when $d \geq 3$ the pinning effect is always dominant and only the state of partial wetting appears for every $J \in \mathbb{R}$.

As a natural index to judge that the system lies in the states of complete or partial wettings, let us introduce the mean density of the dry region

$$\rho_N(J) = \frac{1}{|D_N|} E^{\mu_N^{J,+}}[|\mathcal{A}_N|] \geq 0 .$$

Observing that the limit of $\rho_N(J)$ is 0 or positive, one can probe whether the most region is wet or the dry region substantially survives. Namely, we call the complete wetting if

$$\rho(J) := \lim_{N \to \infty} \rho_N(J) = 0 , \tag{7.14}$$

and the partial wetting if

$$\rho(J) := \liminf_{N \to \infty} \rho_N(J) > 0 . \tag{7.15}$$

Lemma 7.8 below shows that $\rho(J)$ is nondecreasing in J.

The result is summarized in the next theorem. Recall that the potential is taken as $V(\eta) = \frac{1}{2}\eta^2$.

Theorem 7.7. (1) [49] *When $d = 2$, the critical value $J_c \in \mathbb{R}$ exists such that $\rho(J) = 0$ (i.e., complete wetting) for $J < J_c$ and $\rho(J) > 0$ (i.e., partial wetting) for $J > J_c$.*
(2) [30] *When $d \geq 3$, $\rho(J) > 0$ (i.e., partial wetting) for every $J \in \mathbb{R}$.*

Remark 7.6. [49] *The wetting transition exists for the SOS model with the potential $V(\eta) = |\eta|$ for all dimensions $d \geq 1$.*

The proof of Theorem 7.7 is omitted, instead we state two simple lemmas which are needed for the proof of the theorem. Let $Z_N^{J,+}$ be the normalization constant for $\mu_N^{J,+}$:

$$Z_N^{J,+} = \int_{\mathbb{R}_+^{D_N}} e^{-H_N^0(\phi)} \prod_{x \in D_N} \{d\phi(x) + e^J \delta_0(d\phi(x))\} .$$

In particular, for $J = -\infty$

$$Z_N^{-\infty,+} \equiv Z_N^+ = \int_{\mathbb{R}_+^{D_N}} e^{-H_N^0(\phi)} \prod_{x \in D_N} d\phi(x)$$

is the normalization constant when there is no pinning. Then we have the following lemma.

Lemma 7.8. *The dimension $d \geq 1$ is arbitrary. We have that*

$$\frac{d}{dJ} \log Z_N^{J,+} = E^{\mu_N^{J,+}}[|\mathcal{A}_N|] , \qquad (7.16)$$

$$\frac{d^2}{dJ^2} \log Z_N^{J,+} = E^{\mu_N^{J,+}}[|\mathcal{A}_N|; |\mathcal{A}_N|] . \qquad (7.17)$$

Moreover, the limit

$$\tau^{wall}(J) = - \lim_{N\to\infty} \frac{1}{|D_N|} \log \frac{Z_N^{J,+}}{Z_N^+} \in (-\infty, 0] \qquad (7.18)$$

exists.

Proof (Outline). Use the expansion in Lemma 7.6-(1) to show (7.16) and (7.17). The existence of the limit $\tau^{wall}(J)$ is similar to that for the surface tension, Theorem 5.1. □

The constant $\tau^{wall}(J)$ is called the **wall free energy** (or, more precisely, the wall+pinning free energy) which is expressed as the difference

$$\tau^{wall}(J) = - \lim_{N\to\infty} \frac{1}{|D_N|} \log Z_N^{J,+} + \lim_{N\to\infty} \frac{1}{|D_N|} \log Z_N^+ ,$$

between the surface tensions under the hard wall with and without the pinning.

Lemma 7.9. *The dimension $d \geq 1$ is arbitrary. For every large enough $J \in \mathbb{R}$, $\rho(J) > 0$ (i.e., partial wetting) holds.*

Proof. The identity (7.16) shows

$$\int_{-\infty}^{J} \rho_N(J') \, dJ' = \frac{1}{|D_N|} \log \frac{Z_N^{J,+}}{Z_N^+} . \qquad (7.19)$$

Here, the replacement of the region of the integration from $\mathbb{R}_+^{D_N}$ into $\mathbb{R}^{D_N}$ shows $Z_N^+ \leq Z_{D_N}^0$ and this implies that

$$\limsup_{N\to\infty} \frac{1}{|D_N|} \log Z_N^+ \leq \lim_{N\to\infty} \frac{1}{|D_N|} \log Z_{D_N}^0 = -\sigma^*(0) ,$$

where $\sigma^*(0)$ is the unnormalized surface tension at $u = 0$. On the other hand, $Z_N^{J,+}$ can be bounded from below by the integral with respect to $\prod_{x\in D_N} e^J \delta_0(d\phi(x))$ so that we have

$$Z_N^{J,+} \geq e^{J|D_N|} .$$

These two bounds combined with (7.19) show that

$$\liminf_{N\to\infty} \int_{-\infty}^{J} \rho_N(J') \, dJ' \geq J + \sigma^*(0) ,$$

for every $J \in \mathbb{R}$. The right hand side is positive if J is sufficiently large. This proves the conclusion noting that $\rho_N(J)$ is nondecreasing in J. □

When $d \geq 3$, one can prove that $\tau^{wall}(J) < 0$ for all $J \in \mathbb{R}$; the proof is omitted. This physically means that the interfaces always feel the pinning effect at macroscopic level. Then, similarly to the proof of Lemma 7.9, one can deduce $\rho(J) > 0$ from $\tau^{wall}(J) < 0$ and this completes the proof of Theorem 7.7-(2). Velenik [249] discussed the delocalization of interfaces above a wall in a complete wetting regime, and in an external field.

Remark 7.7. *When $d = 1$, Isozaki and Yoshida [154] established the limit theorem at path level for the SOS type model: $\phi(x) \in \mathbb{Z}_+$, which is extended by Deuschel, Giacomin and Zambotti [78] for $\mathbb{R}_+$-valued case including the critical regime; see also [30]. When $d = 1$, the wetting transition for plural interfaces of SOS type was discussed by Tanemura and Yoshida [245].*

Winterbottom Shape

Bolthausen and Ioffe [26, 31] add the pinning effect to the argument for the derivation of the Wulff shape in Sect. 6.2. Assuming $d = 2$ and $V(\eta) = \frac{1}{2}\eta^2$, the measure $\mu_{N,v}^{J,+}$ is introduced from $\mu_N^{J,+}$ by conditioning on the macroscopic volume of droplets as before:

$$\mu_{N,v}^{J,+} = \mu_N^{J,+} \left(\cdot \, \middle| \, \int_D h^N(\theta) \, d\theta \geq v \right).$$

[31] proved that, if $\tau^{wall}(J) < 0$, the LLN holds under $\mu_{N,v}^{J,+}$ and the limits are the minimizers of the functional

$$\Sigma_D^J(h) = \Sigma_D(h) + \tau^{wall}(J)|\{\theta \in D; \, h(\theta) = 0\}|$$

under the conditions that $h \geq 0$ and $\int_D h \, d\theta = v$. The second term represents that the interfacial energy is smaller if the dry region (i.e., the region on which $h(\theta) = 0$) is wider; i.e., if $\tau^{wall}(J) < 0$, it is more favorable for the interface to stay on the wall because of the pinning effect. The minimizer of $\Sigma_D^J(h)$ is called the Winterbottom shape, which is unique except the translation.

Remark 7.8. *(1) De Coninck et al. [61] discussed the above problem for general potential V when $d = 1$, i.e., conditioning μ_N^+ or $\mu_N^{J,+}$ as $\sum_{x=1}^{N-1} \phi(x) = N^2 v$ (constant macroscopic volume condition for the interfaces), they derived the Wulff shape or the Winterbottom shape, respectively. They further obtained the Young's relation for the angle of the interface to the wall at the point it touches the wall.*
(2) De Coninck et al. [62] discussed the generalization of Young's law for the SOS type interfaces on a substrate which is heterogeneous, rough and realized as another SOS interface.

Remark 7.9. *(Ising model) The wetting transition and the derivation of the Winterbottom shape from the two dimensional Ising model were studied by Pfister and Velenik [212]. They impose the + spins on the upper half of the* *boundary of the cube with size N and $-$ spins on the lower half. Moreover, at the lower segment of the boundary of the cube, magnetic field is added and this gives the pinning effect. Changing the strength of the magnetic field, the phase separation curve prefers to stay on the lower boundary of the cube. Three and higher dimensional Ising model was investigated by Bodineau, Ioffe and Velenik* [23].

Remark 7.10. *We refer to* [187, 188, 192, 193] *by Lipowsky et al. for physical motivations to the problem of the wetting transition.*

8 Central Limit Theorem

The long correlations of the φ-field (cf. Theorem 4.13) make the proof of limit theorems like the LDP, the CLT and others nontrivial. This section discusses the central limit theorem (CLT) for an infinite system.

We first assume $d \geq 3$ and consider the φ-Gibbs measure μ on $\mathbb{Z}^d$ constructed by Theorem 4.16. If μ has a very nice mixing property, it is easy to show the CLT under the usual scaling for the fluctuation of the φ-field:

$$\tilde{\Phi}^N = N^{-d/2} \sum_{x \in D_N} \{\phi(x) - E^\mu[\phi(x)]\} \underset{N \to \infty}{\Longrightarrow} N(0, m) \ ,$$

for some $m > 0$, where D is a bounded domain of $\mathbb{R}^d$ and note that $N^{d/2} \approx \sqrt{|D_N|}$. We have seen this in Proposition 3.12 for massive Gaussian systems. However, Theorem 4.13 actually implies that

$$E^\mu[\{\tilde{\Phi}^N\}^2] \approx \sum_{|x| \leq N} |x|^{2-d} \approx N^2 \underset{N \to \infty}{\longrightarrow} \infty \ .$$

Therefore, $\tilde{\Phi}^N$ does not give the right scaling and, as we did in (3.17) or in (3.18), we should scale-down it and consider

$$\Phi^N = N^{-1}\tilde{\Phi}^N = N^{-d/2} \sum_{x \in D_N} N^{-1} \{\phi(x) - E^\mu[\phi(x)]\} \tag{8.1}$$

or the random signed measures

$$\Phi^N(d\theta) = N^{-d/2} \sum_{x \in \mathbb{Z}^d} N^{-1} \{\phi(x) - E^\mu[\phi(x)]\} \delta_{x/N}(d\theta) \ , \tag{8.2}$$

for $\theta \in \mathbb{R}^d$. Since (8.2) is the usual CLT scaling for the $\nabla\varphi$-field (recall Sect. 3.4), it is natural to introduce the fluctuation fields $\Psi_i^N, 1 \leq i \leq d$ for $\nabla\phi = \{\nabla\phi(b); b \in (\mathbb{Z}^d)^*\}$ as

$$\Psi_i^N(d\theta) = N^{-d/2} \sum_{x \in \mathbb{Z}^d} \{\nabla_i\phi(x) - u_i\} \delta_{x/N}(d\theta) \ , \tag{8.3}$$

for $\theta \in \mathbb{R}^d$, where $u_i = E[\nabla_i \phi(x)]$; recall (3.19) for the Gaussian case. In fact, Naddaf and Spencer [202] studied Ψ_i^N under the ergodic $\nabla\varphi$-Gibbs measure μ_u^∇ with mean $u = (u_i)_{i=1}^d \in \mathbb{R}^d$, i.e. $\mu_u^\nabla \in (\text{ext } \mathcal{G}^\nabla)_u$, and established the CLT. The lattice dimension $d \geq 1$ is arbitrary, since the $\nabla\varphi$-field is dealt with. The result is later extended to the dynamic level by Giacomin et al. [135], which actually concludes the static result of Naddaf and Spencer, see Sect.11.

In (8.1) or in (8.2), $\phi(x)$ is divided by N. This may be explained in the following manner: Our real object is the $\nabla\varphi$-field and, from this point of view, $\phi(x)$ is expressed as the sum of $\nabla\phi(b)$'s along a path connecting O and x. Since the typical length of the path is N, it is natural for the φ-field to be divided by N.

We now state the CLT result. We write $\langle \Psi, f \rangle = \int_{\mathbb{R}^d} f \, d\Psi$ or $\langle f, g \rangle = \int_{\mathbb{R}^d} fg \, d\theta$.

Theorem 8.1. [202] *There exists a positive definite $d \times d$ matrix $q = (q_{ij}(u))_{1 \leq i,j \leq d}$ such that*

$$\lim_{N\to\infty} E^{\mu_u^\nabla} \left[e^{\sqrt{-1}\langle \Psi_i^N, f\rangle} \right] = \exp\left\{ -\frac{1}{2} \left\langle \frac{\partial f}{\partial \theta_i}, A \frac{\partial f}{\partial \theta_i} \right\rangle \right\}$$

$$= \exp\left\{ \frac{1}{2} \int_{\mathbb{R}^d} \frac{k_i^2}{k \cdot qk} |\hat{f}(k)|^2 \, dk \right\}$$

holds for every $f = f(\theta) \in C_0^\infty(\mathbb{R}^d)$. Here, A is a positive definite integral operator determined by $A^{-1} = -\sum_{i,j=1}^d q_{ij} \frac{\partial^2}{\partial\theta_i\partial\theta_j}$, $\hat{f}(k)$ is the Fourier transform of f and $k \cdot qk = \sum_{ij} q_{ij} k_i k_j$. The concrete form of the matrix q will be given in Theorem 11.1, (11.1).

We outline the proof of Theorem 8.1. The potential V is always supposed to satisfy the conditions (V1)–(V3) in (2.2). The basic idea is the usage of the Helffer-Sjöstrand representation on $\mathbb{Z}^d$.

Consider the differential operators ∂_x, ∂_x^* and L defined by

$$\partial_x = \frac{\partial}{\partial\phi(x)}, \quad \partial_x^* = -\partial_x + \sum_{y\in\mathbb{Z}^d:|x-y|=1} V'(\phi(x) - \phi(y)),$$

$$L = -\sum_{x\in\mathbb{Z}^d} \partial_x^* \partial_x,$$

acting on the functions $F = F(\phi)$ of $\phi = \{\phi(x); x \in \mathbb{Z}^d\}$, recall Sect. 4.1. ∂_x^* is the dual operator of ∂_x (with respect to the φ-Gibbs measure μ at least when $d \geq 3$) and L is the generator of the SDEs (2.13) on $\mathbb{Z}^d$, i.e., the operator (2.15) with $\Gamma = \mathbb{Z}^d$. We further introduce the operator

$$\mathcal{L} = L + Q$$

acting on $F = F(x, \phi)$, where Q is defined by

$$QF(x, \phi) = -\sum_{i=1}^{d} \left(\nabla_i^* V''(\nabla_i \phi(\cdot)) \nabla_i \right) F(x, \phi) \,.$$

Recall that

$$\nabla_i g(x) = g(x + e_i) - g(x), \quad \nabla_i^* g(x) = g(x - e_i) - g(x) \,.$$

In Sect. 4.1, Q is denoted by $Q_{\mathbb{Z}^d}^{\phi}$. Assuming $u = 0$ for simplicity, set

$$G^N(t) = E^{\mu_0^{\nabla}} \left[e^{t \langle \Psi_i^N, f \rangle} \right] \,.$$

Then, taking

$$v^N(x, \phi) = (-\mathcal{L})^{-1} \{ \nabla_i^* f^N(x) \}, \quad f^N(x) = N^{1 - \frac{d}{2}} f(x/N) \,,$$

we have for every $x \in \mathbb{Z}^d$

$$\partial_x \left(\sum_{y \in \mathbb{Z}^d} \partial_y^* v^N \right) (x, \phi) = -\mathcal{L} v^N(x, \phi) = \nabla_i^* f^N(x) = N \partial_x \langle \Phi^N, \nabla_i^{N*} f \rangle \,,$$

where $\nabla_i^{N*} f(\theta) = N\big(f(\theta - e_i/N) - f(\theta) \big)$. We have used $[\partial_x, \partial_y^*] = \partial_x \partial_y H$ (H is a formal Hamiltonian on $\mathbb{Z}^d$) and $\partial_x v^N(y, \phi) = \partial_y v^N(x, \phi)$ for the first equality. These identities imply

$$N^{-1} \sum_{y \in \mathbb{Z}^d} \partial_y^* v^N = \langle \Phi^N, \nabla_i^{N*} f \rangle \big(= \langle \Psi_i^N, f \rangle \big) \,.$$

We accordingly obtain

$$\begin{aligned}
\frac{d}{dt} G^N(t) &= E^{\mu_0^{\nabla}} \left[\langle \Phi^N, \nabla_i^{N*} f \rangle e^{t \langle \Psi_i^N, f \rangle} \right] \\
&= E^{\mu_0^{\nabla}} \left[\sum_{x \in \mathbb{Z}^d} N^{-1} \partial_x^* v^N e^{t \langle \Psi_i^N, f \rangle} \right] \\
&= \sum_{x \in \mathbb{Z}^d} E^{\mu_0^{\nabla}} \left[N^{-1} v^N \partial_x e^{t \langle \Psi_i^N, f \rangle} \right] \\
&= \sum_{x \in \mathbb{Z}^d} E^{\mu_0^{\nabla}} \left[N^{-2} v^N \, t \nabla_i^* f^N(x) e^{t \langle \Psi_i^N, f \rangle} \right] \\
&= t E^{\mu_0^{\nabla}} \left[\left(\sum_{x \in \mathbb{Z}^d} N^{-2} v^N \nabla_i^* f^N(x) - A_f \right) e^{t \langle \Psi_i^N, f \rangle} \right] + t A_f G^N(t) \,,
\end{aligned}$$

where A_f is an arbitrary constant. The next proposition is essentially a homogenization result for the random walk in random environment:

Proposition 8.2. *Take $A_f = \langle \partial f/\partial \theta_i, A \partial f/\partial \theta_i \rangle$. Then we have*

$$\lim_{N \to \infty} E^{\mu_0^\nabla} \left[\left\{ (\nabla_i^* f^N, (-N^2 \mathcal{L})^{-1} (\nabla_i^* f^N)) - A_f \right\}^2 \right] = 0 \,,$$

where $(\,,\,)$ means the inner product on $\mathbb{Z}^d$.

Once this is shown, the first term in the last line of the above equalities vanishes as $N \to \infty$. Thus we have

$$\lim_{N \to \infty} \frac{d}{dt} \log G^N(t) = tA_f \,,$$

which concludes the proof of Theorem 8.1.   □

Remark 8.1. *In one dimension, $\{\nabla \phi(x); x \in \mathbb{Z}\}$ form i.i.d. for general (nonconvex) potential. Therefore, the CLT is obvious and $q(u)$ coincides with the variance of ν_u, see Remark 4.5.*

9 Characterization of $\nabla \varphi$-Gibbs Measures

In Sect. 4.4, for each average tilt $u \in \mathbb{R}^d$, an infinite volume $\nabla \varphi$-Gibbs measure μ_u^∇ on $(\mathbb{Z}^d)^*$, which is tempered (i.e., square integrable), shift invariant and ergodic under shifts, in other words, $\nabla \varphi$-pure phase was constructed, see Theorem 4.15 and recall Definitions 2.2 and 2.3 for the notion of the $\nabla \varphi$-Gibbs measures, shift invariance and ergodicity. The tightness argument based on the Brascamp-Lieb inequality was applied.

This section addresses the uniqueness problem for the $\nabla \varphi$-pure phase μ_u^∇ for each $u \in \mathbb{R}^d$. The well-known Dobrushin's uniqueness argument [83, 84] does not work here, since if it works the correlation functions of the fields must decay exponentially fast, [179]. But this can not happen as we have seen already. We shall solve the problem based on the **dynamic coupling**, i.e., by characterizing all equilibrium (stationary) measures for the dynamics of gradient fields $\nabla \phi_t$ associated with those of heights fields ϕ_t determined by the SDEs (2.13) by means of the relation (2.6):

$$\eta_t(b) \equiv \nabla \phi_t(b) = \phi_t(x_b) - \phi_t(y_b) \,,$$

where $b = \langle x_b, y_b \rangle \in (\mathbb{Z}^d)^*$ are directed bonds in $\mathbb{Z}^d$. Considering $\nabla \phi_t$ is natural in the sense that the dynamics (2.13) for the φ-field on $\mathbb{Z}^d$ is invariant under the uniform translation $\phi(x) \to \phi(x) + h$.

Since $\nabla \varphi$-Gibbs measures are reversible and therefore stationary for the stochastic processes $\eta_t = \{\eta_t(b); b \in (\mathbb{Z}^d)^*\}$ under the subsidiary assumptions of shift invariance and temperedness (see Proposition 9.4), the study of stationary measures for η_t yields an information for the $\nabla \varphi$-Gibbs measures, see Definition 9.1 below for stationarity and reversibility. Our result can roughly be stated as follows: Under the conditions (V1)-(V3) in (2.2) on the potential

V, for each $u \in \mathbb{R}^d$ there exists a unique tempered, shift invariant, ergodic under shifts and stationary probability measure μ_u^∇ for η_t with mean u (average tilt), cf. Theorem 9.3. Especially, there exists a unique $\nabla\varphi$-Gibbs measure which is tempered, shift invariant, ergodic under shifts and has mean u, see Corollary 9.6. This will play an important role in establishing the hydrodynamic limit later, and has been already applied to show several properties of the surface tension $\sigma = \sigma(u)$, see Sect. 5.4, and also to prove the LDP, see Sect.6.5.

9.1 φ-Dynamics on $\mathbb{Z}^d$ and $\nabla\varphi$-Dynamics on $(\mathbb{Z}^d)^*$

According to (2.13) the dynamics of the height variables $\phi_t = \{\phi_t(x)\} \in \mathbb{R}^{\mathbb{Z}^d}$ is governed by the SDEs

$$d\phi_t(x) = -\sum_{b:x_b=x} V'(\nabla\phi_t(b)) \, dt + \sqrt{2}dw_t(x), \quad x \in \mathbb{Z}^d , \qquad (9.1)$$

where $\{w_t(x); x \in \mathbb{Z}^d\}$ is a family of independent one dimensional standard Brownian motions. The potential V is always assumed to satisfy the conditions (V1)–(V3). The dynamics for height differences $\eta_t = \{\eta_t(b)\} \in \mathbb{R}^{(\mathbb{Z}^d)^*}$ is then determined by the SDEs

$$d\eta_t(b) = -\left\{ \sum_{\bar{b}:x_{\bar{b}}=x_b} V'(\eta_t(\bar{b})) - \sum_{\bar{b}:x_{\bar{b}}=y_b} V'(\eta_t(\bar{b})) \right\} dt + \sqrt{2}dw_t(b) , \qquad (9.2)$$

for $b \in (\mathbb{Z}^d)^*$, where $w_t(b) = w_t(x_b) - w_t(y_b)$. Indeed, writing down the SDEs (9.1) for $\phi_t(x_b)$ and $\phi_t(y_b)$ and then taking their difference, (9.2) are readily obtained. Since η_t fulfills the loop condition, the state space of the process η_t is $\mathcal{X}$ which has been introduced in Sect. 2.3.

The relationship between the solutions of (9.1) and (9.2) is summarized in the next lemma. Recall that the height differences η^ϕ are associated with the heights ϕ by (2.6) and, conversely, the heights $\phi^{\eta,\phi(O)}$ can be constructed from height differences η and the height variable $\phi(O)$ at $x = O$ by (2.7). We always assume $\eta_0 \in \mathcal{X}$ for the initial data of (9.2).

Lemma 9.1. (1) *The solution of (9.2) satisfies* $\eta_t \in \mathcal{X}$ *for all* $t > 0$.
(2) *If* ϕ_t *is a solution of (9.1), then* $\eta_t := \eta^{\phi_t}$ *is a solution of (9.2).*
(3) *Conversely, let* η_t *be a solution of (9.2) and define* $\phi_t(O)$ *through (9.1) for* $x = O$ *and* $\nabla\phi_t(b)$ *replaced by* $\eta_t(b)$ *with arbitrary initial condition* $\phi_0(O) \in \mathbb{R}$. *Then* $\phi_t := \phi^{\eta_t,\phi_t(O)}$ *is a solution of (9.1).*

To discuss the existence and uniqueness of solutions to (9.2), the space $\mathcal{X}$ is rather big and therefore we introduce weighted ℓ^2-spaces on $(\mathbb{Z}^d)^*$ as we have done for φ-field in Sect. 3.2:

$$\ell_{r,*}^2 \equiv \ell_r^2((\mathbb{Z}^d)^*) := \left\{ \eta \in \mathbb{R}^{(\mathbb{Z}^d)^*}; |\eta|_r^2 := \sum_{b \in (\mathbb{Z}^d)^*} \eta(b)^2 e^{-2r|x_b|} < \infty \right\},$$

for $r > 0$. The increasing order of η is controlled exponentially in the space $\ell_{r,*}^2$. We denote $\mathcal{X}_r = \mathcal{X} \cap \ell_{r,*}^2$ equipped with the norm $|\cdot|_r$. Then, the condition (V.3) on V implies global Lipschitz continuity in $\mathcal{X}_r, r > 0$, of the drift term of the SDEs (9.2). Therefore, a standard method of successive approximations yields the following lemma.

Lemma 9.2. *For each $\eta \in \mathcal{X}_r, r > 0$, the SDEs (9.2) have a unique $\mathcal{X}_r$-valued continuous solution η_t starting at $\eta_0 = \eta$.*

We are now in the position to precisely define the stationarity and the reversibility of probability measures under the dynamics η_t. Let $C_{loc,b}^\infty(\mathcal{X})$ denote the family of all (tame) functions F on $\mathcal{X}$ of the form $F(\eta) = \tilde{F}(\{\eta(b); b \in \tilde{\Lambda}^*\})$ for some $\tilde{\Lambda} \Subset \mathbb{Z}^d$ and $\tilde{F} \in C_b^\infty(\mathbb{R}^{\tilde{\Lambda}^*})$.

Definition 9.1. *We say that $\mu^\nabla \in \mathcal{P}(\mathcal{X})$ is stationary under η_t if $E^{\mu^\nabla}[F(\eta_0)] = E^{\mu^\nabla}[F(\eta_t)]$ for all $t \geq 0$ and $F \in C_{loc,b}^\infty(\mathcal{X})$, where $E^{\mu^\nabla}[\cdot]$ means the expectation for η_t with initial distribution μ^∇. We say that μ^∇ is reversible if $E^{\mu^\nabla}[F(\eta_0)G(\eta_t)] = E^{\mu^\nabla}[F(\eta_t)G(\eta_0)]$ for all $t \geq 0$ and $F, G \in C_{loc,b}^\infty(\mathcal{X})$.*

The reversibility implies the stationarity.

9.2 Stationary Measures and $\nabla\varphi$-Gibbs Measures

Let us formulate the results precisely. We shall consider the following classes of probability measures on $\mathcal{X}$:

$$\mathcal{P}_2(\mathcal{X}) = \{\mu^\nabla \in \mathcal{P}(\mathcal{X}); E^{\mu^\nabla}[\eta(b)^2] < \infty \text{ for every } b \in (\mathbb{Z}^d)^*\},$$
$$\mathcal{S} = \{\mu^\nabla \in \mathcal{P}_2(\mathcal{X}); \text{ shift invariant and stationary under } \eta_t\},$$
$$\text{ext } \mathcal{S} = \{\mu^\nabla \in \mathcal{S}; \text{ ergodic under shifts}\},$$
$$(\text{ext } \mathcal{S})_u = \{\mu^\nabla \in \text{ext } \mathcal{S}; \mu^\nabla \text{ has mean } u\}, \quad u = (u_i)_{i=1}^d \in \mathbb{R}^d.$$

Recall Definition 2.3 for the shift invariance and the ergodicity. The last condition "μ^∇ has mean u" means $E^{\mu^\nabla}[\eta(e_i)] = u_i$ for every $1 \leq i \leq d$, where $e_i \in \mathbb{Z}^d$ denotes the i-th unit vector and the bond $\langle e_i, O \rangle$ is also denoted by e_i. The measures $\mu^\nabla \in \mathcal{P}_2(\mathcal{X})$ are called **tempered**. The set $\mathcal{P}(\mathcal{X}_r), r > 0$, is defined correspondingly and $\mathcal{P}_2(\mathcal{X}_r)$ stands for the set of all $\mu \in \mathcal{P}(\mathcal{X}_r)$ such that $E^\mu[|\eta|_r^2] < \infty$. Note that $\mathcal{S} \subset \mathcal{P}_2(\mathcal{X}_r)$ for every $r > 0$.

With these notation, the uniqueness of stationary measures under the $\nabla\varphi$-dynamics is formulated in the next theorem.

Theorem 9.3. [124] *(uniqueness of stationary measures for η_t) For every $u \in \mathbb{R}^d$ there exists at most one $\mu \in (\text{ext } \mathcal{S})_u$.*

The proof of this theorem is given based on a coupling argument; namely, assuming that there exist two different measures $\mu^\nabla, \bar{\mu}^\nabla \in (\text{ext } \mathcal{S})_u$, we construct two solutions ϕ_t and $\bar{\phi}_t$ of the SDEs (2.13) with common Brownian motions $w_t = \{w_t(x); x \in \mathbb{Z}^d\}$ in such a way that the gradients $\nabla\phi_0$ and $\nabla\bar{\phi}_0$ of their initial data are distributed under μ^∇ and $\bar{\mu}^\nabla$, respectively. Then, computing the time derivative of $\sum_{x \in \Lambda} E[(\phi_t(x) - \bar{\phi}_t(x))^2]$ for each $\Lambda \Subset \mathbb{Z}^d$, one can finally conclude $\mu^\nabla = \bar{\mu}^\nabla$ by letting $\Lambda \nearrow \mathbb{Z}^d$ and noting the ergodicity of μ^∇ and $\bar{\mu}^\nabla$. The ergodicity helps to deal with the boundary terms. The strict convexity of V plays an essential role. The details will be discussed in Sect. 9.3.

We now consider the family of the $\nabla\varphi$-Gibbs measures on $(\mathbb{Z}^d)^*$, recall Definition 2.2:

$$\mathcal{G}^\nabla = \{\mu^\nabla \in \mathcal{P}_2(\mathcal{X}); \text{ shift invariant } \nabla\varphi\text{-Gibbs measures}\}.$$

The classes ext $\mathcal{G}^\nabla$ and $(\text{ext } \mathcal{G}^\nabla)_u$ are similarly defined, so that $\mu^\nabla \in \text{ext } \mathcal{G}^\nabla$ is ergodic under shifts and $\mu^\nabla \in (\text{ext } \mathcal{G}^\nabla)_u$ has mean u. Note that, if $\mu \in \mathcal{P}_2(\mathcal{X})$ is shift invariant, then $\mu \in \mathcal{P}_2(\mathcal{X}_r)$ for all $r > 0$. Since the finite volume $\nabla\varphi$-Gibbs measure $\mu_{\Lambda,\xi}^\nabla$ is reversible under the evolution governed by the finite dimensional SDEs for $\nabla\varphi$-field on $\overline{\Lambda^*}$ with boundary condition ξ, one can show the following proposition by letting $\Lambda \nearrow \mathbb{Z}^d$, see Sect. 9.4

Proposition 9.4. *Every* $\mu^\nabla \in \mathcal{G}^\nabla$ *is reversible under the dynamics* η_t *defined by the SDEs (9.2). In particular, we have* $\mathcal{G}^\nabla \subset \mathcal{S}$.

Theorem 9.3 and Proposition 9.4 imply the uniqueness of the tempered, shift invariant and ergodic $\nabla\varphi$-Gibbs measures for each mean tilt u:

Theorem 9.5. [124] *(uniqueness of* $\nabla\varphi$-*Gibbs measures) For every* $u \in \mathbb{R}^d$ *there exists at most one* $\mu^\nabla \in (\text{ext } \mathcal{G}^\nabla)_u$.

Proof. By Proposition 9.4, $\mu^\nabla \in \text{ext } \mathcal{G}^\nabla$ implies $\mu^\nabla \in \text{ext } \mathcal{S}$. Consequently the conclusion follows from Theorem 9.3.   □

Combining this with Theorem 4.15 (existence of $\nabla\varphi$-Gibbs measures) we have the following corollary.

Corollary 9.6. [124] *(characterization of all tempered and shift invariant* $\nabla\varphi$-*Gibbs measures) For every* $u \in \mathbb{R}^d$, $(\text{ext } \mathcal{G}^\nabla)_u = \{\mu_u^\nabla\}$. *In particular, the family* $\mathcal{G}^\nabla$ *is the convex hull of* $\{\mu_u^\nabla; u \in \mathbb{R}^d\}$.

Remark 9.1. *As we saw in Sect. 3.2, the Gaussian measure* $\mu^\psi = N(\psi, G)$ *is a* φ-*Gibbs measure for quadratic potentials if* $d \geq 3$ *and* ψ *is harmonic, where* G *is the Green function. In particular, if* $\phi = \{\phi(x); x \in \mathbb{Z}^d\}$ *is* μ^ψ-*distributed and* $\nabla\psi$ *is not shift invariant, the distribution of* $\nabla\phi$ *is a* $\nabla\varphi$-*Gibbs measure which is not shift invariant. The characterization of unshift invariant* $\nabla\varphi$-*Gibbs measures for general convex potentials* V *is not known.*

9.3 Proof of Theorem 9.3

Energy Inequality

We first prepare an energy inequality for ϕ_t. After computing the time derivative, the proof is essentially due to the rearrangement of the sum, which is a discrete analogue of Green-Stokes' formula.

Lemma 9.7. *Let ϕ_t and $\bar\phi_t$ be two solutions of* (9.1) *and set $\tilde\phi_t(x) := \phi_t(x) - \bar\phi_t(x)$. Then, for every $\Lambda \Subset \mathbb{Z}^d$, we have*

$$\frac{\partial}{\partial t} \sum_{x \in \Lambda} \left(\tilde\phi_t(x) \right)^2 = I_t^\Lambda + B_t^\Lambda , \tag{9.3}$$

where

$$I_t^\Lambda = - \sum_{b \in \Lambda^*} \nabla\tilde\phi_t(b) \left\{ V'(\nabla\phi_t(b)) - V'(\nabla\bar\phi_t(b)) \right\} ,$$

$$B_t^\Lambda = 2 \sum_{b \in \partial\Lambda^*} \tilde\phi_t(y_b) \left\{ V'(\nabla\phi_t(b)) - V'(\nabla\bar\phi_t(b)) \right\} ,$$

and $\partial\Lambda^* = \{ b \in (\mathbb{Z}^d)^*; x_b \notin \Lambda, y_b \in \Lambda \}$. The interior term I_t^Λ and the boundary term B_t^Λ admit the following bounds, respectively,

$$I_t^\Lambda \le -c_- \sum_{b \in \Lambda^*} \left(\nabla\tilde\phi_t(b) \right)^2 , \tag{9.4}$$

$$B_t^\Lambda \le 2c_+ \sum_{b \in \partial\Lambda^*} |\tilde\phi_t(y_b)| \, |\nabla\tilde\phi_t(b)| . \tag{9.5}$$

Proof. From the equation (9.1),

$$\frac{\partial}{\partial t} \left(\tilde\phi_t(x) \right)^2 = -2 \sum_{b: x_b = x} \Phi_t(b) \, \tilde\phi_t(x) = - \left\{ \sum_{b: x_b = x} \Phi_t(b) - \sum_{b: y_b = x} \Phi_t(b) \right\} \tilde\phi_t(x) ,$$

where

$$\Phi_t(b) := V'(\nabla\phi_t(b)) - V'(\nabla\bar\phi_t(b)) .$$

The second equality uses the symmetry of V which implies $V'(\nabla\phi(b)) = -V'(\nabla\phi(-b))$. The right hand side summed over $x \in \Lambda$ becomes

$$- \sum_{x \in \Lambda} \tilde\phi_t(x) \sum_{b: x_b = x} \Phi_t(b) + \sum_{x \in \Lambda} \tilde\phi_t(x) \sum_{b: y_b = x} \Phi_t(b)$$

$$= - \sum_{b \in \Lambda^*} \nabla\tilde\phi_t(b)\Phi_t(b) - \sum_{b: x_b \in \Lambda, y_b \notin \Lambda} \tilde\phi_t(x_b)\Phi_t(b) + \sum_{b: y_b \in \Lambda, x_b \notin \Lambda} \tilde\phi_t(y_b)\Phi_t(b)$$

$$= I_t^\Lambda + B_t^\Lambda ,$$

which proves (9.3). To obtain the term B_t^Λ we again used the symmetry of V. The two bounds (9.4) and (9.5) follow from the condition (V.3) on V.   □

Dynamic Coupling

The proof of Theorem 9.3 is reduced to a proposition which also implies the Lipschitz continuity of the derivative of the surface tension $\sigma(u)$, see Theorem 5.3 above. Suppose that there exist $\mu^\nabla \in (\text{ext}\,\mathcal{S})_u$ and $\bar{\mu}^\nabla \in (\text{ext}\,\mathcal{S})_v$ for $u, v \in \mathbb{R}^d$. Let us construct two independent $\mathcal{X}_r$-valued random variables $\eta = \{\eta(b)\}$ and $\bar{\eta} = \{\bar{\eta}(b)\}$ on a common probability space $(\Omega, \mathcal{F}, P)$ in such a manner that η and $\bar{\eta}$ are distributed by μ^∇ and $\bar{\mu}^\nabla$ under P, respectively. We define $\phi_0 = \phi^{\eta,0}$ and $\bar{\phi}_0 = \phi^{\bar{\eta},0}$ using the notation in (2.7). Let ϕ_t and $\bar{\phi}_t$ be the two solutions of the SDEs (9.1) with common Brownian motions having initial data ϕ_0 and $\bar{\phi}_0$. In view of Lemmas 9.1 and 9.2 such solutions certainly exist. Since $\mu^\nabla, \bar{\mu}^\nabla \in \mathcal{S}$, we conclude that $\eta_t := \eta^{\phi_t}$ and $\bar{\eta}_t := \eta^{\bar{\phi}_t}$ are distributed by μ^∇ and $\bar{\mu}^\nabla$, respectively, for all $t \geq 0$. Our claim is then the following.

Proposition 9.8. *There exists a constant $C > 0$ independent of $u, v \in \mathbb{R}^d$ such that*

$$\limsup_{T \to \infty} \frac{1}{T} \int_0^T \sum_{i=1}^d E^P[(\eta_t(e_i) - \bar{\eta}_t(e_i))^2]\, dt \leq C|u - v|^2 . \tag{9.6}$$

Once this proposition is proved, Theorem 9.3 immediately follows. Indeed, suppose that there exist two measures $\mu^\nabla, \bar{\mu}^\nabla \in (\text{ext}\,\mathcal{S})_u$. Then Proposition 9.8 with $u = v$ implies

$$\lim_{T \to \infty} \int |\eta - \bar{\eta}|_r^2 P_T(d\eta\, d\bar{\eta}) = 0 , \tag{9.7}$$

where P_T is a shift invariant probability measure on $\mathcal{X}_r \times \mathcal{X}_r, r > 0$, defined by

$$P_T(d\eta\, d\bar{\eta}) := \frac{1}{T} \int_0^T P\big(\{\eta_t(b), \bar{\eta}_t(b); b \in (\mathbb{Z}^d)^*\} \in d\eta\, d\bar{\eta}\big)\, dt .$$

The first marginal of P_T is μ^∇ and the second one is $\bar{\mu}^\nabla$. Thus (9.7) implies that the Vaserstein distance between μ^∇ and $\bar{\mu}^\nabla$ vanishes and hence $\mu^\nabla = \bar{\mu}^\nabla$, see, e.g., [109], p.482 for the Vaserstein metric on the space $\mathcal{P}_2(\mathcal{X}_r)$. This concludes the proof of Theorem 9.3.

Proof of Proposition 9.8

Step 1. We apply Lemma 9.7 to the differences $\{\tilde{\phi}_t(x) := \phi_t(x) - \bar{\phi}_t(x)\}$ and obtain, with the choice $\Lambda = \Lambda_\ell$,

$$E^P\left[\sum_{x \in \Lambda_\ell} \big(\tilde{\phi}_T(x)\big)^2\right] + c_- \int_0^T E^P\left[\sum_{b \in \Lambda_\ell^*} \big(\nabla\tilde{\phi}_t(b)\big)^2\right] dt$$

$$\leq E^P\left[\sum_{x \in \Lambda_\ell} \big(\tilde{\phi}_0(x)\big)^2\right] + 2c_+ \int_0^T E^P\left[\sum_{b \in \partial\Lambda_\ell^*} |\tilde{\phi}_t(y_b)|\, |\nabla\tilde{\phi}_t(b)|\right] dt \tag{9.8}$$

for every $T > 0$ and $\ell \in \mathbb{N}$. Set

$$g(t) = \sum_{i=1}^{d} E^P \left[\left(\nabla \tilde{\phi}_t(e_i) \right)^2 \right] .$$

Then, noting that the distribution of $(\eta_t, \bar{\eta}_t) = (\nabla \phi_t, \nabla \bar{\phi}_t)$ on $\mathcal{X}_r \times \mathcal{X}_r$ is shift invariant, the second term on the left hand side of (9.8) coincides with $c_- d^{-1} |\Lambda_\ell^*| \int_0^T g(t)\, dt$. On the other hand, estimating $|\tilde{\phi}_t(y_b)| \, |\nabla \tilde{\phi}_t(b)| \leq \{ \ell \gamma |\nabla \tilde{\phi}_t(b)|^2 + \ell^{-1} \gamma^{-1} |\tilde{\phi}_t(y_b)|^2 \}/2$ for arbitrary $\gamma > 0$, the second term on the right hand side is bounded by

$$c_+ \ell \gamma d^{-1} |\partial \Lambda_\ell^*| \int_0^T g(t)\, dt + c_+ \ell^{-1} \gamma^{-1} |\partial \Lambda_\ell^*| \int_0^T \sup_{y \in \partial \Lambda_\ell} \|\tilde{\phi}_t(y)\|_{L^2(P)}^2 \, dt .$$

Then, choosing $\gamma = c_-/2c_+ c_0$ with $c_0 := \sup_{\ell \geq 1} \{ \ell |\partial \Lambda_\ell^*|/|\Lambda_\ell^*| \} < \infty$, we obtain from (9.8)

$$\int_0^T g(t)\, dt \leq \frac{2d}{c_- |\Lambda_\ell^*|} E^P \left[\sum_{x \in \Lambda_\ell} \left(\tilde{\phi}_0(x) \right)^2 \right]$$
$$+ \frac{(2c_+ c_0)^2 d}{(c_- \ell)^2} \int_0^T \sup_{y \in \partial \Lambda_\ell} \|\tilde{\phi}_t(y)\|_{L^2(P)}^2 \, dt , \qquad (9.9)$$

where we have dropped the nonnegative first term on the left hand side of (9.8).

Step 2. Here we derive the following bound on the boundary term: For each $\epsilon > 0$ there exists an $\ell_0 \in \mathbb{N}$ such that

$$\sup_{y \in \partial \Lambda_\ell} \|\tilde{\phi}_t(y)\|_{L^2(P)}^2 \leq C_1 \left(\epsilon^2 \ell^2 + \ell^2 |u - v|^2 + \ell^{-2} t \int_0^t g(s)\, ds \right) \qquad (9.10)$$

for every $t > 0$ and $\ell \geq \ell_0$, where $C_1 > 0$ is a constant independent of ϵ, ℓ, and t. To this end, as an immediate consequence of the mean ergodic theorem (cf. [24, 175]) applied to $\mu^\nabla \in (\text{ext } \mathcal{S})_u$, we have

$$\lim_{|x| \to \infty} \frac{1}{|x|} \|\phi^{\eta, 0}(x) - x \cdot u\|_{L^2(\mu^\nabla)} = 0 \qquad (9.11)$$

and correspondingly for $\bar{\mu}^\nabla$ with v in place of u. Taking $\Lambda' = \Lambda_{[\ell/2]}$ one obtains

$$\|\tilde{\phi}_t(y)\|_{L^2(P)} \leq \|\phi_t(y) - \frac{1}{|\Lambda'|} \sum_{x \in \Lambda'} \phi_t(x) - y \cdot u\|_{L^2(P)}$$
$$+ \|\bar{\phi}_t(y) - \frac{1}{|\Lambda'|} \sum_{x \in \Lambda'} \bar{\phi}_t(x) - y \cdot v\|_{L^2(P)}$$
$$+ \|\frac{1}{|\Lambda'|} \sum_{x \in \Lambda'} \tilde{\phi}_t(x)\|_{L^2(P)} + \sqrt{d} \ell |u - v|$$

$$=:I_1 + I_2 + I_3 + I_4 \; ,$$

for $y \in \partial \Lambda_\ell$. However, since $\sum_{x \in \Lambda'} x = 0$ and using (9.11),

$$I_1 \le \frac{1}{|\Lambda'|} \sum_{x \in \Lambda'} \|\phi_t(y) - \phi_t(x) - (y - x) \cdot u\|_{L^2(P)}$$

$$= \frac{1}{|\Lambda'|} \sum_{x \in \Lambda'} \|\phi^{\eta,0}(y - x) - (y - x) \cdot u\|_{L^2(\mu^\nabla)} \le \epsilon \ell$$

provided ℓ is sufficiently large; recall that $\nabla \phi_t$ is distributed by μ^∇ for all $t \ge 0$. Similarly, $I_2 \le \epsilon \ell$ for sufficiently large ℓ. Finally, since as in the proof of Lemma 9.7

$$\frac{\partial}{\partial t}\left\{ \sum_{x \in \Lambda'} \tilde{\phi}_t(x) \right\} = -\sum_{x \in \Lambda'} \sum_{b:x_b = x} \Phi_t(b) = \sum_{b \in (\partial \Lambda')^*} \Phi_t(b) \; ,$$

I_3 is bounded as

$$I_3 \le \left\| \frac{1}{|\Lambda'|} \sum_{x \in \Lambda'} \tilde{\phi}_0(x) \right\|_{L^2(P)} + \int_0^t \frac{1}{|\Lambda'|} \sum_{b \in (\partial \Lambda')^*} \|\Phi_s(b)\|_{L^2(P)} \, ds \; .$$

The right hand side can be further estimated as

$$\sum_{b \in (\partial \Lambda')^*} \|\Phi_s(b)\|_{L^2(P)} \le c_+ d^{-1} |(\partial \Lambda')^*| \sum_{i=1}^d \|\nabla \tilde{\phi}_s(e_i)\|_{L^2(P)}$$

and, using again (9.11),

$$\left\| \frac{1}{|\Lambda'|} \sum_{x \in \Lambda'} \tilde{\phi}_0(x) \right\|_{L^2(P)} \le \frac{1}{|\Lambda'|} \sum_{x \in \Lambda'} \{ \|\phi^{\eta,0}(x) - x \cdot u\|_{L^2(\mu^\nabla)}$$

$$+ \|\phi^{\bar{\eta},0}(x) - x \cdot v\|_{L^2(\bar{\mu}^\nabla)} + |x| \cdot |u - v| \}$$

$$\le \epsilon \ell + \sqrt{d} \ell |u - v| \; ,$$

for sufficiently large ℓ. Therefore,

$$I_3 \le \epsilon \ell + \sqrt{d} \ell |u - v| + c_+ d^{-1} |\Lambda'|^{-1} |(\partial \Lambda')^*| \int_0^t \sum_{i=1}^d \|\nabla \tilde{\phi}_s(e_i)\|_{L^2(P)} \, ds$$

for sufficiently large ℓ. This completes the proof of (9.10).

Step 3. Using (9.11), one can choose $\ell_1 \in \mathbb{N}$ such that

$$\frac{1}{|\Lambda_\ell|} \sum_{x \in \Lambda_\ell} E^P \left[\left(\tilde{\phi}_0(x) \right)^2 \right]$$

$$\leq \frac{3}{|\Lambda_\ell|} \sum_{x \in \Lambda_\ell} \left\{ \|\phi^{\eta,0}(x) - x \cdot u\|^2_{L^2(\mu^\nabla)} \right.$$

$$\left. + |x \cdot u - x \cdot v|^2 + \|\phi^{\bar{\eta},0}(x) - x \cdot v\|^2_{L^2(\bar{\mu}^\nabla)} \right\}$$

$$\leq \epsilon^2 \ell^2 + 3d\ell^2 |u - v|^2, \tag{9.12}$$

for every $\ell \geq \ell_1$. Inserting the estimates (9.10) and (9.12) into (9.9), we have

$$\int_0^T g(t)\, dt \leq C_2(\epsilon^2 \ell^2 + \ell^2 |u - v|^2)$$

$$+ C_2 \ell^{-2} \int_0^T \left(\epsilon^2 \ell^2 + \ell^2 |u - v|^2 + \ell^{-2} t \int_0^t g(s)\, ds \right) dt$$

$$\leq C_2(\epsilon^2 + |u - v|^2)(\ell^2 + T) + C_2 \ell^{-4} T^2 \int_0^T g(t)\, dt$$

for every $T > 0$ and $\ell \geq \ell_2 := \max\{\ell_0, \ell_1\}$, which may depend on u, v and $\epsilon > 0$. C_2 is a constant independent of u, v and ϵ. Choosing $\ell = (2C_2 T^2)^{1/4}$ and letting $T \to \infty$, we obtain

$$\limsup_{T \to \infty} \frac{1}{T} \int_0^T g(t)\, dt \leq 2C_2(\sqrt{2C_2} + 1)(\epsilon^2 + |u - v|^2)$$

for every $\epsilon > 0$. Finally, letting $\epsilon \to 0$, the desired estimate (9.6) is shown. $\quad\square$

9.4 Proof of Proposition 9.4

We establish the reversibility of the $\nabla\varphi$-Gibbs measures under the dynamics (9.2). To this end, we need the approximation of the solutions of (9.2) by the corresponding finite volume equations, cf. [89, 231, 257] for related results. For every $\xi \in \mathcal{X}$ and $\Lambda \Subset \mathbb{Z}^d$, let us consider the SDEs

$$\begin{cases} d\eta_t(b) = -\left\{ \displaystyle\sum_{\bar{b}:x_{\bar{b}}=x_b \in \Lambda} V'(\eta_t(\bar{b})) - \sum_{\bar{b}:x_{\bar{b}}=y_b \in \Lambda} V'(\eta_t(\bar{b})) \right\} dt \\ \qquad\qquad + \sqrt{2} dw_t^\Lambda(b), \quad b \in \overline{\Lambda^*}, \\ \eta_t(b) = \xi(b), \qquad b \notin \overline{\Lambda^*}, \\ \eta_0(b) = \xi(b), \qquad b \in (\mathbb{Z}^d)^*, \end{cases} \tag{9.13}$$

where $w_t^\Lambda(b) = 1_{\{x_b \in \Lambda\}} w_t(x_b) - 1_{\{y_b \in \Lambda\}} w_t(y_b)$. The distribution on the space $C([0,T], \mathcal{X})$ of the solution $\eta_t \equiv \eta_t^\Lambda$ is denoted by P_ξ^Λ. The distribution of the solution of the SDEs (9.2) starting at ξ is denoted by P_ξ. Then the next lemma is standard by showing the tightness of $\{P_\xi^\Lambda\}_\Lambda$ and the unique characterization of P_ξ in terms of the martingale problem, see Proposition 2.2 of [124]. Recall Sect. 9.1 for the space $C^\infty_{\text{loc},b}(\mathcal{X})$.

Lemma 9.9. *For every $\xi \in \mathcal{X}_r$ and $F \in C_{\mathrm{loc},b}^{\infty}(\mathcal{X})$,*

$$\lim_{\Lambda \nearrow \mathbb{Z}^d} E^{P_\xi^\Lambda}[F(\eta_t)] = E^{P_\xi}[F(\eta_t)] .$$

We are now in the position to complete the proof of Proposition 9.4. To this end, it suffices to show that every $\mu^\nabla \in \mathcal{G}^\nabla$ satisfies

$$\int_{\mathcal{X}_r} F(\xi) E^{P_\xi}[G(\eta_t)] \, \mu^\nabla(d\xi) = \int_{\mathcal{X}_r} E^{P_\xi}[F(\eta_t)] G(\xi) \, \mu^\nabla(d\xi) \qquad (9.14)$$

for every $t \geq 0$ and $F, G \in C_{\mathrm{loc},b}^{\infty}(\mathcal{X})$, cf. Definition 9.1. However, for every $\psi \in \mathbb{R}^{\mathbb{Z}^d}$ and $\Lambda \Subset \mathbb{Z}^d$, if we consider the SDEs for $\phi_t \in \mathbb{R}^{\mathbb{Z}^d}$:

$$\begin{cases} d\phi_t(x) = -\displaystyle\sum_{b:x_b=x} V'(\nabla\phi_t(b)) \, dt + \sqrt{2}dw_t(x), & x \in \Lambda , \\[2mm] \phi_t(x) = \psi(x), & x \notin \Lambda , \\[2mm] \phi_0(x) = \psi(x), & x \in \mathbb{Z}^d , \end{cases} \qquad (9.15)$$

then the finite volume φ-Gibbs measure μ_Λ^ψ is clearly reversible under (9.15); recall (4.5) taking $\rho = 0$. Therefore, since $\eta_t = \nabla\phi_t$ satisfies (9.13) provided $\psi = \phi^{\xi,0}$, $\mu_{\Lambda,\xi}^\nabla$ is reversible under (9.13), i.e.,

$$\int_{\mathcal{X}_{\overline{\Lambda^*},\xi}} F(\eta) E^{P_{\eta \vee \xi}^\Lambda}[G(\eta_t)] \, \mu_{\Lambda,\xi}^\nabla(d\eta) = \int_{\mathcal{X}_{\overline{\Lambda^*},\xi}} E^{P_{\eta \vee \xi}^\Lambda}[F(\eta_t)] G(\eta) \, \mu_{\Lambda,\xi}^\nabla(d\eta)$$

$$(9.16)$$

for all $\xi \in \mathcal{X}$ if both F and G are supported in Λ. For given $\mu^\nabla \in \mathcal{G}^\nabla$, integrating both sides of (9.16) with respect to $\mu^\nabla(d\xi)$ we have by the DLR equation

$$\int_{\mathcal{X}_r} F(\xi) E^{P_\xi^\Lambda}[G(\eta_t)] \, \mu^\nabla(d\xi) = \int_{\mathcal{X}_r} E^{P_\xi^\Lambda}[F(\eta_t)] G(\xi) \, \mu^\nabla(d\xi) .$$

Hence, (9.14) follows from Lemma 9.9 by letting $\Lambda \nearrow \mathbb{Z}^d$. □

Remark 9.2. (1) *Results similar to Proposition 9.4 together with its converse were obtained for lattice systems by* [89, 231, 258, 225] *and for continuum systems by* [156, 109].
(2) *The dynamic approach might work also to construct the $\nabla\varphi$-Gibbs measures, see* [109], *Proposition 6.2 for the massive continuum field.*

Remark 9.3. *Sheffield* [230] *gives a different proof for Theorem 9.5 in more general setting based on the argument called "cluster swapping".*

Remark 9.4. *Gawedzki and Kupiainen* [127] *considered the $\nabla\varphi$ interface model with $V(\eta) = \frac{1}{2}\eta^2 + \lambda\eta^4$ and proved that, applying the Wilson-Kadanoff's renormalization group repeatedly, the limit becomes the massless Gaussian field with the potential $\frac{1}{2}c(\lambda)\eta^2$ with a proper positive constant $c(\lambda)$ determined depending on λ.*

Remark 9.5. *The nonuniqueness of μ_u^∇ for nonconvex potential V is unknown. The ground states corresponding to such potential are analyzed by* [201] *for continuum field.*

9.5 Uniqueness of φ-Gibbs Measures

The existence of shift invariant and ergodic φ-Gibbs measures is not fully established in general (cf. Theorem 4.16 and Remark 4.6) except the Gaussian case (cf. Sect. 6.6). However, the uniqueness can be shown from Theorem 9.5.

Theorem 9.10. *For every $h \in \mathbb{R}$, the square integrable, shift invariant and ergodic (under the shifts) φ-Gibbs measure μ with mean h (i.e., $E^\mu[\phi(x)] = h$ for every $x \in \mathbb{Z}^d$) is unique; recall Definition 2.3 for shift invariance and ergodicity of φ-fields.*

Proof. Let $\phi = \{\phi(x); x \in \mathbb{Z}^d\}$ be μ-distributed. For each $x \in \Lambda_\ell$, $\phi(x)$ is represented as

$$\phi(x) = \frac{1}{|\partial\Lambda_\ell|} \sum_{y \in \partial\Lambda_\ell} \phi(y) + \frac{1}{|\partial\Lambda_\ell|} \sum_{y \in \partial\Lambda_\ell} \sum_{b \in \mathfrak{C}_{y,x}} \nabla\phi(b) \, ,$$

where $\mathfrak{C}_{y,x}$ are chains connecting y and x. However, letting $\ell \to \infty$, the first term in the right hand side converges to h in $L^1(\mu)$ by the ergodicity of μ. Therefore, from Theorem 9.5, we see that every finite dimensional distribution of ϕ under μ is uniquely determined.   □

For the infinite volume dynamics $\phi_t = \{\phi_t(x); x \in \mathbb{Z}^d\}$ and $\eta_t = \{\eta_t(b); b \in (\mathbb{Z}^d)^*\}$, the convergence rate to the equilibrium or the algebraic decay of correlations:

$$\mathrm{cov}\,(F(\phi_0), G(\phi_t)) \sim ct^{-(d-2)/2}, \quad \mathrm{cov}\,(F(\eta_0), G(\eta_t)) \sim ct^{-d/2}$$

as $t \to \infty$ under equilibrium are proved by [64] for $d \geq 3$ and $d \geq 2$, respectively; compare this with the static results in Sect. 4.3.

10 Hydrodynamic Limit

We now entirely move toward the investigation on the dynamics for the heights. The random time evolution of microscopic height variables $\phi_t = \{\phi_t(x)\}, t \geq 0$ was naturally introduced in Sect. 2.4 from the Hamiltonian $H(\phi)$ by means of the Langevin equations. This section analyzes its macroscopic behavior under the space-time diffusive scaling defined by (2.17). We shall establish the LLN, called hydrodynamic limit, for φ-dynamics on $\mathbb{T}_N^d$ or on D_N. It is shown that the evolutional law of the macroscopic interfaces is governed by the motion by mean curvature with anisotropy in the limit, and described by the nonlinear PDE with diffusion coefficient formally given by the Hessian of the surface tension $\sigma = \sigma(u)$. The corresponding CLT for $\nabla\varphi$-dynamics on $(\mathbb{Z}^d)^*$ which is in equilibrium and the LDP for φ-dynamics on $\mathbb{T}_N^d$ will be studied in Sects. 11 and 12, respectively. The dynamics with the wall effect or those in two media realized by adding weak self potential will be discussed in Sects. 13, 14 and 15.

10.1 Space-Time Diffusive Scaling Limit

Let us consider the SDEs (2.9) on a big but finite lattice domain Γ. We shall take $\Gamma = \mathbb{T}_N^d$ (i.e., we discuss under periodic boundary conditions) or $\Gamma = D_N$ for a bounded domain D in $\mathbb{R}^d$ having piecewise Lipschitz boundary with properly scaled boundary conditions $\psi \in \mathbb{R}^{\partial^+ D_N}$. More exactly saying, when $\Gamma = \mathbb{T}_N^d$, the SDEs have the form

$$d\phi_t(x) = - \sum_{y \in \mathbb{T}_N^d : |x-y|=1} V'(\phi_t(x) - \phi_t(y))dt + \sqrt{2}dw_t(x) , \qquad (10.1)$$

for $x \in \mathbb{T}_N^d$, while, when $\Gamma = D_N$, they have the form

$$d\phi_t(x) = - \sum_{y \in \overline{D_N} : |x-y|=1} V'(\phi_t(x) - \phi_t(y))dt + \sqrt{2}dw_t(x) , \qquad (2.11)'$$

for $x \in D_N$ with the boundary conditions

$$\phi_t(y) = \psi(y), \quad y \in \partial^+ D_N . \qquad (2.12)'$$

(a) *Main Theorem*

Under the space-time diffusive scaling for the evolution of microscopic height variables $\phi_t = \{\phi_t(x)\}$ of the interface, macroscopic height variables $h^N(t) = \{h^N(t, \theta)\}$ are defined as step functions on the torus $\mathbb{T}^d$ or on the domain D by the formula (2.17):

$$h^N(t, \theta) = \frac{1}{N}\phi_{N^2 t}([N\theta]), \quad \theta \in \mathbb{T}^d \text{ or } D .$$

Or, we adopt the definition

$$h^N(t, \theta) = \sum_{x \in \mathbb{T}_N^d (\text{or } D_N)} \frac{1}{N}\phi_{N^2 t}(x)1_{B(x/N, 1/N)}(\theta) , \qquad (10.2)$$

where $B(\theta, a) = \prod_{i=1}^d [\theta_i - a/2, \theta_i + a/2)$ denotes the d dimensional cube (box) with center $\theta = (\theta_i)_{i=1}^d$ and side length $a > 0$.

The goal is to study the behavior of $h^N(t)$ as $N \to \infty$. Two definitions (2.17) and (10.2) coincide if $B(\theta, a)$ is taken as $\prod_{i=1}^d [\theta_i, \theta_i + a)$. The difference is therefore only the componentwise shift in the variable θ by $1/2N$, but this is negligible in the limit. The conditions (V1)-(V3) in (2.2) are always assumed on the potential V.

Theorem 10.1. (Hydrodynamic Limit, *Funaki and Spohn* [124] *on the torus* $\mathbb{T}^d$, *Nishikawa* [204] *on D with boundary conditions) Assume that initial random configuration* $\phi_0 = \{\phi_0(x); x \in \mathbb{T}_N^d\}$ *of the SDEs* (10.1) *converges to some nonrandom* $h_0 \in L^2(\mathbb{T}^d)$ *in the sense that*

$$\lim_{N\to\infty} E[\|h^N(0) - h_0\|^2] = 0 , \qquad (10.3)$$

where $\|\cdot\|$ *denotes the usual* L^2-*norm of the space* $L^2(\mathbb{T}^d)$. *Then, for every* $t > 0$,

$$\lim_{N\to\infty} E[\|h^N(t) - h(t)\|^2] = 0$$

holds and the limit $h(t) = h(t, \theta)$ *is a unique weak solution of the nonlinear PDE*

$$\frac{\partial h}{\partial t}(t, \theta) = \operatorname{div}\left\{\nabla\sigma(\nabla h(t, \theta))\right\}$$

$$\equiv \sum_{i=1}^{d} \frac{\partial}{\partial \theta_i}\left\{\frac{\partial \sigma}{\partial u_i}(\nabla h(t, \theta))\right\}, \quad \theta \in \mathbb{T}^d, \qquad (10.4)$$

having initial data h_0, *where* $\sigma = \sigma(u)$ *is the normalized surface tension defined in Sect. 5.1,* (5.2).

The theorem is only stated for the torus $\mathbb{T}^d$, but a similar result holds on D and in the space $L^2(D)$. In this case, the PDE (10.4) requires a macroscopic boundary condition g at ∂D. The PDE (10.4) describes the **motion by mean curvature** (MMC) with anisotropy, see the next paragraph (b). The limit $h(t)$ is nonrandom and therefore Theorem 10.1 is at the level of the LLN.

Remark 10.1. (1) *If* $\sigma \in C^2(\mathbb{R}^d)$ *which is not yet shown (see Problem 5.1), the diffusion coefficient of the PDE* (10.4) *is given by the Hessian* $(\partial^2\sigma/\partial u_i\partial u_j)_{ij}$ *of* σ.
(2) *In the Gaussian case (i.e.,* $V(\eta) = \frac{1}{2}\eta^2$), $\sigma(u) = \frac{1}{2}|u|^2$ *and the limit equation* (10.4) *is linear heat equation. In fact, this can be directly seen, since the drift term of the SDEs* (10.1) *is* $\Delta\phi_t(x)$ *for such potential* V; *recall the SDEs* (2.20) *and that* Δ *denotes the discrete Laplacian. The space-time diffusive scaling leads the discrete Laplacian to the continuum one.*

(b) *Physical Meaning of the PDE* (10.4)

The total surface tensions $\Sigma_{\mathbb{T}^d}(h)$ on the torus $\mathbb{T}^d$ or $\Sigma_D(h)$ on D of the macroscopic surface $h = \{h(\theta)\}$ were introduced in (6.2). The Fréchet derivatives of $\Sigma = \Sigma_{\mathbb{T}^d}$ or Σ_D are given by

$$\frac{\delta\Sigma}{\delta h(\theta)}(h) = -\operatorname{div}\left\{(\nabla\sigma)(\nabla h(\theta))\right\} .$$

Therefore, the hydrodynamic equation (10.4) can be regarded as a gradient flow for Σ

$$\frac{\partial h}{\partial t}(t) = -\frac{\delta \Sigma}{\delta h}(h(t)) , \qquad (10.5)$$

namely the surface moves relaxing its total surface energy. For isotropic motion by mean curvature one would have $\sigma(u) = \sqrt{1 + |u|^2}$; note that $\Sigma(h)$ is the surface area of h and $\delta \Sigma / \delta h$ is the mean curvature in such case. In our case this is likely to hold for small $|u|$, however $\sigma(u) \simeq |u|^2$ for large $|u|$, which reflects the constraints due to the underlying microscopic lattice structure.

(c) *Formal Derivation of the PDE* (10.4)

We work on $\mathbb{T}^d$ just for fixing the notation. For every test function $f = f(\theta) \in C^\infty(\mathbb{T}^d)$, we have that

$$\langle h^N(t), f \rangle := \int_{\mathbb{T}^d} h^N(t, \theta) f(\theta) \, d\theta$$

$$= \frac{1}{N^{d+1}} \sum_{x \in \mathbb{T}_N^d} \phi_{N^2 t}(x)[f]^N(x/N) ,$$

where

$$[f]^N(x/N) := N^d \int_{B(x/N, 1/N)} f(\theta) \, d\theta .$$

Then, applying Itô's formula and recalling the symmetry of V in our basic conditions (2.2), we have by summation by parts that

$$\langle h^N(t), f \rangle - \langle h^N(0), f \rangle$$

$$= -\int_0^t \frac{1}{N^d} \sum_{i=1}^d \sum_{x \in \mathbb{T}_N^d} V'(\nabla_i \phi_{N^2 s}(x)) \frac{\partial f}{\partial \theta_i}\left(\frac{x}{N}\right) ds + o(1). \qquad (10.6)$$

The last error term $o(1)$ involves those for the replacement of $\nabla_i^N [f]^N$ (macroscopically normalized discrete differential, see Sect. 10.2-(b) or Lemma 6.14) with $\partial f / \partial \theta_i$ and the martingale term:

$$m^N(t, f) = \frac{\sqrt{2}}{N^{d+1}} \sum_{x \in \mathbb{T}_N^d} w_{N^2 t}(x)[f]^N\left(\frac{x}{N}\right) ,$$

which goes to 0 since $E[m^N(t, f)^2] = O(N^{-d})$. Note that the divergent factor N^2 carried under the time change has disappeared, see Problem 10.1 below.

The left hand side of (10.6) would converge to $\langle h(t), f \rangle - \langle h_0, f \rangle$. On the other hand, the right hand side consists of a large scale sum of complex variables. However, one would expect the so-called **local equilibrium states** were realized in the system, i.e., around each macroscopic space-time point (s, θ) the distribution of the gradient field corresponding to the microscopic height variables $\{\phi_{N^2 s}(x); x \sim N\theta\}$ would reach the equilibrium state $\mu_{\nabla h(s,\theta)}^\nabla$, which is the $\nabla\varphi$-Gibbs measure with mean tilt $\nabla h(s, \theta)$. Thus, by means of the **local ergodicity**, i.e., under large sum, $V'(\nabla_i \phi_{N^2 s}(x))$ in the right hand side could be replaced with its ensemble average $E^{\mu^{\nabla}_{\nabla h(s, x/N)}}[V'(\nabla_i \phi(0))]$ which coincides with $\partial\sigma/\partial u_i(\nabla h(s, x/N))$ from (5.14); note that $\nabla_i \phi(0) = \nabla\phi(e_i)$ under different two notation. Therefore, one would obtain in the limit

$$\frac{d}{dt}\langle h(t), f\rangle = -\langle \nabla\sigma(\nabla h(t)), \nabla f\rangle ,$$

for every $f \in C^\infty(\mathbb{T}^d)$, which is a weak form of the nonlinear PDE (10.4). The actual proof given in [124] is slightly different. It is based on the method of entropy production initiated by [141] (see also [170, 239]) and its variant, the H^{-1}-method, by which one can avoid the so-called two blocks' estimate necessary in the standard route for establishing the hydrodynamic limit. The result in Sect. 9 (Corollary 9.6) is substantial to complete the proof.

Remark 10.2. (1) *In one dimension, Theorem 10.1 gives essentially the same result that [141] obtained (without convexity condition on V), since $\eta_t \equiv \nabla\phi_t$ satisfies the same SDEs that [141] considered. However, in higher dimensions, our $\nabla\varphi$-Gibbs measures have long correlations and the situation is very different from [141].*
(2) *The martingale term $m^N(t, f)$ looks simply disappearing in the limit as we have mentioned, but, in fact, this is not really true. For $a > 0$, consider the SDEs (10.1) with the Brownian motions $\sqrt{2}w_t(x)$ replaced by $\sqrt{2a}w_t(x)$. Then, starting from such SDEs, we have the limit equation for the macroscopic heights*

$$\frac{\partial h}{\partial t}(t, \theta) = a \operatorname{div}\{(\nabla\sigma^a)(\nabla h(t, \theta))\} ,$$

where σ^a is the surface tension determined by the potential $a^{-1}V$. Indeed, under the time change $\bar{\phi}_t := \phi_{a^{-1}t}$, one can apply Theorem 10.1 for $\bar{\phi}_t$. When V is quadratic, $a\nabla\sigma^a(u)$ does not depend on a, but this may not be true in general.

Problem 10.1. *In (10.6), we have used the summation by parts formula noting that $V'(\eta) = -V'(-\eta)$ and this makes the right hand side of order $O(1)$ as $N \to \infty$. If the potential V is asymmetric (cf. Remark 2.1), such cancellation does not occur and the right hand side remains to be $O(N)$ at least at first look. Models involving such divergent quantities are called of* **nongradient**. *The hydrodynamic limit for the Ginzburg-Landau dynamics of nongradient type might be established based on Varadhan's argument [247].*

10.2 The Nonlinear PDE (10.4)

We start now the proof of Theorem 10.1 on $\mathbb{T}^d$. This subsection summarizes results on the nonlinear PDE (10.4). We recall that the surface tension σ satisfies the properties stated in Sect. 5.3.

(a) *Existence and Uniqueness of Solutions*

Let us introduce a triple of real separable Hilbert spaces $V \subset H = H^* \subset V^*$ by $H = L^2(\mathbb{T}^d), V = H^1(\mathbb{T}^d) := \{h \in H; |\nabla h| \in H\}$ and $V^* = H^{-1}(\mathbb{T}^d)$. We also denote by H^d the d-fold direct product of H. These three spaces are equipped with their standard norms denoted by $\| \cdot \|, \| \cdot \|_V$ and $\| \cdot \|_{V^*}$, respectively. The duality relation $_V\langle \cdot, \cdot \rangle_{V^*}$ between V and V^* satisfies $_V\langle v, h \rangle_{V^*} = \langle v, h \rangle$ if $v \in V$ and $h \in H$, where $\langle \cdot, \cdot \rangle$ is the scalar product of H. We consider the nonlinear differential operator

$$A(h) = \sum_{i=1}^d \frac{\partial}{\partial \theta_i} \sigma_i'(\nabla h), \quad h \in V,$$

where $\sigma_i'(u) := \partial \sigma / \partial u_i, u \in \mathbb{R}^d$. The next lemma follows from Theorem 5.3 and Corollary 5.4:

Lemma 10.2. *The operator $A : V \to V^*$ has the following properties for all $h, h_1, h_2 \in V$. The constants c_- and C are those appeared in (V3) and Theorem 5.3-(1), respectively.*
(A_1) *(semicontinuity)* $\quad _V\langle h, A(h_1 + \lambda h_2) \rangle_{V^*}$ *is continuous in $\lambda \in \mathbb{R}$,*
(A_2) *(monotonicity)* $\quad _V\langle h_1 - h_2, A(h_1) - A(h_2) \rangle_{V^*} \le -c_- \|\nabla h_1 - \nabla h_2\|^2,$
(A_3) *(coercivity)* $\quad _V\langle h, A(h) \rangle_{V^*} + c_- \|h\|_V^2 \le c_- \|h\|^2,$
(A_4) *(growth condition)* $\quad \|A(h)\|_{V^*} \le C\|h\|_V.$

We call $h(t)$ a solution (or an H-solution) of (10.4) with initial data $h_0 \in H$ if $h(t) \in C([0, T], H) \cap L^2([0, T], V)$ and

$$h(t) = h_0 + \int_0^t A(h(s)) \, ds$$

holds in V^* for a.e. $t \in [0, T]$. The general theory on nonlinear PDEs (e.g., [10, 37, 178, 255]) proves the existence and uniqueness of solutions to (10.4) under the conditions (A_1)–(A_4) of Lemma 10.2.

Proposition 10.3. *For every initial data $h_0 \in H$ the PDE (10.4) has a unique solution $h(t)$. In addition, it admits the uniform bound*

$$\sup_{0 \le t \le T} \|h(t)\|^2 + \int_0^T \|h(t)\|_V^2 \, dt \le K(\|h_0\|^2 + 1), \tag{10.7}$$

where K is a constant depending only on c_- and T.

(b) *Discretization Scheme and Its Convergence*

In order to prove Theorem 10.1, one needs to compare the discrete variable $h^N(t)$ with the continuum one $h(t)$. It is therefore convenient to introduce $h^{\sharp, N}(t)$, a solution of lattice approximated version of the PDE (10.4), and compare $h^N(t)$ with $h^{\sharp,N}(t)$. For this purpose we define the finite difference operators

$$\nabla_i^N f(\theta) = N(f(\theta + e_i/N) - f(\theta)) ,$$
$$\nabla_i^{N*} f(\theta) = -N(f(\theta) - f(\theta - e_i/N)), \quad \theta \in \mathbb{T}^d, \ 1 \le i \le d,$$
$$\nabla^N = (\nabla_1^N, ..., \nabla_d^N) .$$

With these notations the discretized PDE of (10.4) reads

$$\frac{\partial}{\partial t} h^{\sharp,N}(t,\theta) = A^N(h^{\sharp,N}(t))(\theta) := -\sum_{i=1}^d \nabla_i^{N*} \sigma_i'(\nabla^N h^{\sharp,N}(t,\theta)) , \qquad (10.8)$$

for $\theta \in \frac{1}{N}\mathbb{T}_N^d \equiv \{\theta \in \mathbb{T}^d; N\theta \in \mathbb{T}_N^d\} \subset \mathbb{T}^d$. It has to be solved with the initial data

$$h_0^{\sharp,N}(\theta) = [h_0]^N(\theta) := N^d \int_{[\![\theta]\!]_N} h_0(\theta')\, d\theta' \qquad (10.9)$$

where $[\![\theta]\!]_N$ stands for the box with center in $\frac{1}{N}\mathbb{T}_N^d$ of side length $\frac{1}{N}$ containing $\theta \in \mathbb{T}^d$. Denoting by $[\theta]_N$ the center of the box $[\![\theta]\!]_N$, we extend $h^{\sharp,N}(t,\theta)$ to $\mathbb{T}^d$ as a step function,

$$h^{\sharp,N}(t,\theta) := h^{\sharp,N}(t,[\theta]_N), \quad \text{for } \theta \in \mathbb{T}^d . \qquad (10.10)$$

We mention the convergence of the solution $h^{\sharp,N}(t)$ of the discretized PDE (10.8) to $h(t)$ as $N \to \infty$. The monotonicity of the operator is essential for the proof, see Proposition I.2 in [124] for details.

Lemma 10.4. (1) *For every $t > 0$, $h^{\sharp,N}(t)$ converges to $h(t)$ weakly in H as $N \to \infty$, where $h(t)$ is the unique solution of (10.4) with initial data $h_0 \in H$.* (2) *Assume $\sup_{N \in \mathbb{N}} \|\nabla^N h_0^{\sharp,N}\| < \infty$ in addition. Then the above convergence holds strongly in H.*

(c) *Uniform L^p-Bound on $\{\nabla^N h^{\sharp,N}(t); N \in \mathbb{N}\}$*

In the proof of Theorem 10.1, a certain function of $\nabla\phi$-variables diverging quadratically in $\nabla\phi$ arises. Such function can be controlled in the limit, if a uniform L^p-bound on the $\nabla\varphi$-dynamics is available for some $p > 2$, since it implies the uniform L^2-integrability of the function. However, unfortunately, we can only derive a uniform L^2-bound for the $\nabla\varphi$-dynamics, see Sect. 10.3-(a). We shall introduce the notion of coupled local equilibria and show that a uniform L^p-bound for the discretized PDE, which is derived here, compensates with the missing estimate on the $\nabla\varphi$-dynamics.

Let $h^{\sharp,N}(t)$ be the solution of (10.8) with initial data $h_0^{\sharp,N}$ satisfying $\sup_{N \in \mathbb{N}} \|\nabla^N h_0^{\sharp,N}\| < \infty$. We shall derive a uniform L^p-bound on $\nabla^N h^{\sharp,N}(t)$ in N. The norm of the space $L^p(\mathbb{T}^d)$ is denoted by $\|\cdot\|_p$, $1 \le p \le \infty$; recall that $\|\cdot\|_2$ is simply denoted by $\|\cdot\|$.

Lemma 10.5. *We have that*

$$\sup_{N\in\mathbb{N}} \sup_{t\geq 0} \|\nabla^N h^{\sharp,N}(t)\| < \infty \, , \tag{10.11}$$

$$\sup_{N\in\mathbb{N}} \int_0^T \|\nabla^N\nabla^N h^{\sharp,N}(t)\|^2 \, dt < \infty, \quad T > 0 \, , \tag{10.12}$$

where $\|\nabla^N\nabla^N h\|^2 = \sum_{i,j=1}^d \|\nabla_i^N\nabla_j^N h\|^2$ *, and for some* $p > 2$,

$$\sup_{N\in\mathbb{N}} \int_0^T \|\nabla^N h^{\sharp,N}(t)\|_p^p \, dt < \infty \, . \tag{10.13}$$

Proof. The proof is due to an idea quite common in the theory of PDE, e.g., see [180], p.433. Denoting $h = h^{\sharp,N}$ for simplicity, we have from (10.8),

$$\frac{d}{dt}\|\nabla^N h(t)\|^2 = -2N^{d-2} \sum_{\theta\in\frac{1}{N}\mathbb{T}_N^d} \sum_{i=1}^d \{\nabla^N h(t,\theta + e_i/N) - \nabla^N h(t,\theta)\}$$
$$\cdot \{\nabla\sigma(\nabla^N h(t,\theta + e_i/N)) - \nabla\sigma(\nabla^N h(t,\theta))\}$$
$$\leq -2c_-\|\nabla^N\nabla^N h(t)\|^2 \, . \tag{10.14}$$

We have used $\nabla_i^N\nabla_j^{N*} = \nabla_j^{N*}\nabla_i^N$ and $\nabla_j^N\nabla_i^N = \nabla_i^N\nabla_j^N$, and subsequently Corollary 5.4. Hence,

$$\|\nabla^N h(t)\|^2 + 2c_- \int_0^t \|\nabla^N\nabla^N h(s)\|^2 \, ds \leq \|\nabla^N h_0\|^2 \, ,$$

which shows (10.11) and (10.12). To show (10.13), we need Sobolev's lemma for lattice functions,

$$\|f\|_{2^*}^2 \leq C(\|\nabla^N f\|^2 + \|f\|^2), \quad f = \{f(\theta), \theta\in\mathbb{T}_N^d\}, \tag{10.15}$$

for some $C > 0$ independent of the lattice spacing N. Here 2^* is the Sobolev conjugate of 2 defined by $2^* = 2d/(d-2)$ if $d \geq 3$, 2^* is an arbitrary number larger than 1 if $d = 2$ and $2^* = \infty$ if $d = 1$. Given (10.15), the proof of (10.13) can be completed from (10.11) and (10.12) using Hölder's inequality

$$\|f\|_p \leq \|f\|^{1-\tau}\|f\|_q^\tau$$

with the choice of $q = 2^*, p = 4 - 4/2^* (> 2)$ and $\tau = p/2$. $\square$

10.3 Local Equilibria

(a) *Uniform Bound on Second Moments*

As we have mentioned in Sect. 10.1-(c), one would expect that at positive (macroscopic) times the interface has locally a definite tilt u and a statistics as specified by the $\nabla\varphi$-Gibbs measure μ_u^∇. Such a strong property will come out only indirectly. However for the space-time averaged measure we will establish that it is some mixture of $\nabla\varphi$-Gibbs measures. In fact such property will be established for the measure coupled to the solution of a discretized version of the PDE (10.4), see Proposition 10.8 for a precise statement.

Let $\mu_t^{\nabla,N} \in \mathcal{P}(\mathcal{X}_{\mathbb{T}_N^d})$ be the distribution of $\nabla\phi_t$ on $\mathcal{X}_{\mathbb{T}_N^d}$, the state space for the $\nabla\varphi$-field on the torus defined in the proof of Theorem 4.15, and let $\mathrm{Av}_T(\mu^{\nabla,N})$ be its space-time average over $[0, N^2T] \times \mathbb{T}_N^d$:

$$
\mathrm{Av}_T(\mu^{\nabla,N}) = \frac{1}{N^d} \sum_{x\in\mathbb{T}_N^d} \frac{1}{N^2T} \int_0^{N^2T} \mu_t^{\nabla,N} \circ \tau_x^{-1} \, dt ,
$$

for $T > 0$. Here $\tau_x : \mathcal{X}_{\mathbb{T}_N^d} \to \mathcal{X}_{\mathbb{T}_N^d}$ denotes the shift by x on $\mathbb{T}_N^d$ (cf. Definition 2.3 on $\mathbb{Z}^d$) and note that $N^d = |\mathbb{T}_N^d|$. $\nu^\nabla \in \mathcal{P}(\mathcal{X}_{\mathbb{T}_N^d})$ is always regarded as $\nu^\nabla \in \mathcal{P}(\mathcal{X})$ by extending it periodically. We shall simply denote by $\mu_N^\nabla = \mu_{N,0}^\nabla \in \mathcal{P}(\mathcal{X}_{\mathbb{T}_N^d})$ the finite volume $\nabla\varphi$-Gibbs measure with periodic boundary conditions and tilt $u = 0$ (see the proof of Theorem 4.15).

To obtain uniform L^2-bounds, we again use a coupling argument for the SDEs (10.1) on $\mathbb{T}_N^d$. Assume that two initial data ($\mathbb{R}^{\mathbb{T}_N^d}$-valued random variables) $\phi_0 = \{\phi_0(x); x \in \mathbb{T}_N^d\}$ and $\bar\phi_0 = \{\bar\phi_0(x); x \in \mathbb{T}_N^d\}$ are given and let ϕ_t and $\bar\phi_t$ be the corresponding two solutions of the SDEs (10.1) with common Brownian motions. The macroscopic φ-fields obtained from ϕ_t and $\bar\phi_t$ by scaling in space, time and magnitude as in (10.2) are denoted by $h^N(t,\theta)$ and $\bar h^N(t,\theta)$, $\theta \in \mathbb{T}^d$, respectively. Recall that $\|\cdot\|$ denotes the norm of the space $L^2(\mathbb{T}^d)$.

Lemma 10.6. (1) *We have for every $t > 0$*

$$
E[\|h^N(t) - \bar h^N(t)\|^2] \le E[\|h^N(0) - \bar h^N(0)\|^2] .
$$

(2) *Assume the condition (10.3) on the distribution μ_0^N of ϕ_0. Then,*

$$
\sup_{N\in\mathbb{N}} E^{\mathrm{Av}_T(\mu^{\nabla,N})}[\eta(b)^2] < \infty, \quad b \in (\mathbb{Z}^d)^* .
$$

Proof. As in Lemma 9.7 we have

$$
\frac{\partial}{\partial t} \sum_{x\in\mathbb{T}_N^d} \left(\tilde\phi_t(x)\right)^2 \le -c_- \sum_{b\in(\mathbb{T}_N^d)^*} \left(\nabla\tilde\phi_t(b)\right)^2
$$

with $\tilde\phi_t := \phi_t - \bar\phi_t$. On the torus $\mathbb{T}_N^d$ there is no boundary term. Integrating both sides in t and dividing by N^{d+2}, we obtain

$$
E[\|h^N(t) - \bar h^N(t)\|^2] + c_- \int_0^t E\left[\frac{1}{N^{d+2}} \sum_{b\in(\mathbb{T}_N^d)^*} \left(\nabla\tilde\phi_s(b)\right)^2 \right] ds
$$

$$
\le E[\|h^N(0) - \bar h^N(0)\|^2] . \tag{10.16}
$$

This shows (1).

We now take a special $\bar{\phi}_0$: $\bar{\phi}_0(x) = \sum_{b\in\mathfrak{C}_{O,x}} \bar{\eta}(b)$ with the chain $\mathfrak{C}_{O,x}$ connecting O and x and with $\mathcal{X}_{\mathbb{T}_N^d}$-valued random variable $\bar{\eta}$ distributed under μ_N^∇. Then,

$$
\sum_{i=1}^d E^{\mathrm{Av}_T(\mu^{\nabla,N})}[\eta(e_i)^2] = \frac{1}{N^d}\sum_{b\in(\mathbb{T}_N^d)^*}\frac{1}{N^2 T}\int_0^{N^2 T} E\left[(\nabla\phi_t(b))^2\right] dt
$$

$$
\leq \frac{2}{N^d}\sum_{b\in(\mathbb{T}_N^d)^*}\frac{1}{N^2 T}\int_0^{N^2 T} E\left[\left(\nabla\tilde{\phi}_t(b)\right)^2\right] dt + 2\sum_{i=1}^d E^{\mu_N^\nabla}[\eta(e_i)^2]
$$

$$
\leq \frac{4}{Tc_-}\left\{E[\|h^N(0)\|^2] + E[\|\bar{h}^N(0)\|^2]\right\} + 2\sum_{i=1}^d E^{\mu_N^\nabla}[\eta(e_i)^2] .
$$

We used the stationarity of μ_N^∇ under the SDEs (9.2) on $(\mathbb{T}_N^d)^*$ for the second line and then (10.16) in the third. The last term in the right hand side is bounded in N because of the uniform bound (4.18), take $u = 0$. Therefore, since μ_0^N satisfies (10.3), the assertion (2) follows if one can show $\sup_{N\in\mathbb{N}} E[\|\bar{h}^N(0)\|^2] < \infty$. To this end we choose the chain $\mathfrak{C}_{O,x}$ connecting O and x as follows: First we connect O and $(x_1, 0, \ldots, 0)$ through changing only the first coordinate one by one. Then $(x_1, 0, 0, \ldots, 0)$ and $(x_1, x_2, 0, \ldots, 0)$ are connected through changing the second coordinate, etc.. With this choice,

$$
E[\|\bar{h}^N(0)\|^2] = \frac{1}{N^{d+2}}\sum_{x\in\mathbb{T}_N^d} E\left[\left(\sum_{b\in\mathfrak{C}_{0,x}}\bar{\eta}(b)\right)^2\right]
$$

$$
\leq \frac{1}{N^{d+2}}\sum_{x\in\mathbb{T}_N^d} dN E\left[\sum_{b\in\mathfrak{C}_{0,x}}\bar{\eta}(b)^2\right] \leq C\sum_{i=1}^d E^{\mu_N^\nabla}[\eta(e_i)^2] ,
$$

which is bounded in N.   □

(b) *Method of Entropy Production*

To establish the local equilibria we will essentially follow the route of [141]. We first note that, as pointed out in Sect. 4.1 (with $\rho = 0$ and Λ replaced by $\mathbb{T}_N^d$), the generator of the process $\phi_t = \{\phi_t(x); x \in \mathbb{T}_N^d\}$ is given by

$$
L_N = \sum_{x\in\mathbb{T}_N^d} L_x ,
$$

in which L_x are differential operators

$$
L_x := e^{H_N}\frac{\partial}{\partial\phi(x)}e^{-H_N}\frac{\partial}{\partial\phi(x)} = \partial_x^2 - \partial_x H_N \cdot \partial_x ,
$$

where $H_N = H_{\mathbb{T}_N^d}(\phi)$ is the Hamiltonian defined by (2.1) on $\mathbb{T}_N^d$ and $\partial_x = \partial/\partial\phi(x)$. To write down the generator for the corresponding $\nabla\varphi$-dynamics $\eta_t \equiv \nabla\phi_t$, which is the solution of the SDEs (9.2) on $(\mathbb{T}_N^d)^*$, we further note that

$$\partial_x = 2 \sum_{b:x_b=x} \frac{\partial}{\partial\eta(b)} \qquad (10.17)$$

as operators acting on the functions $F = F(\eta) \in C_b^2(\mathcal{X}_{\mathbb{T}_N^d})$ of variables $\nabla\phi$. Replacing ∂_x in the definition of L_x with this formula, we obtain the differential operators

$$L_N^\nabla = \sum_{x \in \mathbb{T}_N^d} L_x^\nabla,$$

$$L_x^\nabla = \sum_{b,\bar{b} \in (\mathbb{T}_N^d)^*:x_b=x_{\bar{b}}=x} \left\{ 4\frac{\partial^2}{\partial\eta(b)\partial\eta(\bar{b})} - 2V'(\eta(\bar{b}))\frac{\partial}{\partial\eta(b)} \right\},$$

for $x \in \mathbb{T}_N^d$. Two operators L_x and L_x^∇ (and therefore L_N and L_N^∇) coincide when they act on $C_b^2(\mathcal{X}_{\mathbb{T}_N^d})$. Thus L_N^∇ is the generator corresponding to η_t. Through integrating by parts its Dirichlet form is given by

$$-\int_{\mathcal{X}_{\mathbb{T}_N^d}} F L_N^\nabla G \, d\mu_N^\nabla$$

$$= \sum_{x \in \mathbb{T}_N^d} \int_{\mathcal{X}_{\mathbb{T}_N^d}} \partial_x F \, \partial_x G \, d\mu_N^\nabla,$$

$$= 4 \sum_{x \in \mathbb{T}_N^d} \int_{\mathcal{X}_{\mathbb{T}_N^d}} \left(\sum_{b:x_b=x} \frac{\partial F}{\partial\eta(b)} \right) \left(\sum_{b:x_b=x} \frac{\partial G}{\partial\eta(b)} \right) d\mu_N^\nabla, \qquad (10.18)$$

for $F, G \in C_b^2(\mathcal{X}_{\mathbb{T}_N^d})$. For $\nu^\nabla \in \mathcal{P}(\mathcal{X}_{\mathbb{T}_N^d})$ let $I_N(\nu^\nabla)$ be the **entropy production** defined by

$$I_N(\nu^\nabla) = -4 \int_{\mathcal{X}_{\mathbb{T}_N^d}} \sqrt{F_N} L_N^\nabla \sqrt{F_N} \, d\mu_N^\nabla,$$

where $F_N(\eta) = d\nu^\nabla/d\mu_N^\nabla$.

In order to apply the argument of [141], it is convenient to extend the differential operators on the whole lattice $\mathbb{Z}^d$ or on $\Lambda \Subset \mathbb{Z}^d$ in the following manner. Let $C_{\text{loc},b}^2(\mathcal{X})$ be the class of all tame functions F on $\mathcal{X}$ of the form $F(\eta) = \tilde{F}(\{\eta(b); b \in \tilde{\Lambda}^*\})$ for some $\tilde{\Lambda} \Subset \mathbb{Z}^d$ and $\tilde{F} \in C_b^2(\mathbb{R}^{\tilde{\Lambda}^*})$. We regard $L_x^\nabla, x \in \mathbb{Z}^d$, the differential operators acting on $C_{\text{loc},b}^2(\mathcal{X})$; the sum in the right hand side should be taken for $b, \bar{b} \in (\mathbb{Z}^d)^* : x_b = x_{\bar{b}} = x$. We further define the differential operator $L_\Lambda^\nabla, \Lambda \Subset \mathbb{Z}^d$, acting on $C_{\text{loc},b}^2(\mathcal{X})$ by

$$L_\Lambda^\nabla = \sum_{x \in \Lambda} L_x^\nabla.$$

The next lemma claims that, if the entropy production per unit volume converges to 0, the limit measure must be a superposition of $\{\mu_u^\nabla; u \in \mathbb{R}^d\}$.

Lemma 10.7. *Let a sequence $\{\tilde{\mu}^{\nabla,N} \in \mathcal{P}(\mathcal{X}_{\mathbb{T}_N^d}); N \in \mathbb{N}\}$ be given, which is tight in $\mathcal{P}(\mathcal{X})$ and satisfies*

$$\lim_{N\to\infty} N^{-d} I_N(\tilde{\mu}^{\nabla,N}) = 0. \tag{10.19}$$

Then, every limit point $\nu^\nabla \in \mathcal{P}(\mathcal{X})$ of $\tilde{\mu}^{\nabla,N}$ is a $\nabla\varphi$-Gibbs measure.

Proof. On the infinite lattice we define the entropy production as follows: For $\nu^\nabla \in \mathcal{P}(\mathcal{X})$ and $\Lambda \Subset \mathbb{Z}^d$,

$$I_\Lambda(\nu^\nabla) = -4 \int_\mathcal{X} \sqrt{F_\Lambda} L_\Lambda^\nabla \sqrt{F_\Lambda} \, d\mu^\nabla,$$

where $F_\Lambda = d\nu^\nabla/d\mu^\nabla|_{\mathcal{F}_{\overline{\Lambda^*}}}$ with the σ-field $\mathcal{F}_{\overline{\Lambda^*}}$ of $\mathcal{X}$ generated by $\{\eta(b); b \in \overline{\Lambda^*}\}$ and $\mu^\nabla = \mu_0^\nabla \in \mathcal{P}(\mathcal{X})$ is the $\nabla\varphi$-Gibbs measure with tilt $u = 0$. Considering $\tilde{\mu}^{\nabla,N} \in \mathcal{P}(\mathcal{X})$, we have

$$I_\Lambda(\tilde{\mu}^{\nabla,N}) = \sup\left\{ -\int_\mathcal{X} \frac{L_\Lambda^\nabla G}{G} \, d\tilde{\mu}^{\nabla,N}; G \text{ is positive and } \mathcal{F}_{\overline{\Lambda^*}}\text{- measurable} \right\}$$

$$\leq \sup\left\{ -\int_{\mathcal{X}_{\mathbb{T}_N^d}} \frac{L_\Lambda^\nabla G}{G} \, d\tilde{\mu}^{\nabla,N}; G \text{ is positive function on } \mathcal{X}_{\mathbb{T}_N^d} \right\}$$

$$= \frac{|\Lambda|}{N^d} I_N(\tilde{\mu}^{\nabla,N}) = |\Lambda| \times o(1)$$

as $N \to \infty$ by assumption, see [141] for the first and the third equalities. Since I_Λ is lower semicontinuous, the above bound implies $I_\Lambda(\nu^\nabla) = 0$ for all weak limits ν^∇ in $\mathcal{P}(\mathcal{X})$ of $\{\tilde{\mu}^{\nabla,N}\}$ as $N \to \infty$. To show that ν^∇ is a $\nabla\varphi$-Gibbs measure, we choose some $\mathcal{F}_{\overline{\Lambda^*}}$-measurable $G \in C_{\text{loc},b}^2(\mathcal{X})$. Then

$$\left| \int_\mathcal{X} L_\Lambda^\nabla G \, d\nu^\nabla \right| = \left| \int_\mathcal{X} L_\Lambda^\nabla G \cdot F_\Lambda \, d\mu^\nabla \right|$$

$$= 4 \left| \sum_{x\in\Lambda} \int_\mathcal{X} \left(\sum_{b:x_b=x} \frac{\partial G}{\partial\eta(b)} \right) \left(\sum_{b:x_b=x} \frac{\partial F_\Lambda}{\partial\eta(b)} \right) d\mu^\nabla \right|$$

$$\leq 2 \sqrt{\sum_{x\in\Lambda} \int_\mathcal{X} \left(\sum_{b:x_b=x} \frac{\partial G}{\partial\eta(b)} \right)^2 d\nu^\nabla} \times \sqrt{I_\Lambda(\nu^\nabla)} = 0.$$

This implies that $\nu^\nabla|_{\mathcal{F}_{\overline{\Lambda^*}}}$ is stationary under L_Λ^∇, the generator for the SDEs (9.13) when the boundary condition ξ is fixed. The dynamics defined by (9.13) is ergodic. This can be seen through the diffeomorphism $J: \mathcal{X}_{\overline{\Lambda^*},\xi} \ni \eta \mapsto \phi = \{\phi(x); x \in \Lambda\} \in \mathbb{R}^\Lambda$ defined by (2.8) and from the fact that the dynamics for $\phi_t = \{\phi_t(x); x \in \Lambda\}$ defined by the SDEs (9.15) is ergodic. Its unique stationary measure is the finite volume $\nabla\varphi$-Gibbs measure $\mu^\nabla_{\Lambda,\xi} \in \mathcal{P}(\mathcal{X}_{\overline{\Lambda^*},\xi})$, cf. Sect. 9.4, which implies the DLR equations for ν^∇,

$$\nu^\nabla(\cdot | \mathcal{F}_{(\mathbb{Z}^d)^* \setminus \overline{\Lambda^*}})(\xi) = \mu^\nabla_{\Lambda,\xi}(\cdot), \quad \nu^\nabla\text{-a.e. } \xi .$$

This proves that ν^∇ is a $\nabla\varphi$-Gibbs measure.   □

(c) *Coupled Local Equilibria*

Set

$$u^{\sharp,N}(t,x) \equiv (u^{\sharp,N}_1(t,x), ..., u^{\sharp,N}_d(t,x)) = \nabla^N h^{\sharp,N}(t, x/N)$$

for $x \in \mathbb{T}^d_N$ and consider the probability measures

$$p^N(d\eta du) = \frac{1}{t} \int_0^t \frac{1}{N^d} \sum_{x \in \mathbb{T}^d_N} 1_{\{u^{\sharp,N}(s,x) \in du\}} \mu^{\nabla,N}_{N^2 s} \circ \tau^{-1}_x (d\eta) \, ds$$

on $\mathcal{X}_{\mathbb{T}^d_N} \times \mathbb{R}^d$ (and therefore on $\mathcal{X} \times \mathbb{R}^d$ by periodic extension). This means that we have coupled the distribution of the stochastic dynamics and the solution of the discrete PDE (10.8). Lemmas 10.5 and 10.6-(2) prove

$$\sup_{N \in \mathbb{N}} \sup_{b \in (\mathbb{Z}^d)^*} \int \{\eta(b)^2 + |u|^p\} \, p^N(d\eta du) < \infty \qquad (10.20)$$

for some $p > 2$. In particular, $\{p^N; N \in \mathbb{N}\}$ is tight in $\mathcal{P}(\mathcal{X} \times \mathbb{R}^d)$ and, consequently, one can choose from an arbitrary sequence $N' \to \infty$ a subsequence $N'' \to \infty$ such that $p^{N''}(d\eta du)$ converges weakly on $\mathcal{X} \times \mathbb{R}^d$ to some $\bar{p}(d\eta du)$ as $N'' \to \infty$.

To characterize $\bar{p}$, the following entropy bound is imposed on the initial distributions $\mu^{\nabla,N}_0$,

$$\lim_{N \to \infty} N^{-(d+2)} \mathcal{H}_N(\mu^{\nabla,N}_0) = 0 . \qquad (10.21)$$

This condition will be removed later. Here $\mathcal{H}_N(\nu^\nabla) \equiv \mathcal{H}(\nu^\nabla | \mu^\nabla_N)$ denotes the relative entropy of $\nu^\nabla \in \mathcal{P}(\mathcal{X}_{\mathbb{T}^d_N})$ with respect to μ^∇_N, recall (5.4).

Proposition 10.8. *Under the condition* (10.21), *there exists* $\bar{\lambda} \in \mathcal{P}(\mathbb{R}^d \times \mathbb{R}^d)$ *such that* $\bar{p}$ *can be represented in the form*

$$\bar{p}(d\eta du) = \int_{v \in \mathbb{R}^d} \mu^\nabla_v(d\eta) \, \bar{\lambda}(dvdu) .$$

Proof. For $G = G(u) \in C_b(\mathbb{R}^d)$ and $p(d\eta du) \in \mathcal{P}(\mathcal{X} \times \mathbb{R}^d)$ we shall denote the integration of G with respect to $p(d\eta du)$ in u by $p(d\eta, G) \in \mathcal{M}(\mathcal{X})$; the class of all signed measures on S having finite total variations is denoted by $\mathcal{M}(S)$. The subsequence N'' is simply denoted by N.

First we note that $\bar{p}(d\eta, G)$ is shift invariant for every $G \in C_b(\mathbb{R}^d)$. In fact, this can be shown by Lemma 10.5 first for $G \in C_b^1(\mathbb{R}^d)$ and then for general G by approximation.

We next show that, for every $G \in C_b(\mathbb{R}^d)$, $\bar{p}(d\eta, G)$ has a representation

$$\bar{p}(d\eta, G) = \int_{\mathbb{R}^d} \mu_v^\nabla(d\eta) \, \lambda(dv, G) \tag{10.22}$$

with some $\lambda(dv, G) \in \mathcal{M}(\mathbb{R}^d)$. Set $\tilde{p}^N(d\eta, G) := p^N(d\eta, G)/\int_{\mathcal{X}} p^N(d\eta, G) \in \mathcal{P}(\mathcal{X})$ for $G \geq c > 0$. Then, since $G > 0$ and the entropy production $I_N(\nu^\nabla)$ is convex in ν^∇, we have

$$
\begin{aligned}
I_N(\tilde{p}^N) &\leq \frac{\|G\|_\infty}{ct} \int_0^t I_N(\mu_{N^2s}^{\nabla,N}) \, ds \\
&= \frac{\|G\|_\infty}{ctN^2} \left\{ \mathcal{H}_N(\mu_0^{\nabla,N}) - \mathcal{H}_N(\mu_{N^2t}^{\nabla,N}) \right\} .
\end{aligned}
$$

The second line is shown by noting that $F_N(t) = d\mu_t^{\nabla,N}/d\mu_N^\nabla$ is the solution of the forward equation $\partial F_N(t)/\partial t = L_N^\nabla F_N(t)$ and then using (10.18). Since $\mathcal{H}_N(\mu_{N^2t}^{\nabla,N}) \geq 0$, we conclude from the assumption (10.21) that $\{\tilde{p}^N; N \in \mathbb{N}\}$ satisfies the condition (10.19) and therefore Lemma 10.7 shows that its weak limit $\tilde{p}(\cdot, G) = \bar{p}(\cdot, G)/\int_{\mathcal{X}} \bar{p}(d\eta, G)$ is a $\nabla\varphi$-Gibbs measure. However, $\tilde{p}(\cdot, G)$ is shift invariant as we have seen above and $\tilde{p}(\cdot, G) \in \mathcal{P}_2(\mathcal{X})$ by using (10.20). Hence $\tilde{p}(\cdot, G) \in \mathcal{G}^\nabla$ and consequently we see from Corollary 9.6

$$\tilde{p}(\cdot, G) = \int_{\mathbb{R}^d} \mu_v^\nabla(\cdot) \, \tilde{\lambda}(dv, G)$$

for some $\tilde{\lambda}(\cdot, G) \in \mathcal{P}(\mathbb{R}^d)$. Thus we have obtained (10.22) for uniformly positive $G \in C_b(\mathbb{R}^d)$ by taking $\lambda(dv, G) = \int_{\mathcal{X}} \bar{p}(d\eta, G) \times \tilde{\lambda}(dv, G)$. It also holds for general G.

The final task is to show that $\lambda(dv, G)$ in (10.22) is represented as

$$\lambda(dv, G) = \int_{\mathbb{R}^d} G(u) \, \bar{\lambda}(dvdu)$$

for every $G \in C_b(\mathbb{R}^d)$ with some $\bar{\lambda} \in \mathcal{P}(\mathbb{R}^d \times \mathbb{R}^d)$. To this end, one can apply Birkhoff's individual ergodic theorem for μ_v^∇, and then Stone-Weierstrass's theorem ([218], p.121) and Riesz-Markov's theorem ([218], p.111). The details are omitted. □

10.4 Proof of Theorem 10.1

We compare the solution of the SDEs with that of a discretized version of the PDE. Recalling the definition of $h^{\sharp,N}(t)$ from Sect. 10.2-(b), we have

$$E\left[\|h(t) - h^N(t)\|^2\right] \leq 2\|h(t) - h^{\sharp,N}(t)\|^2 + 2E\left[\|h^{\sharp,N}(t) - h^N(t)\|^2\right]. \tag{10.23}$$

The first term refers to the PDE only. In Lemma 10.4-(2) we proved that it converges to zero in the limit $N \to \infty$ under additional uniform bound on $\{\nabla^N h_0^{\natural,N}\}$. But this assumption can be easily removed by approximating the initial data $h_0 \in L^2(\mathbb{T}^d)$ with smooth functions. In this subsection only the second term is handled. We first assume the entropy bound (10.21), which is actually removable.

By a straightforward computation of the L^2-norm using Itô's formula

$$E[\|h^{\natural,N}(t) - h^N(t)\|^2]$$

$$= E\left[\frac{1}{N^d} \sum_{x \in \mathbb{T}_N^d} (h^{\natural,N}(t, x/N) - N^{-1}\phi_{N^2 t}(x))^2\right] \tag{10.24}$$

$$= E[\|h^{\natural,N}(0) - h^N(0)\|^2] - 2\int_0^t (I_1^N(s) - I_2^N(s) - I_3^N(s) + I_4^N(s))\, ds \,,$$

where

$$I_1^N(s) = \frac{1}{N^d} \sum_{x \in \mathbb{T}_N^d} \sum_{i=1}^d \sigma_i'(u^{\natural,N}(s, x)) u_i^{\natural,N}(s, x) \,,$$

$$I_2^N(s) = \frac{1}{N^d} \sum_{x \in \mathbb{T}_N^d} \sum_{i=1}^d \sigma_i'(u^{\natural,N}(s, x)) E[\nabla_i \phi_{N^2 s}(x)] \,,$$

$$I_3^N(s) = \frac{1}{N^d} \sum_{x \in \mathbb{T}_N^d} \sum_{i=1}^d u_i^{\natural,N}(s, x) E[V'(\nabla_i \phi_{N^2 s}(x))] \,,$$

$$I_4^N(s) = \frac{1}{N^d} \sum_{x \in \mathbb{T}_N^d} \sum_{i=1}^d E[\nabla_i \phi_{N^2 s}(x) V'(\nabla_i \phi_{N^2 s}(x))] - 1 \,.$$

Recall $u^{\natural,N}(s, x) = \nabla^N h^{\natural,N}(s, x/N)$. With the notation $p^N(d\eta du)$, these terms can be rewritten as

$$\int_0^t I_k^N(s)\, ds = t \int_{\mathcal{X} \times \mathbb{R}^d} f_k(\eta, u)\, p^N(d\eta du) \,,$$

for $k = 1, 2, 3, 4$, where

$$f_1(\eta, u) = \sum_{i=1}^d u_i \sigma_i'(u), \qquad f_2(\eta, u) = \sum_{i=1}^d \eta(e_i) \sigma_i'(u) \,,$$

$$f_3(\eta, u) = \sum_{i=1}^d V'(\eta(e_i)) u_i, \qquad f_4(\eta, u) = \sum_{i=1}^d \eta(e_i) V'(\eta(e_i)) - 1 \,.$$

One can pass to the limit for the first three terms, where the limit $N'' \to \infty$ should be taken along the subsequence $\{N''\}$ chosen in Sect. 10.3-(c),

$$\lim_{N'' \to \infty} \int_0^t I_k^{N''}(s)\, ds = t \int_{\mathcal{X} \times \mathbb{R}^d} f_k(\eta, u)\, \bar{p}(d\eta du) \,, \qquad (10.25)$$

for $k = 1, 2, 3$. Indeed, noting that $|\nabla \sigma(u)| \le C(1 + |u|)$ and $|V'(\eta(e_i))| \le C(1 + |\eta(e_i)|)$, we see

$$|f_1(\eta, u)|^{p/2} \le C(1 + |u|^p), \qquad p/2 > 1 \,,$$

$$|f_2(\eta, u)|^q + |f_3(\eta, u)|^q \le C \left(1 + |u|^p + \sum_{i=1}^d |\eta(e_i)|^2 \right) \,,$$

for $q = 2p/(2+p) > 1$. Therefore $f_k(\eta, u), k = 1, 2, 3$, are uniformly integrable with respect to the probability measures $\{p^N; N \in \mathbb{N}\}$ because of the uniform bound (10.20). Since $p^{N''}$ converges weakly to $\bar{p}$, we obtain (10.25). For the fourth term, since $\eta V'(\eta) \ge c_- \eta^2 \ge 0\, (\eta \in \mathbb{R})$, one can apply Fatou's lemma to obtain

$$\limsup_{N'' \to \infty} \left(-\int_0^t I_4^{N''}(s)\, ds \right) \le -t \int_{\mathcal{X} \times \mathbb{R}^d} f_4(\eta, u)\, \bar{p}(d\eta du). \qquad (10.26)$$

Summarizing (10.25), (10.26) and together with Proposition 10.8, Theorem 5.5, we have proved that

$$\limsup_{N'' \to \infty} \left[-\int_0^t (I_1^{N''}(s) - I_2^{N''}(s) - I_3^{N''}(s) + I_4^{N''}(s))\, ds \right]$$

$$\le t \int_{\mathbb{R}^{2d}} \{-u \cdot \nabla \sigma(u) + v \cdot \nabla \sigma(u) + u \cdot \nabla \sigma(v) - v \cdot \nabla \sigma(v)\}\, \bar{\lambda}(dvdu)$$

$$= -t \int_{\mathbb{R}^{2d}} (u - v) \cdot (\nabla \sigma(u) - \nabla \sigma(v))\, \bar{\lambda}(dvdu) \le 0 \,. \qquad (10.27)$$

The convexity of σ (cf. Corollary 5.4) implies the nonpositivity of the last integral. This holds for some subsequence $\{N'' \to \infty\}$ of arbitrarily taken sequence $\{N' \to \infty\}$. Hence, going back to (10.24), we have without taking subsequence

$$\lim_{N \to \infty} E \left[\|h^{\natural, N}(t) - h^N(t)\|^2 \right] = 0 \,. \qquad (10.28)$$

Finally, from (10.23) we conclude Theorem 10.1 under the auxiliary entropy assumption (10.21).

However, the entropy assumption can be removed. The idea originally observed by Lu [194] is, roughly saying, that after a short time (macroscopically of order $O(N^{-2})$) the system gains the entropy bound. The details are omitted. □

Remark 10.3. (1) *The technique employed for the proof of Theorem 10.1 is called H^{-1}-method, since the L^2-norms for the height variables can be regarded as the H^{-1}-norms for the height differences. This method was proposed by Chang and Yau [56] to establish the nonequilibrium fluctuation for the first* *time. It has an advantage to skip the so-called 2-blocks' estimate. As we have seen, the 1-block estimate follows from the three steps: (a) showing that limit measures have entropy production 0, (b) 0-entropy production implies the $\nabla\varphi$-Gibbs property, and (c) characterization of all $\nabla\varphi$-Gibbs measures. Indeed, in Sect. 9, we have characterized all stationary measures for the infinite volume $\nabla\varphi$-dynamics (Theorem 9.3), which is stronger result than (c). Under such situation, the proof of 1-block estimate can be simplified, in particular, one need not rely on the entropy production, see [114, 220, 240].*

(2) Under different setting, Abraham et al. [2] discussed the dynamics with boundary conditions related to the wetting transition. They observed the interfaces sideways and proved that the wetting transition occurs depending on the strength of the potential V. See also [59]. For lattice gasses, the boundary conditions were discussed by [96, 171].

(3) From the view point of the nonlinear PDE theory, Giga and Giga [136, 137] studied the case where the surface tension has anisotropy and singularity. In particular, they treated the evolution of facets.

(4) The hydrodynamic limit involving phase transitions is not well established except [219].

10.5 Surface Diffusion

So far, we have been considering the dynamics (2.9) (or, equivalently (2.11)–(2.13) or (10.1)) associated with the Hamiltonian H in the sense that they have the φ-Gibbs measures as their equilibrium states. Indeed, one can introduce various types of dynamics having such properties. For instance, let A be a (nonnegative) operator acting on the spatial variable x and consider the SDEs

$$d\phi_t(x) = -A\frac{\partial H}{\partial \phi(x)}(\phi_t)dt + \sqrt{2A}dw_t(x) , \tag{10.29}$$

for $x \in \Gamma$. Then, at least formally, ϕ_t is reversible under the φ-Gibbs measures on Γ. This can be seen by writing down the corresponding generator and checking that it is symmetric under the φ-Gibbs measures. See Hohenberg and Halperin [147] for the physical background for the SDEs (10.29). The dynamics (2.9) is the special case that $A = I$ (identity map).

Let us, in particular, take $A = -\Delta$. Then, (10.29) can be rewritten as

$$d\phi_t(x) = \sum_{y:|x-y|=1}\left\{ \sum_{z:|z-y|=1} V'(\phi_t(y) - \phi_t(z)) \right.$$

$$\left. - \sum_{z:|z-x|=1} V'(\phi_t(x) - \phi_t(z)) \right\} dt$$

$$+\sqrt{2}d\tilde{w}_t(x) , \tag{10.30}$$

for $x \in \Gamma$, where $\{\tilde{w}_t(x); x \in \Gamma\}$ are Gaussian processes with mean 0 and covariance

$$E[\tilde{w}_t(x)\tilde{w}_s(y)] = -\Delta(x,y) \cdot t \wedge s ,$$

which is a precise realization of $\sqrt{-\Delta}w_t(x)$. $\Delta(x,y)$ is the kernel of the discrete Laplacian Δ.

The SDEs (10.30) on the lattice torus $\Gamma = \mathbb{T}_N^d$ have, contrarily to the dynamics (2.9), the **conservation law** (at microscopic level):

$$\sum_{x \in \mathbb{T}_N^d} \phi_t(x) = \sum_{x \in \mathbb{T}_N^d} \phi_0(x) ,$$

for $t > 0$. Indeed, taking the sum in x, the drift term in (10.30) cancels out and, moreover, $\sum_{x \in \mathbb{T}_N^d} \tilde{w}_t(x) = 0$ holds. The sum $\sum_{x \in \mathbb{T}_N^d} \phi_t(x)$ represents the total volume of the phase below the interface. The dynamics determined by (2.9) does not have such property. In this sense, the time evolutions defined by (2.9) and (10.30) may correspond to the Glauber and Kawasaki dynamics, respectively, in particles' systems. The SDEs (10.30) are sometimes adopted as a model for alloys, since the total numbers of atoms of two kinds of metals are preserved, respectively, under the time evolution. Because of such conservation law, one may think of that the atoms move around only over the surface separating two phases. Thus the model is called the surface diffusion, see [240] for details.

Nishikawa [203] introduced the macroscopic scaling

$$h^N(t,\theta) = \sum_{x \in \mathbb{T}_N^d} \frac{1}{N} \phi_{N^4 t}(x) 1_{B(x/N,1/N)}(\theta) , \tag{10.31}$$

for $\theta \in \mathbb{T}^d$ and the solution $\phi_t = \{\phi_t(x); x \in \mathbb{T}_N^d\}$ of (10.30) on the lattice torus and proved the following theorem. The potential V satisfies the basic conditions (V1)-(V3) in (2.2).

Theorem 10.9. *Assume that $h^N(0)$ satisfies two conditions*

$$\lim_{N \to \infty} E[\|h^N(0) - h_0\|_{H^{-1}(\mathbb{T}^d)}^2] = 0 ,$$

$$\sup_N E[\|h^N(0)\|_{L^2(\mathbb{T}^d)}^2] < \infty ,$$

for some $h_0 \in L^2(\mathbb{T}^d)$. Then, for every $t > 0$,

$$\lim_{N \to \infty} E[\|h^N(t) - h(t)\|_{H^{-1}(\mathbb{T}^d)}^2] = 0$$

holds and the limit $h(t) = h(t,\theta)$ is a unique weak solution of the nonlinear PDE

$$\frac{\partial h}{\partial t}(t,\theta) = -\Delta \left[\mathrm{div} \left\{ (\nabla \sigma)(\nabla h(t,\theta)) \right\} \right]$$

$$\equiv - \sum_{i,j=1}^d \frac{\partial^2}{\partial \theta_j^2} \frac{\partial}{\partial \theta_i} \left\{ \frac{\partial \sigma}{\partial u_i}(\nabla h(t,\theta)) \right\}, \quad \theta \in \mathbb{T}^d , \tag{10.32}$$

having initial data h_0, where $H^{-1}(\mathbb{T}^d)$ stands for the Sobolev space over $\mathbb{T}^d$ equipped with the standard norm $\|\cdot\|_{H^{-1}(\mathbb{T}^d)}$ and $\sigma = \sigma(u)$ is the normalized surface tension as before.

Compared with the scaling (10.2) for the case without conservation law, the space-time scaling ratio is $N : N^4$ in (10.31). The limit equation (10.32) is of fourth order. For instance, when the potential is quadratic $V(\eta) = \frac{1}{2}\eta^2$, we easily see that the scaling is proper and the limit equation is of the form

$$\frac{\partial h}{\partial t} = -\Delta^2 h .$$

The space-time scaling ratio is closely related to the spectral gap of the generator corresponding to the process $\phi_t = \{\phi_t(x); x \in \mathbb{T}_N^d\}$. The gap behaves as $O(1/N^2)$ as $N \to \infty$ for the dynamics (10.1), while it behaves as $O(1/N^4)$ for (10.30). In other words, the system with conservation law requires longer time to relax to the equilibrium state. These gaps are seen from the logarithmic Sobolev inequalities, which are obtained based on the theory of Bakry and Emery [8] noting that our potential V is convex, cf. [111]. Compare with the results [126, 183, 195] for particles' systems.

The proof of Theorem 10.9 is similar to that of Theorem 10.1, but based on the H^{-2}-method rather than H^{-1}-method since the basic norms change. The main task is again the characterization of all $\nabla\varphi$-Gibbs measures corresponding to the dynamics (10.30) on $\Gamma = \mathbb{Z}^d$. Under the conservation law, such Gibbs measures should be called the canonical $\nabla\varphi$-Gibbs measures. Unfortunately, the method of energy inequality developed in Sect. 9.3 does not work well for conservative system. Instead, Nishikawa proved that the class of canonical $\nabla\varphi$-Gibbs measures and that of $\nabla\varphi$-Gibbs measures for non-conservative system coincide under the shift invariance. Thus one can apply Corollary 9.6 to characterize the canonical $\nabla\varphi$-Gibbs measures.

In Sect. 10.1-(d), we notified that the macroscopic interface equation (10.4) is nothing but the gradient flow for the total surface tension Σ. The basic space was $L^2(\mathbb{T}^d)$ there and the Fréchet derivatives were computed on this space. For the conservative system, the basic space should be replaced with $H^{-1}(\mathbb{T}^d)$ and the Fréchet derivative $\tilde{\delta}\Sigma/\tilde{\delta}h(\theta)$ of Σ must be computed on this space, i.e.,

$$\frac{d}{d\epsilon}\Sigma(h + \epsilon g)\bigg|_{\epsilon=0} = \left(\frac{\tilde{\delta}\Sigma}{\tilde{\delta}h}, g\right)_{H^{-1}} ,$$

for every $g \in C^\infty(\mathbb{T}^d)$, where the inner product is defined by $(f, g)_{H^{-1}} = ((-\Delta)^{-1}f, g)_{L^2}$. Thus we have

$$\frac{\tilde{\delta}\Sigma}{\tilde{\delta}h(\theta)} = \Delta\left[\text{div}\left\{(\nabla\sigma)(\nabla h(\theta))\right\}\right]$$

and therefore the limit equation (10.32) is again the gradient flow for Σ and has the form

$$\frac{\partial h}{\partial t}(t) = -\frac{\tilde{\delta}\Sigma}{\tilde{\delta}h}(h(t)) \ .$$

The derivation of the motion by mean curvature from bistable reaction-diffusion equations (sometimes called Allen-Cahn equation) via singular limit is extensively studied in recent years in nonlinear PDE theory, see [205] and also Sect. 16.3 (for the equations with noises). The conservative system is described by the fourth order PDE called Cahn-Hilliard equation. The interfacial equation (10.32) derived here might coincide with the equation derived from Cahn-Hilliard type equation via singular limit. See Visintin [251] for various approaches to the problems related to the phase transitions from the view point of PDE theory.

Bertini et al. [17] derived fourth order PDE via hydrodynamic limit. The limit equation is the same as (10.32) in Theorem 10.9 in one dimension. However, in higher dimensions, the equilibrium measures of the model treated by [17] are Bernoulli product measures, while they have long correlations for the model discussed here. Therefore, these two results are essentially different.

11 Equilibrium Fluctuation

Let $\{\phi_t(x); x \in \mathbb{Z}^d\}$ be the solution of the SDEs (2.13) with initial data ϕ_0 whose gradients $\nabla\phi_0$ are distributed according to the $\nabla\varphi$-pure phase μ_u^∇ for some $u \in \mathbb{R}^d$. Note that the process $\eta_t \equiv \nabla\phi_t$ is in equilibrium. Consider an $\mathcal{S}'(\mathbb{R}^d)$-valued process

$$\Psi_i^N(t, d\theta) = N^{-d/2} \sum_{x \in \mathbb{Z}^d} (\nabla_i \phi_{N^2 t}(x) - u_i) \, \delta_{x/N}(d\theta), \quad \theta \in \mathbb{R}^d \ ,$$

for each $1 \leq i \leq d$, where $\mathcal{S}'(\mathbb{R}^d)$ stands for the class of Schwartz distributions on $\mathbb{R}^d$. The potential V satisfies the conditions (V1)-(V3) in (2.2).

Theorem 11.1. (Equilibrium fluctuation, [135]) *The process $\Psi_i^N(t)$ weakly converges as $N \to \infty$ in the space $C([0, \infty), \mathcal{S}'(\mathbb{R}^d))$ to an equilibrium solution $\Psi_i(t)$ of the stochastic PDE (stochastic partial differential equation)*

$$\frac{\partial \Psi_i}{\partial t}(t, \theta) = -\tilde{A}\Psi_i(t, \theta) + \sqrt{2}\frac{\partial \dot{B}}{\partial \theta_i}(t, \theta) \ ,$$

where $\dot{B}(t, \theta)$ is the space-time white noise and

$$\tilde{A} = -\sum_{i,j=1}^d q_{ij}\frac{\partial^2}{\partial \theta_i \partial \theta_j} \ .$$

The positive definite $d \times d$ matrix $(q_{ij}) \equiv (q_{ij}(u))_{1 \leq i,j \leq d}$ is characterized by the variational formula:

$$
v \cdot qv = 2 \inf_{F} \left\{ \sum_{i=1}^{d} E^{\mu_u^{\nabla}} \left[(u_i - D_i F(\eta))^2 V''(\eta(e_i)) \right] \right.
$$

$$
\left. + \sum_{x \in \mathbb{Z}^d} E^{\mu_u^{\nabla}} [(\partial_x F)^2] \right\}, \tag{11.1}
$$

where $v \in \mathbb{R}^d$, the infimum is taken over all $F = F(\eta) \in C^{\infty}_{\text{loc},b}(\mathcal{X})$, $D_i F(\eta) = F(\tau_{e_i} \eta) - F(\eta)$, $\tau_{e_i} : \mathcal{X} \to \mathcal{X}$ is the shift (cf. Definition 2.3) and $\partial_x F(\eta)$ is defined by (10.17).

The proof relies on the Helffer-Sjöstrand type representation for correlation functions of the process $\eta_t \equiv \nabla \phi_t$ in terms of random walk in moving random environment. Such representation is also used in [202] (see Sect. 8) for static correlation functions. Then the problem is reduced to establishing the homogenization for this random walk, cf. [173].

In the stochastic PDE appearing in Theorem 11.1, the strength of the noise (diffusion coefficient), the drift term (indicating frictional drag or dissipation of energy) and the variance of the equilibrium measure (called fluctuation) automatically satisfy a certain relation, which is called the **fluctuation-dissipation theorem**. Einstein relation for the Langevin equation is the typical and simplest example.

Remark 11.1. *Theorem 11.1 covers the static CLT discussed in Sect. 8.*

Problem 11.1. [135] *If $\sigma \in C^2(\mathbb{R}^d)$ (see Problem 5.1), does the covariance matrix $q(u) = (q_{ij}(u))$ appearing in the CLT coincide with the Hessian of $\sigma(u)$? Recall that $\sigma(u)$ arises in the LLN and the LDP in various manner. Such identity is ordinary in statistical mechanics, for instance if the Gibbs measures have exponentially fast mixing property.*

12 Dynamic Large Deviation

12.1 Dynamic LDP

The hydrodynamic limit is the LLN for the macroscopic height variables $h^N(t)$ defined in (2.17). In this section we study the corresponding LDP on the lattice torus $\mathbb{T}_N^d$. We always assume the conditions (V1)-(V3) in (2.2) on the potential V. The result is the following: Assume that the initial data $\{\phi_0(x) \equiv \phi_0^N(x); x \in \mathbb{T}_N^d\}$ of the SDEs (10.1) are deterministic and satisfy

$$
\sup_{N} \left\{ |\phi_0^N(O)| + \frac{1}{N^d} \sum_{b \in (\mathbb{T}_N^d)^*} (\nabla \phi_0^N(b))^2 \right\} < \infty. \tag{12.1}
$$

We also assume the condition (10.3) without taking expectations for the corresponding $h^N(0)$ and some $h_0 \in L^2(\mathbb{T}^d)$.

Theorem 12.1. (Dynamic LDP, *Funaki and Nishikawa* [121] *on* $\mathbb{T}^d$) *The LDP holds for* $\{h^N(t); t \in [0, T]\}$ *with speed* N^d *and rate functional* $I_T(h)$:

$$P\left(h^N(t) \sim h(t),\ t \leq T\right) \underset{N \to \infty}{\asymp} \exp\{-N^d I_T(h)\}\,,$$

where $h(t) = h(t, \theta)$ *is a given motion of surface. More precisely, for every closed set* $\mathcal{C}$ *and open set* $\mathcal{O}$ *of* $C([0, T], L^2_w(\mathbb{T}^d))$, *we have that*

$$\limsup_{N \to \infty} \frac{1}{N^d} \log P(h^N(\cdot) \in \mathcal{C}) \leq -\inf_{h \in \mathcal{C}} I_T(h)\,, \qquad (12.2)$$

$$\liminf_{N \to \infty} \frac{1}{N^d} \log P(h^N(\cdot) \in \mathcal{O}) \geq -\inf_{h \in \mathcal{O}} I_T(h)\,, \qquad (12.3)$$

where $L^2_w(\mathbb{T}^d)$ *is the space* $L^2(\mathbb{T}^d)$ *equipped with the weak topology and* $C([0, T], L^2_w(\mathbb{T}^d))$ *stands for the class of all continuous functions* $h : [0, T] \to L^2_w(\mathbb{T}^d)$. *The precise form of the rate functional* $I_T(h) \equiv I_T(h(\cdot))$ *is stated in the subsequent section.*

12.2 Dynamic Rate Functional

For each $h = h(t, \theta)$ which is differentiable,

$$I_T(h) = \frac{1}{4} \int_0^T dt \int_{\mathbb{T}^d} \left\{ \frac{\partial h}{\partial t}(t, \theta) - \text{div}[(\nabla \sigma)(\nabla h(t, \theta))] \right\}^2 d\theta\,, \qquad (12.4)$$

if $h(0) = h_0$ and $I_T(h) = +\infty$ if $h(0) \neq h_0$, where σ is the normalized surface tension. More precisely saying, for $h \in C([0, T], L^2_w(\mathbb{T}^d))$ satisfying $h(t) \in H^1(\mathbb{T}^d)$ for a.e. $t \in [0, T]$

$$I_T(h) = \sup_{J = J(t, \theta) \in C^1([0, T] \times \mathbb{T}^d)} I_T(h; J)\,,$$

where

$$I_T(h; J) = \int_{\mathbb{T}^d} J(T, \theta) h(T, \theta)\, d\theta - \int_{\mathbb{T}^d} J(0, \theta) h_0(\theta)\, d\theta$$

$$- \int_0^T dt \int_{\mathbb{T}^d} \frac{\partial J}{\partial t}(t, \theta) h(t, \theta)\, d\theta$$

$$+ \int_0^T dt \int_{\mathbb{T}^d} \nabla J(t, \theta) \cdot \nabla \sigma(\nabla h(t, \theta))\, d\theta - \int_0^T dt \int_{\mathbb{T}^d} J^2(t, \theta)\, d\theta\,,$$

and $H^1(\mathbb{T}^d)$ denotes the Sobolev space on $\mathbb{T}^d$.

The upper bound in Theorem 12.1 is shown based on the exponential Chebyshev's inequality, while for the lower bound the hydrodynamic limit for a weakly perturbed system is established and then Girsanov's formula is applied. These ideas are rather standard. Essential role is played by the superexponential estimate, namely the probability of replacing sample mean over box of side length $N\epsilon$ with ensemble mean is superexponentially small as $N \to \infty$ and then $\epsilon \downarrow 0$. The H^{-1}-method is effectively used to prove such estimate.

Remark 12.1. *In one dimension, Theorem 12.1 was proved by Donsker and Varadhan* [88]. *See also* [172, 184] *for the LDPs corresponding to the hydrodynamic limit.*

12.3 Relation to the Static LDP

Let $\mu_N^{\psi} \equiv \mu_{D_N}^{\psi}$ be the finite volume φ-Gibbs measure (2.4) on D_N with boundary condition ψ satisfying the conditions (6.7). Then, Theorem 6.1 with $U \equiv 0$ shows the LDP for macroscopic height variables $h^N = \{h^N(\theta); \theta \in D\}$ distributed under μ_N^{ψ} and the (unnormalized) rate functional is given by the total surface tension $\Sigma_D(h)$ in (6.2).

Going back to the torus, since the distribution of $\nabla^N h^N(T)$ weakly converges as $T \to \infty$ to the macroscopically scaled $\nabla \varphi$-field under the finite volume $\nabla \varphi$-Gibbs measure on $\mathbb{T}_N^d$, one would expect that the static LDP could be recovered from the dynamic LDP. An affirmative answer is not known at present, however one can at least recover the static rate functional from the dynamic one as $T \to \infty$. In fact, denoting the distribution of $h^N(T)$ by $\mu_N(T)$, the contraction principle implies the LDP for $\{\mu_N(T)\}_N$ with rate functional

$$S_T(\bar{h}) = \inf \left\{ I_T(h); \ h = h(t, \theta) \ \text{s.t.} \ h(T, \theta) = \bar{h}(\theta) \right\}$$

for $\bar{h} = \bar{h}(\theta) \in H^1(\mathbb{T}^d)$. The relationship between $S_T(\bar{h})$ and the total surface tension $\Sigma_{\mathbb{T}^d}(\bar{h})$ on the torus (defined by (6.2) with $D = \mathbb{T}^d$) is stated in the following proposition.

Proposition 12.2.
$$\lim_{T \to \infty} S_T(\bar{h}) = \Sigma_{\mathbb{T}^d}(\bar{h}) \ .$$

The limit in the left hand side is called a quasi-potential and the corresponding classical flow is a gradient flow for the potential Σ as pointed out in Sect. 10.1. In such case the quasi-potential coincides with the potential Σ itself. The infimum is attained by the reversed trajectory of the gradient flow.

13 Hydrodynamic Limit on a Wall

Under the static situation, several modifications were made to the Hamiltonian and the corresponding φ-Gibbs measures. We have considered, for instance, wall effect by conditioning $\phi \geq 0$, two media system by introducing weak self potentials and pinning effect near the height level 0. The associated dynamics can be constructed in such a manner that it is reversible under the modified

φ-Gibbs measures. In the following sections, we discuss the problems of hydrodynamic limit and fluctuations for such dynamics. Entropic repulsion is also studied.

13.1 Dynamics on a Wall

The dynamics for the microscopic interfaces $\phi_t = \{\phi_t(x); x \in \Gamma\}$ on a wall is introduced by **SDEs of Skorokhod type**:

$$d\phi_t(x) = -\frac{\partial H}{\partial \phi(x)}(\phi_t)\,dt + \sqrt{2}dw_t(x)$$
$$+ \frac{1}{N}f\left(\frac{t}{N^2}, \frac{x}{N}, \frac{1}{N}\phi_t(x)\right)dt + d\ell_t(x), \quad x \in \Gamma, \qquad (13.1)$$

subject to the conditions

$$\phi_t(x) \geq 0, \quad \ell_t(x) \nearrow, \quad \ell_0(x) = 0 \quad \text{and} \quad \int_0^\infty \phi_t(x)\,d\ell_t(x) = 0, \qquad (13.2)$$

for each $x \in \Gamma$, where $\partial H/\partial \phi(x)$ is defined as in (2.10), $w_t = \{w_t(x); x \in \Gamma\}$ is a family of independent one dimensional standard Brownian motions and $f = f(t, \theta, h)$ is a given macroscopic external force, for instance, a mild pinning effect on the interfaces from the wall. The interfaces can move over the wall settled at height level 0 so that the height variables always satisfy $\phi_t(x) \geq 0$. The condition "$\ell_t(x) \nearrow$" means that $\ell_t(x)$ called local time of $\phi_t(x)$ at 0 is nondecreasing in t and the last condition in (13.2) implies that $\ell_t(x)$ increases only when $\phi_t(x) = 0$. In particular, $d\ell_t(x) = 0$ if $\phi_t(x) > 0$ and $d\ell_t(x)$ represents a strong repelling force from the wall when the interfaces touch it. The external force f is microscopically scaled in the equation (13.1) to have nontrivial macroscopic limit, see Theorem 13.1. If $f = f(t, \theta, h)$ is jointly continuous in these three variables and Lipschitz continuous in h, then the SDEs (13.1) subject to (13.2) (and with boundary conditions (2.12)) have a unique solution, see [191, 244].

The unique stationary measure of the dynamics determined by (13.1)–(13.2) when $f = 0$ and $\Gamma = D_N$ with 0-boundary conditions is given by $\mu_N^+ \equiv \mu_{D_N}^0(\cdot | \phi \geq 0)$, the conditional probability of the finite volume φ-Gibbs measure $\mu_{D_N}^0$. This measure is reversible under the dynamics.

13.2 Hydrodynamic Limit

In the next theorem, we work on the lattice torus $\Gamma = \mathbb{T}_N^d$ and assume the following conditions on f

(E1) $f \in C^1([0, \infty) \times \mathbb{T}^d \times [0, \infty))$,

(E2)   there exist constants $C > 0$ and $\kappa \in (0, 1)$ such that

$$|f(t, \theta, h)| + \left|\frac{\partial f}{\partial \theta_i}(t, \theta, h)\right| + \left|\frac{\partial f}{\partial t}(t, \theta, h)\right| \leq C(1 + |h|^\kappa) ,$$

for $1 \leq i \leq d$ and

$$-C \leq \frac{\partial f}{\partial h}(t, \theta, h) \leq 0,$$

for every $(t, \theta, h) \in [0, \infty) \times \mathbb{T}^d \times [0, \infty)$.

The condition (E2) means the sublinear growth and nonincreasing property of f in h.

Theorem 13.1. (Hydrodynamic limit, *Funaki* [117] *on* $\mathbb{T}^d$**)** *As* $N \to \infty$, *the macroscopic height variables* $h^N(t, \theta)$ *defined by* (2.17) *or* (10.2) *converge to* $h(t, \theta)$ *in probability, i.e., for every* $\varphi \in H = L^2(\mathbb{T}^d)$ *and* $t, \delta > 0$,

$$\lim_{N \to \infty} P\left(|\langle h^N(t), \varphi\rangle - \langle h(t), \varphi\rangle| > \delta\right) = 0 ,$$

if this condition holds at $t = 0$ *and if* $\sup_N E[\|h^N(0)\|_H^2] < \infty$. *The limit* $h(t, \theta)$ *is a unique solution of the* **evolutionary variational inequality**(*MMC with reflection (obstacle)*):

$$h \in L^2([0, T], V), \frac{\partial h}{\partial t} \in L^2([0, T], V'), \quad {}^\forall T > 0 , \tag{a}$$

$$\left\langle \frac{\partial h}{\partial t}(t), h(t) - v \right\rangle + \langle \nabla \sigma(\nabla h(t)), \nabla h(t) - \nabla v\rangle$$

$$\leq \langle f(t, h(t)), h(t) - v\rangle , \quad a.e.\, t, \quad {}^\forall v \in V : v \geq 0 , \tag{b}$$

$$h(t, \theta) \geq 0, \quad a.e. , \tag{c}$$

$$h(0, \theta) = h_0(\theta) , \tag{d}$$

where $V = H^1(\mathbb{T}^d), V' = H^{-1}(\mathbb{T}^d)$ *and* $\langle \cdot, \cdot \rangle$ *denotes the inner product of* H (*or* H^d) *or the duality between* V' *and* V.

Note that, if $h(t, \theta) > 0$, the condition (b) implies that $h(t)$ satisfies the PDE (10.4) with external force f (roughly saying, by taking $v = h(t) + \epsilon \tilde{v}$ for sufficiently small ϵ and $\tilde{v} \in V$):

$$\frac{\partial h}{\partial t}(t, \theta) = \text{div} \{(\nabla \sigma)(\nabla h(t, \theta))\} + f(t, \theta, h(t, \theta)) .$$

The evolutionary variational inequality (EVI) describes the strong repelling effect from the wall when the macroscopic interfaces touch it, i.e., $h(t, \theta) = 0$.

The proof of Theorem 13.1 is completed based on the penalty method, comparison theorem on SDEs, superexponential 1-block and 2-blocks' estimates, tightness argument from energy inequality and results on the EVI due to Bensoussan and Lions [16].

Remark 13.1. *Rezakhanlou* [221, 222] *derived a Hamilton-Jacobi equation under hyperbolic scaling from growing SOS dynamics* ($\phi(x) \in \mathbb{Z}$) *with constraints on the gradients (e.g.,* $\nabla\phi(x) \leq v$). *Related results were obtained by Evans and Rezakhanlou* [95] *and Seppäläinen* [229].

14 Equilibrium Fluctuation on a Wall and Entropic Repulsion

The first two subsections discuss the fluctuation problem for equilibrium φ-dynamics on a wall in one dimension with boundary conditions. The third subsection briefly summarizes recent results on dynamic entropic repulsion.

14.1 The Case Attached to the Wall

Let us consider the equilibrium dynamics ϕ_t on the wall in one dimension, i.e., ϕ_t is a solution of the SDEs (13.1)–(13.2) with $d = 1, \Gamma = D_N \equiv \{1, 2, \dots, N-1\}, D = (0,1), f = 0$ under the 0-boundary conditions:

$$\phi_t(0) = \phi_t(N) = 0 \tag{14.1}$$

and with an initial distribution $\mu_N^+ \equiv \mu_{D_N}^0(\,\cdot\,|\phi \geq 0)$. **Macroscopic fluctuation field** around the hydrodynamic limit $h(t,\theta) = 0$ is defined by

$$
\begin{aligned}
\Phi^N(t,\theta) &= \sqrt{N}h^N(t,\theta) \\
&= \sum_{x \in D_N} \frac{1}{\sqrt{N}}\phi_{N^2t}(x)1_{[\frac{x}{N}-\frac{1}{2N},\frac{x}{N}+\frac{1}{2N})}(\theta), \quad \theta \in \bar{D} = [0,1]\,.
\end{aligned}
$$

Since $\Phi^N(t,\theta) \geq 0$, the limit is certainly non-Gaussian if it exists and the result must be different from the usual CLT. In fact, we have the following theorem.

Theorem 14.1. (Equilibrium fluctuation, *Funaki and Olla* [122]) *As $N \to \infty$, $\Phi^N(t,\theta)$ weakly converges to $\Phi(t,\theta)$ on the space $C([0,T], H^{-\alpha}([0,1])) \cap L_w^2([0,T] \times [0,1])$ for every $T > 0$ and $\alpha > 1/2$. The limit $\Phi(t,\theta)$ is a unique weak stationary solution of the* **stochastic PDE with reflection of Nualart-Pardoux type** (*cf.* [208]):

$$\frac{\partial\Phi}{\partial t}(t,\theta) = q\frac{\partial^2\Phi}{\partial\theta^2}(t,\theta) + \sqrt{2}\dot{B}(t,\theta) + \xi(t,\theta), \quad \theta \in (0,1)\,,$$

$$\Phi(t,\theta) \geq 0, \quad \int_0^\infty\int_0^1 \Phi(t,\theta)\,\xi(dtd\theta) = 0\,,$$

$$\Phi(t,0) = \Phi(t,1) = 0, \quad \xi: \text{ random measure}\,,$$

where $H^{-\alpha}([0,1])$ is the Sobolev space on $[0,1]$ determined from Dirichlet 0-boundary conditions, $\dot{B}(t,\theta)$ is the space-time white noise, $q = 1/\langle\eta^2\rangle_{\hat{\nu}_0}(= 1/u'(0))$ and $\hat{\nu}_0 \in \mathcal{P}(\mathbb{R})$ is defined by (5.25) with $\lambda = 0$.

The proof is based on the penalization, i.e., we replace the terms of the local times $\ell_t(x)$ with strong positive drifts when the interfaces try to move toward the negative side. This replacement gives the lower bound for the SDEs (13.1)–(13.2) by comparison theorems. Therefore the equilibrium fluctuation result for the SDEs with penalization, which is established through the so-called Boltzmann-Gibbs principle, gives the lower estimate for the limit $\Phi(t, \theta)$. On the other hand, to obtain the upper bound for $\Phi(t, \theta)$, since we are concerned with the equilibrium situation, once the dynamic lower bound is established, one only needs to show the static upper bound. Indeed, it is not difficult to prove that the stationary measure μ_N^+ weakly converges under the scaling of our interest to the distribution of (properly time changed) **three dimensional pinned Bessel process**, which is the stationary measure of the stochastic PDE of Nualart-Pardoux type. Zambotti [256] has some extension.

14.2 The Case Away from the Wall

In the last subsection, we have discussed under the 0-boundary conditions (14.1). Then the macroscopic heights $h^N(t, \theta)$ converges to 0 and, in this sense, the interfaces are attached to the wall at the macroscopic level.

Here, taking $a, b > 0$, let us consider the SDEs (13.1)–(13.2) under the positive boundary conditions

$$\phi_t(0) = aN, \quad \phi_t(N) = bN . \tag{14.2}$$

The initial distribution is taken as $\mu_{D_N}^{+,aN,bN} \equiv \mu_{D_N}^{aN,bN}(\cdot|\phi \geq 0)$, the finite volume φ-Gibbs measure on $D_N = \{1, 2, \ldots, N-1\}$ with boundary conditions aN and bN at $x = 0$ and N, respectively, conditioned to be $\phi \geq 0$. Then the system is stationary and reversible. Since the limit of $h^N(t, \theta)$ as $N \to \infty$ is given by

$$h(\theta) = a + (b - a)\theta, \quad \theta \in [0, 1] ,$$

the fluctuation field is defined by

$$\Phi^N(t, \theta) = \sqrt{N} \left(h^N(t, \theta) - h(\theta) \right) .$$

Theorem 14.2. ([122]) *As $N \to \infty$, $\Phi^N(t, \theta)$ weakly converges to $\Phi(t, \theta)$ on the space $C([0, T], H^{-\alpha}([0, 1])) \cap L_w^2([0, T] \times [0, 1])$ for every $T > 0$ and $\alpha > 1/2$. The limit $\Phi(t, \theta)$ is a unique weak stationary solution of the stochastic PDE:*

$$\frac{\partial \Phi}{\partial t}(t, \theta) = q_{b-a} \frac{\partial^2 \Phi}{\partial \theta^2}(t, \theta) + \sqrt{2}\dot{B}(t, \theta), \quad \theta \in (0, 1) ,$$
$$\Phi(t, 0) = \Phi(t, 1) = 0 ,$$

where $q_u = 1/u'(\lambda)$ with $\lambda = \lambda(u)$, see Sect. 5.5.

Intuitively saying, since the fluctuation of the interfaces is of $O(\sqrt{N})$, they do not feel the effect from the wall under the boundary conditions (14.2) so that the limit of the fluctuation fields becomes Gaussian contrary to the case studied in Theorem 14.1.

Problem 14.1. *In higher dimensions, the expected scaling for the fluctuation field might be*

$$N^{d/2} \left(h^N(t, \theta) - E[h^N(t, \theta)] \right) ,$$

i.e., the order of the fluctuation for the φ-field is expected to be of $O(N^{-d/2+1})$. Compare this with the results in entropic repulsion and then you see that the fluctuation is much smaller than the order of the mean $E[h^N(\theta)]$ except when $d = 1$. The fluctuation field accordingly may not feel the wall and the limit might be Gaussian even under the 0-boundary conditions when $d \geq 2$.

14.3 Dynamic Entropic Repulsion

The problem of entropic repulsion (see Sect. 7.1 for static results) can be investigated under the time evolutions.

In fact, Deuschel and Nishikawa [80] discussed the φ-dynamics on a wall on $\mathbb{Z}^d$, i.e. the dynamics governed by the SDEs of Skorokhod type (13.1)–(13.2) with $f \equiv 0$ and $\Gamma = \mathbb{Z}^d$. They proved that, starting from an *i.i.d.* initial distribution with finite variance, the solution behaves as $\phi_t(x) = O(\sqrt{\log_d(t)})$ as $t \to \infty$ for $d \geq 2$.

Dunlop et al. [91] studied the SOS type dynamics on a wall on $\mathbb{Z}$ (under the constraint $|\nabla \phi_t(x)| = 1$) and proved that $c_1 t^{1/4} \leq E[\phi_t(x)] \leq c_2 t^{1/4} \log t$. Ferrari et al. [98] and Fontes et al. [103] discussed the serial harness process (introduced by Hammersley). In particular, the latter was concerned with the dynamic entropic repulsion on a random wall, cf. [18] under the static situation.

15 Dynamics in Two Media and Pinning Dynamics on a Wall

15.1 Dynamics in Two Media

Let us consider the microscopic dynamics associated with the Hamiltonian (6.3) having a weak self potential in one dimension, i.e., let $\phi_t = \{\phi_t(x); 0 \leq x \leq N\}$ be the solution of the SDEs

$$d\phi_t(x) = -\frac{\partial H}{\partial \phi(x)}(\phi_t)\, dt + \sqrt{2}dw_t(x) - f(\phi_t(x))dt ,$$

for $1 \leq x \leq N - 1$ with the boundary conditions

$$\phi_t(0) = aN, \quad \phi_t(N) = bN ,$$

where $f(r) = W'(r)$; we assume $Q \equiv 1$ so that $U(\theta, r) = W(r)$ for simplicity. Then the following result is expected on its hydrodynamic behavior.

*Assume $a \geq 0 \geq b$ and $A = \int_{\mathbb{R}} f(r)\,dr \geq 0$. Then, as $N \to \infty$, the macroscopically scaled height variable $h^N(t, \theta)$ converges to $h(t, \theta)$ in probability. The limit $h(t, \theta)$ is a solution of the **free boundary problem** for the nonlinear PDE* (10.4)

$$\frac{\partial h}{\partial t}(t, \theta) = \frac{\partial}{\partial \theta}\left\{\sigma'\left(\frac{\partial h}{\partial \theta}(t, \theta)\right)\right\} \quad on \quad \{(t, \theta); h(t, \theta) \neq 0\},$$

$$\Psi(h'_+(t, \theta)) - \Psi(h'_-(t, \theta)) = A \quad on \quad \{(t, \theta); h(t, \theta) = 0\},$$

$$h(t, 0) = a, \quad h(t, 1) = b,$$

where $\Psi(u) = \sigma'(u)u - \sigma(u)$ and, $h'_+(\theta)$ and $h'_-(\theta)$ are derivatives of h at θ from the positive and negative sides of h, respectively.

The above mentioned free boundary problem was studied by Caffarelli et al. [43, 44, 45].

15.2 Pinning Dynamics on a Wall

This subsection is taken from unpublished notes based on a discussion with J.-D. Deuschel. We construct the dynamics of microscopic interfaces under the effects of pinning and repulsion, and discuss its reversibility.

Dynamics Without Volume Conservation Law

Let $\Lambda \Subset \mathbb{Z}^d$ be given. For nonnegative height variables $\phi = \{\phi(x); x \in \Lambda\} \in \mathbb{R}^\Lambda_+$ on Λ, the Hamiltonian $H(\phi) \equiv H^\psi_\Lambda(\phi)$ with boundary conditions $\phi(x) = \psi(x) \geq 0, x \in \partial^+ \Lambda$ was introduced in (2.1). We consider the SDEs for $\phi_t = \{\phi_t(x); x \in \Lambda\}$

$$d\phi_t(x) = -1_{(0,\infty)}(\phi_t(x))\frac{\partial H}{\partial \phi(x)}(\phi_t)dt$$

$$+ 1_{(0,\infty)}(\phi_t(x)) \cdot \sqrt{2}dw_t(x) + d\ell_t(x), \quad x \in \Lambda, \quad (15.1)$$

subject to the conditions:

(a) $\phi_t(x) \geq 0, \quad \ell_t(x) \nearrow, \quad \ell_0(x) = 0,$

(b) $\displaystyle\int_0^\infty \phi_t(x)\,d\ell_t(x) = 0,$ (15.2)

(c) $\displaystyle c\ell_t(x) = \int_0^t 1_{\{0\}}(\phi_s(x))\,ds,$

for every $x \in \Lambda$. We shall choose $c = e^J (\geq 0)$ for $J \in [-\infty, \infty)$. The boundary conditions (2.12) at $y \in \partial^+ \Lambda$ is automatically imposed through the Hamiltonian H_Λ^ψ.

The first basic problems we should address are (1) construction and uniqueness of dynamics and (2) identification of invariant or reversible measures. For the problem (1) we refer to [149, 242, 243]. The case of $\Lambda = \mathbb{Z}^d$ should also be considered.

Reversibility

Set $\Omega^+(\Lambda) = \mathbb{R}_+^\Lambda$ and let $\mu_\Lambda^{\psi, J, +} \in \mathcal{P}(\Omega^+(\Lambda))$ be the finite volume φ-Gibbs measure with hard wall and δ-pinning defined by (7.13) (with D_N replaced by Λ and with boundary condition ψ), i.e.,

$$\mu_\Lambda^{\psi, J, +}(d\phi) = \frac{1}{Z_\Lambda^{\psi, J, +}} e^{-H_\Lambda^\psi(\phi)} \prod_{x \in \Lambda} \left(c \delta_0(d\phi(x)) + d\phi^+(x) \right) ,$$

where $c = e^J$ and $d\phi^+(x)$ stands for the Lebesgue measure on $\mathbb{R}_+$. We denote

$$\mathcal{A}(\phi) = \{x \in \Lambda; \phi(x) = 0\},$$

and

$$\mathcal{B}(\phi) = \{x \in \Lambda; \phi(x) > 0\} ,$$

for $\phi \in \Omega^+(\Lambda)$. The sets $\mathcal{A}(\phi)$ and $\mathcal{B}(\phi)$ represent dry and wet regions associated with the height variables ϕ, respectively. Then the space $\Omega^+(\Lambda)$ can be decomposed in two ways as

$$\Omega^+(\Lambda) = \bigcup_{A \subset \Lambda} \Omega_A^0 = \bigcup_{B \subset \Lambda} \Omega_B^+$$

where $\Omega_A^0 = \{\phi \in \Omega^+(\Lambda); \mathcal{A}(\phi) = A\}$ and $\Omega_B^+ = \{\phi \in \Omega^+(\Lambda); \mathcal{B}(\phi) = B\}$, respectively.

Let us return to the SDEs (15.1)–(15.2). The corresponding generators when ϕ_t moves on the region Ω_B^+ are given by

$$\mathcal{L}_B = \sum_{x \in B} \mathcal{L}_x$$

where

$$\mathcal{L}_x F(\phi) = e^{H(\phi)} \frac{\partial}{\partial \phi(x)} \left(e^{-H(\phi)} \frac{\partial F}{\partial \phi(x)} \right), \quad \phi \in \Omega^+(\Lambda) ,$$

for $F = F(\phi) \in C_b^2(\Omega^+(\Lambda))$. We simply denote $H(\phi)$ for $H_\Lambda^\psi(\phi)$. We set

$$\mathcal{L} = \mathcal{L}_\Lambda ,$$

which is the free generator without pinning nor repulsion. In order to glue Ω_B^+ to $\Omega_{B\cup\{x\}}^+$ at $\phi(x) = 0, x \notin B$, we need to introduce the boundary operator L by

$$LF(\phi, x) = \frac{1}{c}\frac{\partial F}{\partial \phi(x)} - \mathcal{L}_x F(\phi), \quad \phi \in \Omega^+(\Lambda), \ x \in \mathcal{A}(\phi) .$$

Note that $LF(\phi, x)$ can be rewritten as

$$LF(\phi, x) = \mathcal{L}_{\mathcal{B}(\phi)}F(\phi) + \frac{1}{c}\frac{\partial F}{\partial \phi(x)} - \mathcal{L}_{\mathcal{B}(\phi)\cup\{x\}}F(\phi) ,$$

and compare this expression with the boundary operator $Lf(x)$ defined by the formula (7.2) of [149], p.204. The gluing operators for $\Omega_{B\setminus C}^+$ with Ω_B^+ for $C \subset B$ is unnecessary if $|C| \geq 2$, since the direct transitions between such two sets never occur (more precisely, occur with probability 0) for ϕ_t.

Lemma 15.1. *For $F \in C_b^2(\Omega^+(\Lambda))$,*

$$F(\phi_t) - F(\phi_0) - \int_0^t \mathcal{L}F(\phi_t)\,dt - \sum_{x \in \Lambda}\int_0^t cLF(\phi_t, x)\,d\ell_t(x)$$

is a martingale.

Proof. Applying Itô's formula, we have

$$
\begin{aligned}
dF(\phi_t) &= \sum_{x \in \Lambda}\frac{\partial F}{\partial \phi(x)}(\phi_t)\,d\phi_t(x) + \sum_{x \in \Lambda}\frac{\partial^2 F}{\partial \phi(x)^2}(\phi_t)1_{(0,\infty)}(\phi_t(x))\,dt \\
&= \sum_{x \in \Lambda}1_{(0,\infty)}(\phi_t(x))\mathcal{L}_x F(\phi_t)\,dt + \sum_{x \in \Lambda}\frac{\partial F}{\partial \phi(x)}(\phi_t)\,d\ell_t(x) + dm_t \\
&= \mathcal{L}F(\phi_t)\,dt + \sum_{x \in \Lambda}cLF(\phi_t, x)\,d\ell_t(x) + dm_t ,
\end{aligned}
$$

where

$$m_t = \sum_{x \in \Lambda}\int_0^t 1_{(0,\infty)}(\phi_s(x)) \cdot \sqrt{2}\frac{\partial F}{\partial \phi(x)}(\phi_s)\,dw_s(x)$$

is a martingale. Note that $d\ell_t(x) = 0$ if $x \in \mathcal{B}(\phi_t)$ and $dt = cd\ell_t(x)$ if $x \in \mathcal{A}(\phi_t)$, which follow from the conditions (15.2)-(b) and (c), respectively. $\square$

Lemma 15.2. *Let $F, G \in C_{b,0}^2(\Omega^+(\Lambda))$ and assume that F satisfies the "boundary conditions" $LF(\phi, x) = 0$ for every $\phi \in \Omega^+(\Lambda)$ and $x \in \mathcal{A}(\phi)$. Then, we have*

$$\int_{\Omega^+(\Lambda)} G\mathcal{L}F\,d\mu_\Lambda^{\psi,J,+} = -\int_{\Omega^+(\Lambda)}\sum_{x \in \mathcal{B}(\phi)}\frac{\partial F}{\partial \phi(x)}\frac{\partial G}{\partial \phi(x)}\,d\mu_\Lambda^{\psi,J,+} . \qquad (15.3)$$

In particular, $\mu_\Lambda^{\psi,J,+}$ is reversible for $(\mathcal{L}, L)$-diffusion, cf. [149], p.204.

Proof. The probability measure $\mu_\Lambda^{\psi,J,+}$ admits a decomposition (recall Lemma 7.6):

$$\mu_\Lambda^{\psi,J,+}(\cdot) = \sum_{A \subset \Lambda} \nu(A) \mu_A^0(\cdot) \,,$$

where

$$\mu_A^0(d\phi) \equiv \mu_{A^c}^+(d\phi) = \frac{1}{Z_A^0} e^{-H(\phi)} \prod_{x \in A^c} d\phi^+(x) \prod_{x \in A} \delta_0(d\phi(x)) \,,$$

$$\nu(A) = \frac{c^{|A|} Z_A^0}{Z_\Lambda^{\psi,J,+}} = \mu_\Lambda^{\psi,J,+}(\mathcal{A}(\phi) = A) \,.$$

If $x \in A^c (\equiv \Lambda \setminus A)$, by the integration by parts,

$$\int_{\Omega^+(\Lambda)} G\mathcal{L}_x F \, \mu_A^0(d\phi) = -\int_{\Omega^+(\Lambda)} \frac{\partial G}{\partial \phi(x)} \frac{\partial F}{\partial \phi(x)} \mu_A^0(d\phi)$$

$$- \int_{\Omega^+(\Lambda)} G \cdot e^{-H} \left.\frac{\partial F}{\partial \phi(x)}\right|_{\phi(x)=0} \frac{1}{Z_A^0} \prod_{y \in A^c} d\phi^+(y) \prod_{y \in A} \delta_0(d\phi(y))$$

and the second term can be further rewritten as

$$- \int_{\Omega^+(\Lambda)} G \frac{\partial F}{\partial \phi(x)} \frac{Z_{A \cup \{x\}}^0}{Z_A^0} \mu_{A \cup \{x\}}^0(d\phi) \,.$$

Therefore,

$$\int_{\Omega^+(\Lambda)} G\mathcal{L}F \, \mu_\Lambda^{\psi,J,+}(d\phi) = \sum_{x \in \Lambda} \sum_{A \subset \Lambda} \nu(A) \int_{\Omega^+(\Lambda)} G\mathcal{L}_x F \, \mu_A^0(d\phi)$$

$$= -\sum_{x \in \Lambda} \sum_{A : x \in A^c} \nu(A) \int_{\Omega^+(\Lambda)} \frac{\partial G}{\partial \phi(x)} \frac{\partial F}{\partial \phi(x)} \mu_A^0(d\phi)$$

$$- \sum_{x \in \Lambda} \sum_{A : x \in A^c} \nu(A) \int_{\Omega^+(\Lambda)} G \frac{\partial F}{\partial \phi(x)} \frac{Z_{A \cup \{x\}}^0}{Z_A^0} \mu_{A \cup \{x\}}^0(d\phi)$$

$$+ \sum_{x \in \Lambda} \sum_{A : x \in A} \nu(A) \int_{\Omega^+(\Lambda)} G\mathcal{L}_x F \, \mu_A^0(d\phi) \,.$$

The first term coincides with the right hand side of (15.3). Setting $A' := A \cup \{x\}$ first and then writing A' by A again, the second term can be rewritten as

$$- \sum_{x \in \Lambda} \sum_{A : x \in A} \nu(A \setminus \{x\}) \int_{\Omega^+(\Lambda)} G \frac{\partial F}{\partial \phi(x)} \frac{Z_A^0}{Z_{A \setminus \{x\}}^0} \mu_A^0(d\phi)$$

$$= - \sum_{x \in \Lambda} \sum_{A : x \in A} c^{-1} \nu(A) \int_{\Omega^+(\Lambda)} G \frac{\partial F}{\partial \phi(x)} \mu_A^0(d\phi) \,.$$

Therefore, the sum of the second and the third terms becomes

$$\sum_{x \in \Lambda} \sum_{A:x \in A} \nu(A) \int_{\Omega^+(\Lambda)} G\left\{\mathcal{L}_x F - c^{-1}\frac{\partial F}{\partial \phi(x)}\right\} \mu_A^0(d\phi)$$

$$= -\int_{\Omega^+(\Lambda)} G \sum_{x \in \mathcal{A}(\phi)} LF(\phi, x)\, \mu_\Lambda^{\psi, J, +}(d\phi) = 0$$

by the boundary conditions.   □

Dynamics With Volume Conservation Law

Mixing the ideas behind the SDEs (10.30) and (15.1), one can introduce dynamics with conservation law by means of other SDEs

$$d\phi_t(x) = 1_{(0,\infty)}(\phi_t(x))\Delta\frac{\partial H}{\partial \phi(x)}(\phi_t)dt$$

$$+ 1_{(0,\infty)}(\phi_t(x)) \cdot \sqrt{2}dw_t^\Lambda(x) - \Delta d\ell_t^\Lambda(x), \quad x \in \Lambda, \tag{15.4}$$

subject to the conditions (15.2) with $\ell_t(x)$ replaced by $\ell_t^\Lambda(x)$. The operator Δ in the first and the last terms of the right hand side is the discrete Laplacian acting on the variable x and the Brownian motions $\{w_t^\Lambda(x); x \in \Lambda\}$ has a covariance structure

$$E[w_t^\Lambda(x)w_s^\Lambda(y)] = -\Delta_\Lambda(x,y) \cdot t \wedge s,$$

where $\Delta_\Lambda(x,y), x, y \in \Lambda$ is the kernel of the discrete Laplacian.

The first fundamental questions are the same as before, i.e., construction, uniqueness of the dynamics and the identification of all reversible measures. Then, an interesting question for the dynamics is the derivation of the motion of the Winterbottom shape, cf. Sect. 7.3.

Let us consider the SDEs (15.4) taking $\Lambda = \mathbb{T}_N^d$. The corresponding macroscopic height variables are defined by

$$h^N(t,\theta) = \frac{1}{N}\sum_{x \in \mathbb{T}_N^d} \phi_{N^\alpha t}(x)1_{B(\frac{x}{N}, \frac{1}{N})}(\theta), \quad \theta \in \mathbb{T}^d.$$

To pick up the correct scaling N^α in time, we take a test function $f = f(\theta) \in C^\infty(\mathbb{T}^d)$ and consider $\langle h^N(t), f \rangle$ as in Sect. 10.1-(c). Then, its martingale term is given (with small error) by

$$\frac{\sqrt{2}}{N^{d+1}}\sum_{x \in \mathbb{T}_N^d} f\left(\frac{x}{N}\right)1_{(0,\infty)}(\phi_{N^\alpha t}(x))w_{N^\alpha t}^\Lambda(x)$$

whose quadratic variational process is

$$= 2N^{\alpha-2d-2}t \sum_{x,y\in\mathbb{T}_N^d} f\left(\frac{x}{N}\right) f\left(\frac{y}{N}\right)$$

$$\times 1_{(0,\infty)}(\phi_{N^\alpha t}(x))\,(-\Delta(x,y))\,1_{(0,\infty)}(\phi_{N^\alpha t}(y))$$

$$\sim 2N^{\alpha-d-4}t\langle f1_{D_t}, (-\Delta)f1_{D_t}\rangle$$

where $D_t :=$ the support of the Winterbottom shape (arising in the limit) at time t and the last Δ is the continuum Laplacian on $\mathbb{T}^d$. Therefore, one can expect that the correct time scaling should be $\alpha = d+4$.

16 Other Dynamic Models

16.1 Stochastic Lattice Gas and Free Boundary Problems

Particle Systems on $\mathbb{Z}^d$

At sufficiently low temperature, physical systems exhibit phase transition phenomena. Suppose that more than one phase coexist in the initial state of the system. Then a boundary will separate the phases, and this phase boundary would move according to a proper evolutional rule. Determination of the motion of the phase boundary is called the problem of phase separation, dynamic phase transition, pattern formation, etc., and is analyzed in various kinds of situations. So far, we have been mostly concerned with the $\nabla\varphi$ interface model. This subsection briefly summarizes approaches from the particle systems.

Infinitely many particles are scattered over $\mathbb{Z}^d$ and each of them performs a random walk with jump rate determined from the surrounding configuration under the exclusion rule that at most one particle can occupy each site at each time. Such particle system is called the (stochastic) lattice gas or the Kawasaki dynamics. The model for generation and extinction of particles at each site is called the Glauber dynamics. The system taking all these effects (i.e., jumps, generations and extinctions) into account is called the Glauber-Kawasaki dynamics. Liggett [189, 190] are basic references for the interacting particle systems.

Starting from the Glauber dynamics and others, Spohn [240] studied a pattern formed after proper scaling and derived the motion by mean curvature. The argument there is rather heuristic, but contains several suggestive conjectures. Presutti and others [69, 129, 163], derived (isotropic) motion by mean curvature for the interfaces from the Glauber-Kawasaki dynamics. Since one can derive the reaction-diffusion equation from the Glauber-Kawasaki dynamics under proper hydrodynamic scaling limit (cf. [66, 70]) and the motion by mean curvature can be obtained from the reaction-diffusion equation under singular limit (cf. Sect. 16.3), these results are thought of as the two scalings are accomplished at once. In [67, 68, 164], the motion by mean curvature was derived from the Glauber dynamics corresponding to the Kac's type potential with long range interaction. The arising limits are nonrandom, while [69] treated the case that several random phase separation points appear. See a review paper by Giacomin et al. [134].

Free Boundary Problems

One phase or two phases Stefan free boundary problems are derived from the systems with two types of particles (e.g., A/B types) in which each type performs the Kawasaki dynamics possibly with different jump rates depending on the types and, if different type of particles meet, both of them disappear. One dimensional case was discussed by Chayes and Swindle [57], and Funaki [114] extended their results to higher dimensions taking the effect of latent heat into account. See Quastel [217], Landim et al. [181], Ben Arous and Ramírez [14], Gravner and Quastel [139] for investigations of systems with two types of particles. Komoriya [174] dealt with the case that, if different type of particles meet, they reflect. The result in [181] is applicable to the system of three types of particles called Potts model. See Ben Arous et al. [13] for internal DLA in a random environment.

16.2 Interacting Brownian Particles at Zero Temperature

The Wulff shape of interfaces for crystals is derived from the ferromagnetic Ising model as we have mentioned in Sect. 1.1; see also Remark 6.7. This is a static result so that a natural question arises: Can one derive the motion of the Wulff shape from the corresponding dynamic system? The Kawasaki dynamics, a system of interacting random walks on the lattice $\mathbb{Z}^d$, is indeed the stochastic evolutions which have the canonical Gibbs measure for the Ising model as an equilibrium state. Unfortunately, the Kawasaki dynamics has a certain technical difficulty because of their discrete nature at this moment. Instead, one can consider its continuum version, a system of interacting Brownian particles in $\mathbb{R}^d$. As we have seen (in Sect. 10), for analyzing the system, the structure of the Gibbs measures corresponding to such dynamics (with infinitely many particles) needs to be clarified first. However, if $d \geq 2$, this (and therefore the Wulff shape for this system) is not known for long except the simplest situation that the temperature of the system is zero. We shall therefore study the system under the zero temperature limit hoping that this would serve as the first stage toward a deeper analysis on the motion of the Wulff shape. This subsection reviews the results of [118, 119]. The model was suggested by D. Ioffe.

Model and Problems

The time evolution of the positions of interacting Brownian N particles in $\mathbb{R}^d$, denoted by $\mathbf{x}(t) = (x_i(t))_{i=1}^N \in (\mathbb{R}^d)^N$, is prescribed by the SDEs

$$dx_i(t) = -\frac{\beta}{2}\nabla_{x_i}H(\mathbf{x}(t))\,dt + dw_i(t), \quad 1 \le i \le N , \qquad (16.1)$$

where $\beta > 0$ represents the inverse temperature of the system and $(w_i(t))_{i=1}^N$ is a family of independent d dimensional Brownian motions. The Hamiltonian is the sum of pairwise interactions between particles

$$H(\mathbf{x}) = \sum_{1 \le i < j \le N} U(x_i - x_j)$$

and the gradient $\nabla_{x_i}H(\mathbf{x}) \equiv \sum_{j\ne i}\nabla U(x_i - x_j)$ is taken in the variable x_i. The potential $U(x) = U(|x|), x \in \mathbb{R}^d$ is radially symmetric and satisfies the following three conditions:

(U1) $U \in C_0^3(\mathbb{R})$, where $U(-r) := U(r)$,

(U2) U attains a unique minimum at $r = a > 0$ such that $U(a) = \min_{r>0} U(r)$,

(U3) $\check{c} = U''(a) > 0$.

The range of U is defined by $b = \inf\{r > 0; U(s) = 0 \text{ for every } s > r\}$.

  The problem is to study the zero temperature limit for the system (16.1). One can expect that the configurations $\mathbf{x}$ are crystallized (i.e., frozen) in an equal distance a as $\beta \to \infty$. More precisely, we shall study the following properties.

- Microscopic behavior: The structure of crystallization is kept under the time evolution except the isometric movement.
- Macroscopic behavior: The limits of translational and rotational motions are characterized.
- Coagulation of several crystals for one dimensional system.

Rigid Crystals

A configuration $\mathbf{z} = (z_i)_{i=1}^N \in (\mathbb{R}^d)^N$ is called a **crystal** if

$$|z_i - z_j| = a \quad \text{or} \quad |z_i - z_j| > b$$

for every $i \ne j$. (A certain condition is required on b for such $\mathbf{z}$ to exist.) For $\theta \in SO(d)$ and $\eta \in \mathbb{R}^d$, let $\varphi_{\theta,\eta}$ be an isometry on $\mathbb{R}^d$ or on $(\mathbb{R}^d)^N$ defined by $\varphi_{\theta,\eta}(y) = \theta y + \eta$ for $y \in \mathbb{R}^d$ and $\varphi_{\theta,\eta}(\mathbf{x}) = (\varphi_{\theta,\eta}(x_i))_{i=1}^N$ for $\mathbf{x} = (x_i)_{i=1}^N$. A crystal $\mathbf{z}$ is called **rigid** if the energy H increases under any perturbative transformations except isometries, i.e., if there exists $\delta > 0$ such that

$$H(\mathbf{x}) > H(\mathbf{z}) \quad \text{for every } \mathbf{x} \in \mathcal{M}(\delta) \setminus \mathcal{M} ,$$

where $\mathcal{M} \equiv \mathcal{M}_\mathbf{z} = \{\varphi_{\theta,\eta}(\mathbf{z}); \theta \in SO(d), \eta \in \mathbb{R}^d\}$ and $\mathcal{M}(\delta)$ is a δ-neighborhood of $\mathcal{M}$ in $(\mathbb{R}^d)^N$. The rigidity means that $\mathbf{z}$ has no internal degree of freedom except for the isometry. For example in 2 dimension, the three vertices of equilateral triangle form a rigid crystal, but the four vertices of square do not. The rigid crystal is a local minimum of H by definition, but not necessarily a global one.

We further introduce the notion of **infinitesimal rigidity**. Tangent space to $\mathcal{M}$ at $\mathbf{z}$ is defined by

$$\mathcal{H}_{\mathbf{z}} = \{X\mathbf{z} + h; X \in \mathfrak{so}(d), h \in \mathbb{R}^d\} \subset (\mathbb{R}^d)^N ,$$

where $X\mathbf{z}+h = (Xz_i+h)_{i=1}^N$ and $\mathfrak{so}(d) = \{X : d\times d \text{ real matrices}; X+{}^tX = 0\}$ is the Lie algebra of $SO(d)$. Let $\mathcal{H}_{\mathbf{z}}^{\perp}$ be the orthogonal subspace to $\mathcal{H}_{\mathbf{z}}$ in $(\mathbb{R}^d)^N$. Note that, if $\delta > 0$ is sufficiently small, $\mathbf{x} \in \mathcal{M}(\delta)$ admits a unique decomposition $\mathbf{x} = \mathbf{z}(\mathbf{x}) + \mathbf{h}(\mathbf{x})$ such that $\mathbf{z}(\mathbf{x}) \in \mathcal{M}$ and $\mathbf{h}(\mathbf{x}) \in \mathcal{H}_{\mathbf{z}(\mathbf{x})}^{\perp}$. The Hessian of H on $\mathcal{M}$ is given by

$$\mathcal{E}(\mathbf{h}) \equiv \mathcal{E}_{\mathbf{z}}(\mathbf{h}) = \frac{\check{c}}{a^2} \sum_{\langle i,j\rangle} (h_i - h_j, z_i - z_j)^2 ,$$

for $\mathbf{h} = (h_i)_{i=1}^N \in (\mathbb{R}^d)^N$. The sum $\langle i,j\rangle$ is taken over all pairs $\{i,j\}$ such that $|z_i - z_j| = a$. A crystal $\mathbf{z}$ is called infinitesimally rigid if

$$\mathcal{E}(\mathbf{h}) = 0 \Longleftrightarrow \mathbf{h} \in \mathcal{H}_{\mathbf{z}}$$

i.e., the Hessian is nondegenerate to the orthogonal direction.

Note that "$\mathbf{h} \in \mathcal{H}_{\mathbf{z}} \Rightarrow \mathcal{E}(\mathbf{h}) = 0$" is obvious, in fact, the translational invariance of H implies $\mathcal{E}(h) = 0$, while its rotational invariance implies $\mathcal{E}(X\mathbf{z}) = 0$. The infinitesimal rigidity implies the rigidity.

Microscopic Shape Theorem

The basic scaling parameter is the ratio $\epsilon =$ (microscopic spatial unit length)/(macroscopic spatial unit length) so that $(\epsilon x_i)_{i=1}^N$ is the macroscopic correspondence to the microscopic configuration $\mathbf{x} = (x_i)_{i=1}^N$. The number of particles and the inverse temperature will be rescaled in ϵ: $N = N(\epsilon), \beta = \beta(\epsilon)$. Let a sequence of infinitesimally rigid crystals $\{\mathbf{z}^{(\epsilon)} = (z_i^{(\epsilon)})_{i=1}^N; 0 < \epsilon < 1\}$ be given. For sufficiently small $c > 0$, the c-neighborhood of $\mathcal{M} = \mathcal{M}_{\mathbf{z}^{(\epsilon)}}$ is defined by $\mathcal{M}^{\nabla}(c) \equiv \mathcal{M}^{\nabla,N}(c) = \{\mathbf{x} \in (\mathbb{R}^d)^N; \|\nabla\mathbf{h}(\mathbf{x})\|_{\infty} \le c\}$, where $\mathbf{h}(\mathbf{x}) \in \mathcal{H}_{\mathbf{z}(\mathbf{x})}^{\perp}$ was determined by decomposing $\mathbf{x}$ and $\|\nabla\mathbf{h}\|_{\infty} = \sup_{\langle i,j\rangle} |h_i - h_j|$. We introduce the macroscopic time change for the solution $\mathbf{x}(t)$ of the SDEs (16.1)

$$\mathbf{x}^{(\epsilon)}(t) = \mathbf{x}(\epsilon^{-\kappa}t), \quad \kappa = d + 2 .$$

Theorem 16.1. [118] *Let $c(\epsilon) \downarrow 0$ (as $\epsilon \downarrow 0$) be given and assume that $\mathbf{x}^{(\epsilon)}(0) \in \mathcal{M}^{\nabla}(c'(\epsilon))$ for some $c'(\epsilon) \ll c(\epsilon)$ (i.e., the initial configuration is nearly an infinitesimally rigid crystal) and $\beta = \beta(\epsilon) \to \infty$ sufficiently fast as $\epsilon \downarrow 0$. Then, we have that*

$$\lim_{\epsilon \downarrow 0} P(\sigma^{(\epsilon)} \geq t) = 1 \,,$$

for every $t > 0$, where $\sigma^{(\epsilon)} = \inf\{t \geq 0; \mathbf{x}^{(\epsilon)}(t) \notin \mathcal{M}^{\nabla}(c(\epsilon))\}$. In other words, asymptotically with probability one $\mathbf{x}^{(\epsilon)}(t)$ keeps its rigidly crystallized shape within fluctuations $c(\epsilon)$.

For the proof of the theorem, Lyapunov type argument is applied combining with a spectral gap estimate for $\mathcal{E}(\mathbf{h})$.

Motion of a Macroscopic Body

We say that a sequence $\mathbf{x}^{(\epsilon)} = (x_i^{(\epsilon)})_{i=1}^N, N = N(\epsilon)$ of configurations has a **macroscopic density function** $\rho(y), y \in \mathbb{R}^d$ if

$$\epsilon^d \sum_{i=1}^N \delta_{\epsilon x_i^{(\epsilon)}}(dy) \Rightarrow \rho(y)\, dy$$

weakly as $\epsilon \downarrow 0$. The initial configurations $\mathbf{x}^{(\epsilon)}(0) = \mathbf{z}^{(\epsilon)}(= (z_i^{(\epsilon)})_{i=1}^N)$ are assumed to be infinitesimally rigid crystals with macroscopic density function $\rho(y)$ and $|z_i^{(\epsilon)}| \leq R\epsilon^{-1}$ for all i and ϵ and for some $R > 0$. The particles' number behaves as $N = N(\epsilon) \sim \bar{\rho}\epsilon^{-d}$, where $\bar{\rho} = \int_{\mathbb{R}^d} \rho(y)\, dy$. We may assume, by shifting the system if necessary, that the body is centered: $\int_{\mathbb{R}^d} y\rho(y)\, dy = 0$. Let $\bar{Q} = (\bar{q}^{\alpha\beta})_{1 \leq \alpha,\beta \leq d}$ be the matrix defined by

$$\bar{q}^{\alpha\beta} = \int_{\mathbb{R}^d} y^\alpha y^\beta \rho(y)\, dy \,,$$

where y^α denotes the αth component of y. We may assume by rotating the system that $\bar{Q}$ is diagonal. The sum

$$v^{\alpha\beta} = \bar{q}^{\alpha\alpha} + \bar{q}^{\beta\beta}$$

is called moments of inertia.

Theorem 16.2. [118] *Assume that $\beta = \beta(\epsilon) \to \infty$ sufficiently fast as $\epsilon \downarrow 0$. Then, $\mathbf{x}^{(\epsilon)}(t)$ has a macroscopic density function $\rho_t(y)$, which is congruent to $\rho(y)$, i.e., $\rho_t(y) = \rho\left(\varphi_{\theta(t),\eta(t)}^{-1}(y)\right)$. The translational and rotational motions $(\eta(t), \theta(t))$ of the limit body are random and characterized as follows:*

(1) *$\eta(t)$ and $\theta(t)$ are mutually independent.*
(2) *$\eta(t) = (d \text{ dimensional Brownian motion})/\sqrt{\bar{\rho}}$*
(3) *$\theta(t)$ is a Brownian motion on $SO(d)$ which is a solution of the SDE of Stratonovich's type*

$$d\theta(t) = \theta(t) \circ dm(t), \quad \theta(0) = I$$

where $m(t) = (m^{\alpha\beta}(t))$ is an $\mathfrak{so}(d)$-valued Brownian motion such that the components $\{m^{\alpha\beta}(t); \alpha < \beta\}$ in the upper half of the matrix $m(t)$ are mutually independent and $m^{\alpha\beta}(t) = (\text{one dimensional Brownian motion})/\sqrt{v^{\alpha\beta}}$.

Coagulation in One Dimension

Consider the SDEs (16.1) in $\mathbb{R}$ taking $\beta = \epsilon^{-\gamma}, \gamma > 0$:

$$dx_i(t) = -\frac{\epsilon^{-\gamma}}{2} \sum_{j \neq i} U'(x_i(t) - x_j(t))\, dt + dw_i(t), \quad 1 \leq i \leq N .$$

Theorems 16.1 and 16.2 have dealt with motion of single macroscopic body. In one dimension one can establish coagulation of several bodies (bodies are called rods in one dimension). We need additional assumptions on U such that the well at a is deep and located away from 0.

Since the rods evolve independently until they meet (i.e., until the time when the microscopic distance between two rods becomes b) and since the analysis of multiple rods can be essentially reduced to the two rods case, we assume the following two conditions on the initial configuration $\mathbf{x}(0)$:

(1) $\mathbf{x}(0) = \mathbf{x}^{(1)}(0) \cup \mathbf{x}^{(2)}(0)$ consists of two nearly rigid crystals (called chains in one dimension) with particles' numbers $N_1 = [\rho_1 \epsilon^{-1}]$, $N_2 = [\rho_2 \epsilon^{-1}]$ and fluctuation $\epsilon^\nu, \nu > 1/2$, i.e., $\mathbf{x}^{(\ell)}(0) \in \mathcal{M}^{\nabla, N_\ell}(\epsilon^\nu), \ell = 1, 2$, where $\rho_1, \rho_2 > 0$.
(2) The distance of these two chains (that between the right most particle of $\mathbf{x}^{(1)}(0)$ and the left most one of $\mathbf{x}^{(2)}(0)$) is b.

Theorem 16.3. [119] *Let* $\mathbf{x}^{(\epsilon)}(t) = \mathbf{x}(\epsilon^{-3}t)$ *be the macroscopically time changed process and assume the above condition on* $\mathbf{x}(0)$. *Take another* $\nu' > 0$ *and suppose that* $\gamma > \max\{4, 2\nu' + 3\}$. *Then, for every* $\delta > 0$, *we have*

$$\lim_{\epsilon \downarrow 0} P\left\{ \mathbf{x}^{(\epsilon)}(t) \in \mathcal{M}^{\nabla, N}(\epsilon^{\nu'}) \text{ for some } t \leq \epsilon^{1-\delta} \right\} = 1 ,$$

where $N = N_1 + N_2$.

This theorem claims that two rods in $\mathbf{x}^{(\epsilon)}(t)$ coagulate and form a single rod within a very short time $\epsilon^{1-\delta}$. After the coagulation $\mathbf{x}^{(\epsilon)}(t)$ moves as a single rod as we have seen in Theorem 16.1.

16.3 Singular Limits for Stochastic Reaction-Diffusion Equations

In this subsection we refer to the results concerning a reaction-diffusion equation with additive noise, i.e., (16.2) below. See the survey paper [113] for more details. The equation (16.2) involves a small parameter $\epsilon > 0$ representing the temperature of the system; $\epsilon = \beta^{-1}$ in the last subsection. The spatial variable x is already macroscopic. Taking the zero temperature limit $\epsilon \downarrow 0$ again, or a singular limit in mathematical terminology, we expect that the solution of the equation converges under an appropriate time change to a point at which the potential appearing in the nonlinear reaction term is minimized. Points at which the potential attains its minimum, simply called minimal points or bottoms, correspond to phases in the physical context. If we have more than one phase, phase separation will occur.

Reaction-diffusion equations have been used to describe various kinds of phenomena, including dynamical phase transitions. In connection with microscopic particle systems, we note that the equation treated here is considered to describe an intermediate (mesoscopic) level, between microscopic and macroscopic. Indeed, the reaction-diffusion equation can be derived from the Glauber-Kawasaki dynamics by means of the hydrodynamic limit (as we have referred in Sect. 16.1), and the noise term naturally appears in the fluctuation problem.

Stochastic Reaction-Diffusion Equations

The following reaction-diffusion equation with noise is considered in this subsection:

$$\frac{\partial u}{\partial t} = \Delta u + \frac{1}{\epsilon} f(u) + \dot{w}^\epsilon(t, x), \quad t > 0, \, x \in D , \qquad (16.2)$$

where $\dot{w}^\epsilon(t, x)$ is a noise depending on $\epsilon > 0$ and D is a domain in $\mathbb{R}^d$. We assume that the reaction term $f \in C^\infty(\mathbb{R})$ is bistable:

$$\exists u_* \in (-1, 1) \text{ such that } f(\pm 1) = f(u_*) = 0, \ f'(\pm 1) < 0, \ f'(u_*) > 0 ,$$

and fulfills a technical condition:

$$\exists C, p > 0 \text{ such that } |f(u)| \leq C(1 + |u|^p) \text{ and } \sup_u f'(u) < \infty ,$$

which ensures the existence and uniqueness of the solution of (16.2) for noises we shall treat. Two values ± 1 are stable points of f and u_* is an unstable point. A function F satisfying $f = -F'$ is the associated potential. Our goal is to study the behavior of the solution $u = u^\epsilon(t, x)$ of (16.2) as $\epsilon \downarrow 0$. When the reaction term dominates the noise term, multiplying both sides by ϵ yields

$$\lim_{\epsilon \downarrow 0} f(u^\epsilon) = 0$$

formally, while the unstable solution is not considered to appear in the limit. So we guess that

$$\lim_{\epsilon \downarrow 0} u^\epsilon(t, x) = +1 \text{ or } -1$$

depending on (t, x). In other words, a random boundary separating $+1$ and -1 might appear. To find the motion of the boundary is the main problem which we discuss here. The results in the absence of noise (i.e., in the case $\dot{w}^\epsilon = 0$) are summarized in [113].

Singular Limits

(a) *The Case where $d = 1, D = \mathbb{R}$ and $\dot{w}^\epsilon(t, x) = \epsilon^\gamma a(x)\dot{w}_h(t, x)$*

Here $\gamma > 0$, and $a \in C_0^2(\mathbb{R})$ is a function representing the magnitude of the noise at each point. The condition that a has a compact support is introduced so that the problem can be localized and the boundary conditions $u^\epsilon(t, \pm\infty) = \pm 1$ of the equation (16.2) at $x = \pm\infty$ hold. $\dot{w}_h, 1/2 \leq h \leq 1$, is a self-similar Gaussian noise, i.e., the covariance structure of the noise is formally given by

$$\langle \dot{w}_h(t, x)\dot{w}_h(s, y)\rangle = \delta_0(t - s)Q_h(x - y),$$

where δ_0 is the δ-function at 0 and Q_h is the Riesz potential of order $2h - 1$:

$$Q_h(x) = \begin{cases} h(2h - 1)|x|^{2h-2}, & 1/2 < h \leq 1, \\ \delta_0(x) & , & h = 1/2. \end{cases}$$

In particular, $\dot{w}_{1/2}(t, x)$ is the space-time white noise and $\dot{w}_1(t, x) = \dot{w}(t)$ is the one parameter white noise independent of the spatial variable. Under the above conditions, the stochastic PDE (16.2) has a unique solution $u^\epsilon(t, x)$ in the sense of mild solutions or generalized functions. Although the solution is not differentiable, it is Hölder continuous:

$$u^\epsilon(t, x) \in \cap_{\delta>0} C^{\frac{h}{2}-\delta, h-\delta}((0, \infty) \times \mathbb{R}), \quad \text{a.s.}$$

Under suitable assumptions on the initial condition and an assumption of symmetry of the reaction term, i.e., f is odd: $f(u) = -f(-u)$, we can prove the following theorem. The function $m = m(y)$ is a standing wave solution to the reaction-diffusion equation, i.e. a solution of the ODE (ordinary differential equation) on $\mathbb{R}$

$$m'' + f(m) = 0, \quad y \in \mathbb{R},$$

satisfying $m(\pm\infty) = \pm 1$. Since m is determined uniquely up to parallel displacement, we normalize it as $m(0) = 0$.

Theorem 16.4. [110, 112] *There exists $\bar{\gamma}(h) > 0$ such that for all $\gamma \geq \bar{\gamma}(h)$*

$$\bar{u}^\epsilon(t, x) := u^\epsilon(\epsilon^{-2\gamma-h}t, x) \underset{\epsilon\downarrow 0}{\Longrightarrow} \chi_{\xi_t}(x),$$

where $\chi_\xi(x) = 1_{\{x>\xi\}} - 1_{\{x<\xi\}}$. The motion of the phase separating point ξ_t that appeared in the limit is governed by the SDE

$$d\xi_t = \alpha_1 a(\xi_t)dB_t + \alpha_2 a(\xi_t)a'(\xi_t)dt, \tag{16.3}$$

where B_t is a one dimensional Brownian motion and the constants $\alpha_1 = \alpha_1(h)$ and $\alpha_2 = \alpha_2(h)$ are given by

$$\alpha_1^2 = \frac{1}{\|m'\|_{L^2(\mathbb{R})}^4} \int_{\mathbb{R}^2} m'(x)m'(y)Q_h(x-y)\,dxdy\,,$$

$$\alpha_2 = -\frac{1}{\|m'\|_{L^2(\mathbb{R})}^2} \int_0^\infty dt \int_{\mathbb{R}^3} xp(t,x,z)p(t,y,z)$$

$$\times f''(m(z))m'(z)Q_h(x-y)\,dxdydz\,,$$

respectively. The function $p(t,x,y)$ is the fundamental solution of the linearized operator $\partial/\partial t - \{\partial^2/\partial y^2 + f'(m(y))\}$.

A related result was obtained in the microscopic situation by Presutti et al. [36]; see also [35]. In physics, the case $h = 1/2$ (that is, the space-time white noise) is important, and it is of particular interest to identify the diffusion coefficient governing the motion of the random boundary in the limit. We see from Theorem 16.4 that the diffusion coefficient (also called mobility) of the limit SDE is

$$\{\alpha_1(1/2)\}^2 = \|m'\|_{L^2(\mathbb{R})}^{-2}$$

if $h = 1/2$. Here $\|m'\|_{L^2(\mathbb{R})}^2$ is called the surface tension in this model. This result is consistent with conjectures of Kawasaki and Ohta [166] and Spohn [240].

(b) *The Case where $d = 2$ and D is a Bounded Domain in $\mathbb{R}^2$ with Smooth Boundary*

In (a) we considered the one dimensional case, while results for higher dimensional case, especially the two dimensional case, are described here. Consider (16.2) with Neumann boundary condition: $\partial u/\partial n = 0$ $(x \in \partial D)$. Here the assumption that f is an odd function is not necessary; instead, we merely assume $A(f) \equiv \int_{-1}^1 f(u)\,du = 0$. For simplicity the noise is assumed to be independent of the spatial variable, depending on only the time variable: $\dot{w}^\epsilon(t,x) = \xi_t^\epsilon/\sqrt{\epsilon}$, where $\xi_t^\epsilon = \epsilon^{-\gamma}\xi(\epsilon^{-2\gamma}t), 0 < \gamma < 1/3$, and $\xi(t) \in C^1(\mathbb{R}_+)$, a.s. is a mean 0 stationary process with the strong mixing property. Roughly speaking, $\xi_t^\epsilon \Rightarrow \alpha \dot{w}_t$ as $\epsilon \downarrow 0$, where w_t is a one dimensional Brownian motion and α is a constant given by the Green-Kubo formula

$$\alpha = \sqrt{2 \int_0^\infty E[\xi(0)\xi(t)]\,dt}\,.$$

Then, under suitable initial conditions, the following holds for the solution $u^\epsilon(t,x)$ of (16.2).

Theorem 16.5. [115] *As long as the phase separation curve Γ_t in the limit is strictly convex and does not touch the boundary ∂D, we have*

$$u^\epsilon(t,x) \underset{\epsilon \downarrow 0}{\Longrightarrow} \chi_{\Gamma_t}(x)\,,$$

where the motion of the curve Γ_t is given by a random perturbation of the curvature flow:

$$V = \kappa + (c_0 \alpha) \dot{w}_t . \tag{16.4}$$

Here, V is the inner normal speed of Γ_t, κ denotes the curvature of Γ_t, and c_0 is given by

$$c_0 = \sqrt{2} \Big/ \int_{-1}^{1} du \sqrt{\int_{u}^{1} f(v)\, dv} .$$

Remark 16.1. *In the case where $\dot{w}^\epsilon(t,x)$ is the space-time white noise and $D = \mathbb{R}^3$, Kawasaki and Ohta [166, 209] considered (16.2) and derived a random mean curvature flow describing the motion of the phase boundary, calling the limit the* **drumhead model.** *However, equation (16.2) does not have a mathematical meaning in dimensions greater than one, and the proof of existence of the solution is impossible. Other physical papers on the derivation of kinetic equations of interfaces or analysis of their equilibrium states include [4, 82, 211]. Among the literatures concerning interface curves we mention [142], and its probabilistic version is discussed in [162].*

Remark 16.2. (1) *Problems discussed in Sects. 16.2 and 16.3 can be viewed as convergence of solutions of SDEs in an infinite (or large) dimensional space toward a submanifold where the minimum energy is attained. Such problems are treated in [120, 165] in finite dimensional spaces.*
(2) *A related problem of a large deviation principle is studied in [25], etc., especially for the case where minimal points of the rate functional are not unique, and [97] discussed the case of stochastic PDEs for which exactly two minimal points exist.*

Remark 16.3. *The zero temperature limit and the metastable behavior of the Glauber dynamics (under the periodic boundary conditions) or the Kawasaki dynamics (in an infinite gas reservoir) were investigated by Ben Arous and Cerf [11] and by den Hollander et al. [65], respectively.*

16.4 Limit Shape of Random Young Diagrams

An asymptotic shape of typical Young diagrams is studied by Vershik [250], cf. Kerov [169]. The similarity of this approach to the Wulff problem in the Ising model is discussed by Shlosman [233]. The related stochastic dynamics is analyzed by several authors; especially, a randomly growing Young diagram by Johansson.

Model

For each $N \in \mathbb{N}$, an array $p = (n_1, n_2, \ldots, n_k)$ of positive integers is called a partition of N if $n_1 \geq n_2 \geq \cdots \geq n_k$ and $\sum_{i=1}^{k} n_i = N$ for some $1 \leq k \leq N$.

The number of groups of size i in the partition p is defined by $r_i = \sharp\{j; n_j = i\}$, $1 \leq i \leq N$. Then a Young diagram is associated with p by

$$\varphi_p(x) = \sum_{i > x} r_i, \quad x \geq 0,$$

which is the number of groups with size larger than x. The function φ_p is a nonincreasing step function satisfying $\varphi_p(0) = k, \varphi_p(x) = 0$ for $x \geq n_1$ and $\int_0^\infty \varphi_p(x)\,dx = N$. More precisely, the region $A = A_p$ surrounded by the step function φ_p and the boundary $\partial\mathbb{R}_+^2$ of quadrant is called the Young diagram; note that each A is built by piling up unit cubes in $\mathbb{R}_+^2$.

Let $\mathcal{P}_N$ be the set of all partitions p of N. Vershik [250] introduced several kinds of statistics (like finite volume Gibbs measures) on the set $\mathcal{P}_N$, which leads to random Young diagrams. The most typical and natural one is the uniform probability on it, sometimes called Bose statistics, defined by $\mu_B^N(\{p\}) = 1/\sharp\mathcal{P}_N$ for every $p \in \mathcal{P}_N$. Fermi statistics is the uniform probability μ_F^N on the set of strict partitions p such that $n_1 > n_2 > \cdots > n_k$.

Scaling and Law of Large Numbers

Introduce the scaling for Young diagrams in two dimensions as

$$\tilde{\varphi}_p^N(x) = \frac{1}{\sqrt{N}}\varphi_p(\sqrt{N}x), \quad x \geq 0,$$

so that $\int_0^\infty \tilde{\varphi}_p^N(x)\,dx = 1$. Then, as $N \to \infty$, the LLN holds and the limit shapes of $\tilde{\varphi}_p^N(x)$ are given by curves in $\mathbb{R}_+^2$:

$$\Phi_B(x) = -\frac{1}{\alpha}\log\left(1 - e^{-\alpha x}\right), \quad \alpha = \frac{\pi}{\sqrt{6}},$$

$$\Phi_F(x) = \frac{1}{\beta}\log\left(1 + e^{-\beta x}\right), \quad \beta = \frac{\pi}{\sqrt{12}},$$

under μ_B^N and μ_F^N, respectively; see [250]. Three dimensional case called skyscraper problem is discussed by Cerf and Kenyon [51]; see also [167]. The limit shape is a surface in $\mathbb{R}_+^3$, which has flat pieces (called facets) embedded on $\partial\mathbb{R}_+^3$ (being consisted of three quadrants $\mathbb{R}_+^2$).

Large Deviation

The corresponding LDP is established by Dembo, Vershik and Zeitouni [71] for μ_B^N and μ_F^N in two dimension. The speed of the LDP is $\sqrt{N}$ and its rate functionals I_B and I_F are explicitly specified. In particular, the limit shapes Φ_B and Φ_F are the solutions of the variational problems for I_B and I_F, respectively. Shlosman [233] discussed the analogy between the Wulff problem arising from the Ising model and the variational problem for I_B including higher dimensional case. Especially, he observes that the latter problem has a similar solution to the Wulff construction.

For the proof, the conditioning approach of Fristedt is effectively used. For instance, for Bose statistics, we have the following representation. Let $\{Z_j\}_{j=1,2,\ldots,N}$ be a sequence of independent $\mathbb{Z}_+$-valued random variables with geometric distributions of parameter q^j, respectively. Then, under conditioning that $\sum_{j=1}^{N} jZ_j = N$, $\{Z_j\}_j$ has the same distribution as $\{r_j\}_j$ distributed under μ_B^N for every $q \in (0,1)$. This corresponds to the equivalence of ensemble in statistical mechanics.

Central Limit Theorem

The corresponding CLT is established by Pittel [214]. In fact, under μ_B^N,

$$V_N(x) = N^{-1/4} x \left\{ \varphi_p \left(\sqrt{N} \alpha^{-1} \log 1/(1-x) \right) - \sqrt{N} \alpha^{-1} \log 1/x \right\}, \quad x \in [0,1]$$

weakly converges to a certain Gaussian process $V(x)$.

Dynamic Approach

Let Ω_N be the set of all $r = (r_1, r_2, \ldots, r_N)$ determined from partitions $p \in \mathcal{P}_N$. Durrett, Granovsky and Gueron [94] introduced a stochastic evolution of $r_t, t \geq 0$ called the coagulation-fragmentation process for each N. It is a Markov chain on Ω_N determined by the rates of coagulation of two groups of sizes i and j into a single group of size $i+j$ and those of fragmentation of a group into two groups. The reversibility of the process r_t is characterized under a proper assumption on the rates (see also [140]) and an explicit formula of the invariant measure is found. It has a conditioning representation of Fristedt's type but the distributions of random variables $\{Z_j\}_j$ should be replaced from geometric to Poissonian.

Norris [206, 207] proved the LLN (a kind of mean field limit) for coagulation process (without fragmentation) under a suitable scaling as $N \to \infty$. The limit is described by the Smoluchowski's coagulation equation, which is a nonlinear equation associated with an infinite system. From the view point of the Young diagrams, his scaling is in a different regime, i.e., the vertical size behaves as $O(N)$, while the horizontal one is kept at $O(1)$ as $N \to \infty$.

Randomly Growing Young Diagram

Let $\mathcal{Y} = \cup_{N=1}^{\infty} \{\varphi_p; p \in \mathcal{P}_N\}$ be the set of all Young diagrams. A randomly growing Young diagram (sometimes called the corner growth model) is a Markov chain $\varphi(t), t = 1, 2, \ldots$ on $\mathcal{Y}$ defined by the following transition rule. We call a unit cube $Q \subset \mathbb{R}_+^2$ a corner outside of $\varphi \in \mathcal{Y}$ if the left and lower sides of Q are both on φ or on $\partial \mathbb{R}_+^2$. Then, $\varphi(t+1)$ is obtained by picking each corner outside of $\varphi(t)$ independently with probability $p \in (0, 1)$ and adding those corners to $\varphi(t)$. The explicit form of the asymptotic shape φ_0 of $\frac{1}{t}\varphi(t)$ as $t \to \infty$ is known. This is a result of LLN type. Johansson [158] (see also [7] and a review paper [159]) further studied the fluctuation of $\varphi(t)$ around $t\varphi_0$ and found that the limit distribution is described by the Tracy-Widom largest eigenvalue distribution arising in the random matrix theory. The model with continuous time parameter $\{\varphi(t); t \geq 0\}$ can be discussed in a parallel way.

The randomly growing Young diagram is equivalent to the totally asymmetric one dimensional exclusion process (first observed by Rost, see also [90, 215]), the last-passage directed percolation model and it is related to the Hammersley's model [157], domino tilings and dimer model [160]. The distribution of the last-passage time in the directed percolation model, in particular, behaves asymptotically as $N \to \infty$ as

$$c_1 N + c_2 N^{1/3} X \ ,$$

where c_1 and c_2 are certain positive constants, N is the size of the system and X is a random variable with Tracy-Widom distribution; [99, 158]. In recent years, it is realized that this type of asymptotic behavior arises universally in several related models. Prähofer and Spohn [216] found a connection in the so-called polynuclear growth (PNG) model studying the fluctuations from its limit semi-circular shape.

Results for the (undirected) first-passage percolation, including an asymptotic shape and fluctuations of percolated region, are reviewed by [148]. See [197] for the directed first-passage and last-passage percolation models.

16.5 Growing Interfaces

Krug and Spohn [177] give a nice review on growing interfaces, including Eden models, SOS models, ballistic deposition models, KPZ equation, directed polymer, percolations, DLA (nonlocal model) and others.

SOS Dynamics

The SOS type dynamics is the φ-dynamics $\phi = \{\phi_t(x)\}$ with values in $\mathbb{Z}$. If $\phi_t(x)$ is nondecreasing in t for each x, it defines the growing interfaces or sometimes called marching soldiers. In one dimension, under the constraint that $\nabla\phi(x) = 0$ or -1 on the height differences, the associated $\nabla\varphi$-dynamics is the totally asymmetric exclusion processes as we have pointed out in the last subsection. For more general case of height differences $\nabla\phi(x) \in \mathbb{Z}$, $\nabla\varphi$-dynamics is the zero-range processes. The hyperbolic space-time scaling limit has been studied by Seppäläinen and Rezakhanlou; cf. Remark 13.1. Fritz [104, 105] extended the Tartar-Murat theory of compensated compactness to prove the hydrodynamic limit (LLN); cf. [182]. Varadhan [248] established the LDP for the asymmetric exclusion process.

KPZ Equation

A model for interfaces growing by depositions of particles, which randomly fall from the ambient atmosphere on the surface, was introduced by Kardar, Parisi and Zhang [161]. The dynamics of the surface is described by a stochastic PDE called KPZ equation:

$$\frac{\partial}{\partial t} h(t, x) = \Delta h(t, x) + |\nabla h(t, x)|^2 + \dot{w}(t, x) \ ,$$

where $h(t, x)$ is the height of the surface and $\dot{w}(t, x)$ is the space-time white noise; see also [176].

References

1. D. ABRAHAM, *Surface structures and phase transitions – exact results*, In: Phase Transitions and Critical Phenomena, **10** (1986), pp. 1–74, Academic Press.
2. D. ABRAHAM, P. COLLET, J. DE CONINCK AND F. DUNLOP, *Langevin dynamics of an interface near a wall*, J. Statis. Phys., **61** (1990), pp. 509–532.
3. K.S. ALEXANDER, *Cube-root boundary fluctuations for droplets in random cluster models*, Commun. Math. Phys., **224** (2001), pp. 733–781.
4. S.M. ALLEN AND J.W. CAHN, *A microscopic theory for antiphase boundary motion and its application to antiphase domain coarsening*, Acta Metall., **27** (1979), pp. 1085–1095.
5. H.W. ALT AND L.A. CAFFARELLI, *Existence and regularity for a minimum problem with free boundary*, J. Reine Angew. Math., **325** (1981), pp. 105–144.
6. H.W. ALT, L.A. CAFFARELLI AND A. FRIEDMAN, *Variational problems with two phases and their free boundaries*, Trans. Amer. Math. Soc., **282** (1984), pp. 431–461.
7. J. BAIK, P. DEIFT AND K. JOHANSSON, *On the distribution of the length of the longest increasing subsequence of random permutations*, J. Amer. Math. Soc., **12** (1999), pp. 1119–1178.
8. B. BAKRY AND M. EMERY, *Diffusions hypercontractives*, In: Séminaire de Probabilités XIX, Lect. Notes Math., **1123** (1985), Springer, pp. 179–206.
9. J.M. BALL, *A version of the fundamental theorem for Young measures*, in *PDEs and continuum models of phase transitions (Nice, 1988)*, Springer, Lecture Notes in Physics, **344** (1989), pp. 207–215.
10. V. BARBU, *Nonlinear Semigroups and Differential Equations in Banach Spaces*, Noordhoff, 1976.
11. G. BEN AROUS AND R. CERF, *Metastability of the three-dimensional Ising model on a torus at very low temperatures*, Electron. J. Probab., **1** (1996), approx. 55 pp. (electronic).
12. G. BEN AROUS AND J.-D. DEUSCHEL, *The construction of the d + 1-dimensional Gaussian droplet*, Commun. Math. Phys., **179** (1996), pp. 467–488.
13. G. BEN AROUS, J. QUASTEL AND A.F. RAMÍREZ, *Internal DLA in a random environment*, Ann. Inst. Henri Poincaré, **39** (2003), pp. 301–324.

14. G. BEN AROUS AND A.F. RAMÍREZ, *Asymptotic survival probabilities in the random saturation process*, Ann. Probab., **28** (2000), pp. 1470–1527.
15. G. BENFATTO, E. PRESUTTI AND M. PULVIRENTI, *DLR measures for one-dimensional harmonic systems*, Z. Wahr. verw. Gebiete, **41** (1978), pp. 305–312.
16. A. BENSOUSSAN AND J.L. LIONS, *Applications of Variational Inequalities in Stochastic Control*, North-Holland, 1982.
17. L. BERTINI, C. LANDIM AND S. OLLA, *Derivation of Cahn-Hilliard equations from Ginzburg-Landau models*, J. Statis. Phys., **88** (1997), pp. 365–381.
18. D. BERTACCHI AND G. GIACOMIN, *Enhanced interface repulsion from quenched hard-wall randomness*, Probab. Theory Relat. Fields, **124** (2002), pp. 487–516.
19. D. BERTACCHI AND G. GIACOMIN, *Wall repulsion and mutual interface repulsion: an harmonic crystal model in high dimensions*, Stoch. Proc. Appl., **110** (2004), pp. 45–66.
20. T. BODINEAU, *The Wulff construction in three and more dimensions*, Commun. Math. Phys., **207** (1999), pp. 197–229.
21. T. BODINEAU, G. GIACOMIN AND Y. VELENIK, *On entropic reduction of fluctuations*, J. Statist. Phys., **102** (2001), pp. 1439–1445.
22. T. BODINEAU, D. IOFFE AND Y. VELENIK, *Rigorous probabilistic analysis of equilibrium crystal shapes*, J. Math. Phys., **41** (2000), pp. 1033–1098.
23. T. BODINEAU, D. IOFFE AND Y. VELENIK, *Winterbottom construction for finite range ferromagnetic models: an $\mathbb{L}_1$-approach*, J. Statist. Phys., **105** (2001), pp. 93–131.
24. D. BOIVIN AND Y. DERRIENNIC, *The ergodic theorem for additive cocycles of $\mathbb{Z}^d$ or $\mathbb{R}^d$*, Ergod. Th. Dynam. Sys., **11** (1991), pp. 19–39.
25. E. BOLTHAUSEN, *Laplace approximations for sums of independent random vectors, Part II. Degenerate maxima and manifolds of maxima*, Probab. Theory Relat. Fields, **76** (1987), pp. 167–206.
26. E. BOLTHAUSEN, *Large deviations and perturbations of random walks and random surfaces*, European Congress of Mathematics, Vol. I (Budapest, 1996), pp. 108–120, Progr. Math., **168**, Birkhäuser, 1998.
27. E. BOLTHAUSEN AND J.-D. DEUSCHEL, *Critical large deviations for Gaussian fields in the phase transition regime I*, Ann. Probab., **21** (1993), pp. 1876–1920.
28. E. BOLTHAUSEN, J.-D. DEUSCHEL AND G. GIACOMIN, *Entropic repulsion and the maximum of two dimensional harmonic crystal*, Ann. Probab., **29** (2001), pp. 1670–1692.
29. E. BOLTHAUSEN, J.-D. DEUSCHEL AND O. ZEITOUNI, *Entropic repulsion of the lattice free field*, Commun. Math. Phys., **170** (1995), pp. 417–443 (Erratum: ibid, **209** (2000), pp. 547–548).
30. E. BOLTHAUSEN, J.-D. DEUSCHEL AND O. ZEITOUNI, *Absence of a wetting transition for a pinned harmonic crystal in dimensions three and larger*, J. Math. Phys., **41** (2000), pp. 1211–1223.
31. E. BOLTHAUSEN AND D. IOFFE, *Harmonic crystal on the wall: a microscopic approach*, Commun. Math. Phys., **187** (1997), pp. 523–566.
32. E. BOLTHAUSEN AND Y. VELENIK, *Critical behavior of the massless free field at the depinning transition*, Commun. Math. Phys., **223** (2001), pp. 161–203.
33. H.J. BRASCAMP AND E.H. LIEB, *On extensions of the Brunn-Minkowski and Prékopa-Leindler theorems, including inequalities for log concave functions, and with an application to the diffusion equation*, J. Funct. Anal., **22** (1976), pp. 366–389.

34. H.J. BRASCAMP, E.H. LIEB AND J.L. LEBOWITZ, *The statistical mechanics of anharmonic lattices*, Bull. Int. Statis. Inst., 1975, pp. 393–404.
35. S. BRASSESCO AND P. BUTTA, *Interface fluctuations for the D = 1 stochastic Ginzburg-Landau equation with nonsymmetric reaction term*, J. Statist. Phys., **93** (1998), pp. 1111–1142.
36. S. BRASSESCO, A. DE MASI AND E. PRESUTTI, *Brownian fluctuations of the instanton in the d = 1 Ginzburg-Landau equation with noise*, Ann. Inst. Henri Poincaré, **31** (1995), pp. 81–118.
37. H. BRÉZIS, *Operateurs Maximaux Monotones et Semi-Groupes de Contractions dans les Espaces de Hilbert*, North-Holland, Amsterdam, 1973.
38. J. BRICMONT, A. EL MELLOUKI AND J. FRÖHLICH, *Random surfaces in statistical mechanics: roughening, rounding, wetting,...*, J. Statist. Phys., **42** (1986), pp. 743–798.
39. J. BRICMONT, J.-R. FONTAINE, J.L. LEBOWITZ AND T. SPENCER, *Lattice systems with a continuous symmetry, I. Perturbation theory for unbounded spins*, Commun. Math. Phys., **78** (1980), pp. 281–302.
40. J. BRICMONT, J.-R. FONTAINE, J.L. LEBOWITZ AND T. SPENCER, *Lattice systems with a continuous symmetry, II. Decay of correlations*, Commun. Math. Phys., **78** (1981), pp. 363–371.
41. D. BRYDGES, J. FRÖHLICH AND T. SPENCER, *The random walk representation of classical spin systems and correlation inequalities*, Commun. Math. Phys., **83** (1982), pp. 123–150.
42. D. BRYDGES AND H.-T. YAU, *Grad φ perturbations of massless Gaussian fields*, Commun. Math. Phys., **129** (1990), pp. 351–392.
43. L.A. CAFFARELLI AND J.L. VÁZQUEZ, *A free-boundary problem for the heat equation arising in flame propagation*, Trans. Amer. Math. Soc., **347** (1995), pp. 411–441.
44. L.A. CAFFARELLI, C. LEDERMAN AND N. WOLANSKI, *Uniform estimates and limits for a two phase parabolic singular perturbation problem*, Indiana Univ. Math. J., **46** (1997), pp. 453–489.
45. L.A. CAFFARELLI, C. LEDERMAN AND N. WOLANSKI, *Pointwise and viscosity solutions for the limit of a two phase parabolic singular perturbation problem*, Indiana Univ. Math. J., **46** (1997), pp. 719–740.
46. P. CAPUTO AND J.-D. DEUSCHEL, *Large deviations and variational principle for harmonic crystals*, Commun. Math. Phys., **209** (2000), pp. 595–632.
47. P. CAPUTO AND J.-D. DEUSCHEL, *Critical large deviations in harmonic crystals with long-range interactions*, Ann. Probab., **29** (2001), pp. 242–287.
48. P. CAPUTO AND D. IOFFE, *Finite volume approximation of the effective diffusion matrix: the case of independent bond disorder*, Ann. Inst. H. Poincaré, **39** (2003), pp. 505–525.
49. P. CAPUTO AND Y. VELENIK, *A note on wetting transition for gradient fields*, Stoch. Proc. Appl., **87** (2000), pp. 107–113.
50. E.A. CARLEN, S. KUSUOKA AND D.W. STROOCK, *Upper bounds for symmetric Markov transition functions*, Ann. Inst. H. Poincaré, **23** (1987), pp. 245–287.
51. R. CERF AND R. KENYON, *The low-temperature expansion of the Wulff crystal in the 3D Ising model*, Commun. Math. Phys., **222** (2001), pp. 147–179.
52. R. CERF AND A. PISZTORA, *On the Wulff crystal in the Ising model*, Ann. Probab., **28** (2000), pp. 947–1017.
53. R. CERF AND A. PISZTORA, *Phase coexistence in Ising, Potts and percolation models*, Ann. Inst. H. Poincaré, **37** (2001), pp. 643–724.

54. F. CESI AND F. MARTINELLI, *On the layering transition of an SOS surface interacting with a wall, I. equilibrium results*, J. Statis. Phys., **82** (1996), pp. 823–913.

55. F. CESI AND F. MARTINELLI, *On the layering transition of an SOS surface interacting with a wall, II. the Glauber dynamics*, Commun. Math. Phys., **177** (1996), pp. 173–201.

56. C.C. CHANG AND H.-T. YAU, *Fluctuations of one dimensional Ginzburg-Landau models in nonequilibrium*, Commun. Math. Phys., **145** (1992), pp. 209–239.

57. L. CHAYES AND G. SWINDLE, *Hydrodynamic limits for one-dimensional particle systems with moving boundaries*, Ann. Probab., **24** (1996), pp. 559–598.

58. H. COHN, R. KENYON AND J. PROPP, *A variational principle for domino tilings*, J. Amer. Math. Soc., **14** (2001), pp. 297–346.

59. P. COLLET, F. DUNLOP, J. FRITZ AND T. GOBRON, *Langevin dynamics of a semi-infinite interface*, Markov Processes Rel. Fields, **3** (1997), pp. 261–274.

60. O. DAVIAUD, *Extremes of the discrete two-dimensional Gaussian free field*, math.PR/0406609, preprint, 2004.

61. J. DE CONINCK, F. DUNLOP AND V. RIVASSEAU, *On the microscopic validity of the Wulff construction and of the generalized Young equation*, Commun. Math. Phys., **121** (1989), pp. 401–419.

62. J. DE CONINCK, S. MIRACLE-SOLÉ AND J. RUIZ, *Rigorous generalization of Young's law for heterogeneous and rough substrates*, J. Statis. Phys., **111** (2003), pp. 107–127.

63. T. DELMOTTE AND J.-D. DEUSCHEL, *On estimating the derivatives of symmetric diffusions in stationary random environment*, preprint, 2002.

64. T. DELMOTTE AND J.-D. DEUSCHEL, *Algebraic L^2 convergence rates for $\nabla\phi$ interface models*, preprint, 2003.

65. F. DEN HOLLANDER, F.R. NARDI, E. OLIVIERI AND E. SCOPPOLA, *Droplet growth for three-dimensional Kawasaki dynamics*, Probab. Theory Relat. Fields, **125** (2003), pp. 153–194.

66. A. DE MASI, P.A. FERRARI AND J.L. LEBOWITZ, *Reaction diffusion equations for interacting particle systems*, J. Statis. Phys., **44** (1986), pp. 589–644.

67. A. DE MASI, E. ORLANDI, E. PRESUTTI AND L. TRIOLO, *Motion by curvature by scaling nonlocal evolution equations*, J. Statis. Phys., **73** (1993), pp. 543–570.

68. A. DE MASI, E. ORLANDI, E. PRESUTTI AND L. TRIOLO, *Glauber evolution with Kac potentials, I. Mesoscopic and macroscopic limits, interface dynamics*, Nonlinearity, **7** (1994), pp. 633–696.

69. A. DE MASI, A. PELLEGRINOTTI, E. PRESUTTI AND M.E. VARES, *Spatial patterns when phases separate in an interacting particle system*, Ann. Probab., **22** (1994), pp. 334–371.

70. A. DE MASI AND E. PRESUTTI, *Mathematical Methods for Hydrodynamic Limits*, Lect. Notes Math., **1501** (1991), Springer.

71. A. DEMBO, A. VERSHIK AND O. ZEITOUNI, *Large deviations for integer partitions*, Markov Processes Relat. Fields, **6** (2000), pp. 147–179.

72. A. DEMBO AND O. ZEITOUNI, *Large Deviations Techniques and Applications*, 2nd edition, Applications of Mathematics **38**, Springer, 1998.

73. X. DESCOMBES AND E. PECHERSKY, *Droplet shapes for a class of models in $\mathbb{Z}^2$ at zero temperature*, J. Statis. Phys., **111** (2003), pp. 129–169.

74. J.-D. DEUSCHEL, *Entropic repulsion of the lattice free field, II. The 0-boundary case*, Commun. Math. Phys., **181** (1996), pp. 647–665.

75. J.-D. DEUSCHEL AND G. GIACOMIN, *Entropic repulsion for the free field: path-wise characterization in $d \geq 3$*, Commun. Math. Phys., **206** (1999), pp. 447–462.

76. J.-D. DEUSCHEL AND G. GIACOMIN, *Entropic repulsion for massless fields*, Stoch. Proc. Appl., **89** (2000), pp. 333–354.

77. J.-D. DEUSCHEL, G. GIACOMIN AND D. IOFFE, *Large deviations and concentration properties for $\nabla\varphi$ interface models*, Probab. Theory Relat. Fields, **117** (2000), pp. 49–111.

78. J.-D. DEUSCHEL, G. GIACOMIN AND L. ZAMBOTTI, *Scaling limits of equilibrium wetting models in (1+1)-dimension*, preprint, 2004.

79. J.-D. DEUSCHEL AND D.W. STROOCK, *Large Deviations*, Academic Press, 1989.

80. J.-D. DEUSCHEL AND T. NISHIKAWA, The dynamic of entropic repulsion, in preparation, 2004.

81. J.-D. DEUSCHEL AND Y. VELENIK, *Non-Gaussian surface pinned by a weak potential*, Probab. Theory Relat. Fields, **116** (2000), pp. 359–377.

82. H.W. DIEHL, D.M. KROLL AND H. WAGNER, *The interface in a Ginsburg-Landau-Wilson model: Derivation of the drumhead model in the low-temperature limit*, Z. Physik B, **36** (1980), pp. 329–333.

83. R.L. DOBRUSHIN, *The description of a random field by means of conditional probabilities and conditions of its regularity*, Theor. Probab. Appl., **13** (1968), pp. 197–224.

84. R.L. DOBRUSHIN, *Prescribing a system of random variables by conditional distributions*, Theor. Probab. Appl., **15** (1970), pp. 458–486.

85. R.L. DOBRUSHIN AND O. HRYNIV, *Fluctuations of shapes of large areas under paths of random walks*, Probab. Theory Relat. Fields, **105** (1996), pp. 423–458.

86. R.L. DOBRUSHIN, R. KOTECKÝ AND S. SHLOSMAN, *Wulff Construction: a Global Shape from Local Interaction*, AMS translation series, **104** (1992).

87. R.L. DOBRUSHIN, S.B. SHLOSMAN, *Thermodynamic inequalities for the surface tension and the geometry of the Wulff construction*, "Ideas and methods in quantum and statistical physics" (Oslo, 1988, eds. Albeverio, Holden and Lindstom), Cambridge Univ. Press, 1992, pp. 461–483.

88. M.D. DONSKER AND S.R.S. VARADHAN, *Large deviations from a hydrodynamic scaling limit*, Commun. Pure Appl. Math., **42** (1989), pp. 243–270.

89. H. DOSS AND G. ROYER, *Processus de diffusion associé aux mesures de Gibbs sur $\mathbb{R}^{\mathbb{Z}^d}$*, Z. Wahr. verw. Gebiete, **46** (1978), pp. 107–124.

90. F. DUNLOP, *Stationary states and scaling shapes of one-dimensional interfaces*, J. Statis. Phys., **111** (2003), pp. 433–442.

91. F. DUNLOP, P.A. FERRARI AND L.R.G. FONTES, *A dynamic one-dimensional interface interacting with a wall*, J. Statis. Phys., **107** (2002), pp. 705–727.

92. F. DUNLOP, J. MAGNEN AND V. RIVASSEAU, *Mass generation for an interface in the mean field regime*, Ann. Inst. H. Poincaré Phys. Théor., **57** (1992), pp. 333–360.

93. F. DUNLOP, J. MAGNEN, V. RIVASSEAU AND P. ROCHE, *Pinning of an interface by a weak potential*, J. Statis. Phys., **66** (1992), pp. 71–98.

94. R. DURRETT, B.L. GRANOVSKY AND S. GUERON, *The equilibrium behavior of reversible coagulation-fragmentation processes*, J. Theoret. Probab., **12** (1999), pp. 447–474.

95. L.C. EVANS AND F. REZAKHANLOU, *A stochastic model for growing sandpiles and its continuum limit*, Commun. Math. Phys., **197** (1998), pp. 325–345.

96. G. EYINK, J.L. LEBOWITZ AND H. SPOHN, *Lattice gas models in contact with stochastic reservoirs: local equilibrium and relaxation to the steady state*, Commun. Math. Phys., **140** (1991), pp. 119–131.

97. W.G. FARIS AND G. JONA-LASINIO, *Large fluctuations for a nonlinear heat equation with noise*, J. Phys. A: Math. Gen., **15** (1982), pp. 3025–3055.

98. P.A. FERRARI, L.R.G. FONTES, B. NIEDERHAUSER AND M. VACHKOVSKAIA, *The serial harness interacting with a wall*, Stoch. Proc. Appl., **114** (2004), pp. 175–190.

99. P.L. FERRARI AND H. SPOHN, *Last branching in directed last passage percolation*, Markov Processes Relat. Fields, **9** (2003), pp. 323–339.

100. R. FERNÁNDEZ, J. FRÖHLICH AND A.D. SOKAL, *Random Walks, Critical Phenomena, and Triviality in Quantum Field Theory*, Springer, Berlin, 1992.

101. M.E. FISHER, *Walks, walls, wetting, and melting*, J. Statis. Phys., **34** (1984), pp. 667–729.

102. J.R. FONTAINE, *Non-perturbative methods for the study of massless models*, in "Scaling and Self-similarity in Physics" (ed. Fröhlich), 1983, pp. 203–226.

103. L.R.G. FONTES, M. VACHKOVSKAIA AND A. YAMBARTSEV, *Repulsion of an evolving surface on random walls*, math.PR/0405186, preprint, 2004.

104. J. FRITZ, *An Introduction to the Theory of Hydrodynamic Limits*, Lectures in Mathematical Sciences **18**, The University of Tokyo, Tokyo 2001.

105. J. FRITZ, *Entropy pairs and compensated compactness for weakly asymmetric systems*, Proceedings of Shonan/Kyoto meetings "Stochastic Analysis on Large Scale Interacting Systems" (2002, eds. Funaki and Osada), Adv. Stud. Pure Math., **39**, Math. Soc. Japan, 2004, pp. 143–171.

106. J. FRÖHLICH, C.E. PFISTER AND T. SPENCER, *On the statistical mechanics of surfaces*, Stochastic processes in quantum theory and statistical physics (Marseille, 1981), Lect. Notes Phys., **173**, pp. 169–199, Springer, 1982.

107. J. FRÖHLICH AND T. SPENCER, *The Kosterlitz-Thouless transition in two-dimensional abelian spin systems and the Coulomb gas*, Commun. Math. Phys., **81** (1981), pp. 527–602.

108. T. FUNAKI, *Random motion of strings and related stochastic evolution equations*, Nagoya Math. J., **89** (1983), pp. 129–193.

109. T. FUNAKI, *The reversible measures of multi-dimensional Ginzburg-Landau type continuum model*, Osaka J. Math., **28** (1991), pp. 463–494.

110. T. FUNAKI, *The scaling limit for a stochastic PDE and the separation of phases*, Probab. Theory Relat. Fields, **102** (1995), pp. 221–288.

111. T. FUNAKI, *SPDE approach and log-Sobolev inequalities for continuum field with two-body interactions*, Proceedings of Japan-German Seminar, 1996, Hiroshima, unpublished manuscript.

112. T. FUNAKI, *Singular limit for reaction-diffusion equation with self-similar Gaussian noise*, Proceedings of Taniguchi symposium "New Trends in Stochastic Analysis" (eds. Elworthy, Kusuoka and Shigekawa), World Sci., 1997, pp. 132–152.

113. T. FUNAKI, *Stochastic models for phase separation and evolution equations of interfaces*, Sugaku Expositions, **16** (2003), pp. 97–116 (In Japanese: **50** (1998), pp. 68–85).

114. T. FUNAKI, *Free boundary problem from stochastic lattice gas model*, Ann. Inst. H. Poincaré, **35** (1999), pp. 573–603.

115. T. FUNAKI, *Singular limit for stochastic reaction-diffusion equation and generation of random interfaces*, Acta Math. Sinica, **15** (1999), pp. 407–438.

116. T. FUNAKI, *Recent results on the Ginzburg-Landau* $\nabla\phi$ *interface model*, In: "Hydrodynamic Limits and Related Topics", edited by S. Feng, A.T. Lawniczak and S.R.S. Varadhan, Fields Institute Communications and Monograph Series, 2000, pp. 71–81.

117. T. FUNAKI, *Hydrodynamic limit for* $\nabla\phi$ *interface model on a wall*, Probab. Theory Relat. Fields, **126** (2003), pp. 155–183.

118. T. FUNAKI, *Zero temperature limit for interacting Brownian particles, I. Motion of a single body*, Ann. Probab., **32** (2004), pp. 1201–1227.

119. T. FUNAKI, *Zero temperature limit for interacting Brownian particles, II. Coagulation in one dimension*, Ann. Probab., **32** (2004), pp. 1228–1246.

120. T. FUNAKI AND H. NAGAI, *Degenerative convergence of diffusion process toward a submanifold by strong drift*, Stochastics, **44** (1993), pp. 1–25.

121. T. FUNAKI AND T. NISHIKAWA, *Large deviations for the Ginzburg-Landau* $\nabla\phi$ *interface model*, Probab. Theory Relat. Fields, **120** (2001), pp. 535–568.

122. T. FUNAKI AND S. OLLA, *Fluctuations for* $\nabla\phi$ *interface model on a wall*, Stoch. Proc. Appl., **94** (2001), pp. 1–27.

123. T. FUNAKI AND H. SAKAGAWA, *Large deviations for* $\nabla\varphi$ *interface model and derivation of free boundary problems*, Proceedings of Shonan/Kyoto meetings "Stochastic Analysis on Large Scale Interacting Systems" (2002, eds. Funaki and Osada), Adv. Stud. Pure Math., **39**, Math. Soc. Japan, 2004, pp. 173–211.

124. T. FUNAKI AND H. SPOHN, *Motion by mean curvature from the Ginzburg-Landau* $\nabla\phi$ *interface model*, Commun. Math. Phys., **185** (1997), pp. 1–36.

125. T. FUNAKI AND K. UCHIYAMA, *From Micro to Macro 1, Mathematical Theory of Interface Models*, in Japanese. Springer-Verlag Tokyo, 2002, 283+xi pages.

126. T. FUNAKI, K. UCHIYAMA AND H.T. YAU, *Hydrodynamic limit for lattice gas reversible under Bernoulli measures*, "Nonlinear Stochastic PDE's: Hydrodynamic Limit and Burgers' Turbulence" (eds. Funaki and Woyczynski), IMA volume (Univ. Minnesota) **77** (1995), pp. 1–40, Springer.

127. K. GAWEDZKI AND A. KUPIAINEN, *Renormalization group study of a critical lattice model I, II*, Commun. Math. Phys., **82** (1981), pp. 407–433 and **83** (1982), pp. 469–492.

128. H.-O. GEORGII, *Gibbs Measures and Phase Transitions*, Walter, Berlin New York, 1988.

129. G. GIACOMIN, *Onset and structure of interfaces in a Kawasaki+Glauber interacting particle system*, Probab. Theory Relat. Fields, **103** (1995), pp. 1–24.

130. G. GIACOMIN, *Anharmonic lattices, random walks and random interfaces*, in Recent research developments in statistical physics, Transworld research network, **1** (2000), pp. 97–118.

131. G. GIACOMIN, *On stochastic domination in the Brascamp-Lieb framework*, Math. Proc. Cambridge Philos. Soc., **134** (2003), pp. 507–514.

132. G. GIACOMIN, *Limit theorems for random interfaces of Ginzburg-Landau type*, Stochastic partial differential equations and applications (Trento, 2002), pp. 235–253, Lecture Notes in Pure and Appl. Math., **227**, Dekker, New York, 2002.

133. G. GIACOMIN, *Aspects of statistical mechanics of random surfaces*, Notes of the lectures given at IHP, fall 2001, preprint, 2002.

134. G. GIACOMIN, J.L. LEBOWITZ AND E. PRESUTTI, *Deterministic and stochastic hydrodynamic equations arising from simple microscopic model systems*, In: Stochastic Partial Differential Equations: Six Perspectives (eds R. Carmona and B. Rozovskii), pp. 107–152, Math. Surveys Monogr., **64**, AMS, 1999.

135. G. GIACOMIN, S. OLLA AND H. SPOHN, *Equilibrium fluctuations for* $\nabla\varphi$ *interface model*, Ann. Probab., **29** (2001), pp. 1138–1172.

136. M.-H. GIGA AND Y. GIGA, *Evolving graphs by singular weighted curvature*, Arch. Rational Mech. Anal., **141** (1998), pp. 117–198.

137. M.-H. GIGA AND Y. GIGA, *A PDE Approach for Motion of Phase-Boundaries by a Singular Interfacial Energy*, Proceedings of Shonan/Kyoto meetings "Stochastic Analysis on Large Scale Interacting Systems" (2002, eds. Funaki and Osada), Adv. Stud. Pure Math., **39**, Math. Soc. Japan, 2004, pp. 213–232.

138. J. GLIMM AND A. JAFFE, *Quantum Physics, a Functional Integral Point of View*, Springer, 1981.

139. J. GRAVNER AND J. QUASTEL, *Internal DLA and the Stefan problem*, Ann. Probab., **28** (2000), pp. 1528–1562.

140. S. GUERON AND Y. RUBINSTEIN, *The minimal reversible coagulation-fragmentation process having no factorized coagulation and fragmentation rates*, Markov Processes Relat. Fields, **6** (2000), pp. 257–264.

141. M.Z. GUO, G.C. PAPANICOLAOU AND S.R.S. VARADHAN, *Nonlinear diffusion limit for a system with nearest neighbor interactions*, Commun. Math. Phys., **118** (1988), pp. 31–59.

142. M.E. GURTIN, *Thermomechanics of Evolving Phase Boundaries in the Plane*, Oxford Press, 1993.

143. B. HELFFER, *Remarks on decay of correlations and Witten Laplacians, Brascamp-Lieb inequalities and semiclassical limit*, J. Funct. Anal., **155** (1998), pp. 571–586.

144. B. HELFFER AND J. SJÖSTRAND, *On the correlation for Kac-like models in the convex case*, J. Statis. Phys., **74** (1994), pp. 349–409.

145. C. HIERGEIST AND R. LIPOWSKY, *Local contacts of membranes and strings*, Physica A, **244** (1997), pp. 164–175.

146. Y. HIGUCHI, J. MURAI AND J. WANG, *The Dobrushin-Hryniv theory for the two-dimensional lattice Widom-Rowlinson model*, Proceedings of Shonan/Kyoto meetings "Stochastic Analysis on Large Scale Interacting Systems" (2002, eds. Funaki and Osada), Adv. Stud. Pure Math., **39**, Math. Soc. Japan, 2004, pp. 233–281.

147. P.C. HOHENBERG AND B.I. HALPERIN, *Theory of dynamic critical phenomena*, Rev. Mod. Phys., **49** (1977), pp. 435–475.

148. C.D. HOWARD, *Models of first-passage percolation*, in "Probability on Discrete Structures" (ed. Kesten), Encyclopaedia Math. Sci., **110** (2004), Springer, pp. 125–173.

149. N. IKEDA AND S. WATANABE, *Stochastic Differential Equations and Diffusion Processes*, North Holland/Kodansha, 1981 (2nd edition 1989).

150. D. IOFFE, *Large deviations for the 2D Ising model: a lower bound without cluster expansions*, J. Statis. Phys., **74** (1994), pp. 411–432.

151. D. IOFFE, *Exact large deviation bounds up to* T_c *for the Ising model in two dimensions*, Probab. Theory Relat. Fields, **102** (1995), pp. 313–330.

152. D. IOFFE AND R.H. SCHONMANN, *Dobrushin-Kotecký-Shlosman theorem up to the critical temperature*, Commun. Math. Phys., **199** (1998), pp. 117–167.

153. D. IOFFE AND Y. VELENIK, *A note on the decay of correlations under* δ-*pinning*, Probab. Theory Relat. Fields, **116** (2000), pp. 379–389.

154. Y. ISOZAKI AND N. YOSHIDA, *Weakly pinned random walk on the wall: pathwise descriptions of the phase transition*, Stoch. Proc. Appl., **96** (2001), pp. 261–284.

155. K. ITÔ AND H.P. MCKEAN, *Potentials and the random walk*, Illinois J. Math., **4** (1960), pp. 119–132.

156. K. IWATA, *Reversible measures of a $P(\phi)_1$-time evolution*, "Probabilistic Methods in Mathematical Physics" (eds. Itô and Ikeda), Proceedings, Katata Kyoto 1985, pp. 195–209, Kinokuniya, Tokyo, 1987.

157. K. JOHANSSON, *Transversal fluctuations for increasing subsequences on the plane*, Probab. Theory Relat. Fields, **116** (2000), pp. 445–456.

158. K. JOHANSSON, *Shape fluctuations and random matrices*, Commun. Math. Phys., **209** (2000), pp. 437–476.

159. K. JOHANSSON, *Random growth and random matrices*, European Congress of Mathematics, Vol. I (Barcelona, 2000), pp. 445–456, Progr. Math., **201**, Birkhäuser, Basel, 2001.

160. K. JOHANSSON, *Non-intersecting paths, random tilings and random matrices*, Probab. Theory Relat. Fields, **123** (2002), pp. 225–280.

161. M. KARDAR, G. PARISI AND Y.-C. ZHANG, *Dynamical scaling of growing interfaces*, Phys. Rev. Lett., **56** (1986), pp. 889–892.

162. T. KASAI, *Curvature flow added random forces*, Nagoya Univ. Master Thesis, 1994.

163. M.A. KATSOULAKIS AND P.E. SOUGANIDIS, *Interacting particle systems and generalized evolution of fronts*, Arch. Rat. Mech. Anal., **127** (1994), pp. 133–157.

164. M.A. KATSOULAKIS AND P.E. SOUGANIDIS, *Generalized motion by mean curvature as a macroscopic limit of stochastic Ising models with long range interactions and Glauber dynamics*, Commun. Math. Phys., **169** (1995), pp. 61–97.

165. G.S. KATZENBERGER, *Solutions of a stochastic differential equation forced onto a manifold by a large drift*, Ann. Probab., **19** (1991), pp. 1587–1628.

166. K. KAWASAKI AND T. OHTA, *Kinetic drumhead model of interface I*, Prog. Theoret. Phys., **67** (1982), pp. 147–163.

167. R. KENYON, *The planar dimer model with boundary: a survey*, "Directions in Mathematical Quasicrystals" (eds. Baake and Moody), Amer. Math. Soc., Providence, RI 2000, pp. 307–328.

168. R. KENYON, *Dominos and the Gaussian free field*, Ann. Probab., **29** (2001), pp. 1128–1137.

169. S.V. KEROV, *Asymptotic Representation Theory of the Symmetric Group and its Applications in Analysis*, Translations of Math. Monographs, **219**, Amer. Math. Soc., Providence, RI, 2003, xvi+201 pp.

170. C. KIPNIS AND C. LANDIM, *Scaling Limits of Interacting Particle Systems*, Springer, 1999.

171. C. KIPNIS, C. LANDIM AND S. OLLA, *Macroscopic properties of a stationary non-equilibrium distribution for a non-gradient interacting particle system*, Ann. Inst. H. Poincaré, **31** (1995), pp. 191–221.

172. C. KIPNIS, S. OLLA AND S.R.S. VARADHAN, *Hydrodynamics and large deviation for simple exclusion processes*, Commun. Pure Appl. Math., **42** (1989), pp. 115–137.

173. C. KIPNIS AND S.R.S. VARADHAN, *Central limit theorem for additive functionals of reversible Markov processes and applications to simple exclusions*, Commun. Math. Phys., **104** (1986), pp. 1–19.

174. K. KOMORIYA, *Free boundary problem from two component system on $\mathbb{Z}$*, to appear in J. Math. Sci. Univ. Tokyo, **12** (2005), pp. 141–163.

175. S.M. KOZLOV, *The method of averaging and walks in inhomogeneous environments*, Russian Math. Surveys, **40** (1985), pp. 73–145.

176. J. KRUG AND H. SPOHN, *Universality classes for deterministic surface growth*, Phys. Rev. A, **38** (1988), pp. 4271–4283.

177. J. KRUG AND H. SPOHN, *Kinetic roughening of growing surfaces*, in "Solids Far from Equilibrium" (ed. Godrèche), Cambridge Univ. Press, 1991, pp. 479–582.

178. N.V. KRYLOV AND B.L. ROZOVSKI, *Stochastic evolution equations*, J. Soviet Math., **16** (1981), pp. 1233–1277.

179. H. KÜNSCH, *Decay of correlations under Dobrushin's uniqueness condition and its applications*, Commun. Math. Phys., **84** (1982), pp. 207–222.

180. O.A. LADYZENSKAYA, V.A. SOLONNIKOV AND N.N. URAL'CEVA, *Linear and quasilinear equations of parabolic type*, Translations of Math. Monographs, **23**, Amer. Math. Soc., Providence, RI, 1968.

181. C. LANDIM, S. OLLA AND S. VOLCHAN, *Driven tracer particle in one dimensional symmetric simple exclusion*, Commun. Math. Phys., **192** (1998), pp. 287–307.

182. C. LANDIM, S. OLLA AND H.T. YAU, *First order correction for the hydrodynamic limit of asymmetric simple exclusion processes in dimension $d \geq 3$*, Commun. Pure Appl. Math., **50** (1997), pp. 149–203.

183. C. LANDIM, S. SETHURAMAN AND S. VARADHAN, *Spectral gap for zero-range dynamics*, Ann. Probab., **24** (1996), pp. 1871–1902.

184. C. LANDIM AND H.-T. YAU, *Large deviations of interacting particle systems in infinite volume*, Commun. Pure Appl. Math., **48** (1995), pp. 339–379

185. G.F. LAWLER, *Intersections of Random Walks*, Birkhäuser, Boston, 1991.

186. J.L. LEBOWITZ AND C. MAES, *The effect of an external field on an interface, entropic repulsion*, J. Statis. Phys., **46** (1987), pp. 39–49.

187. P. LENZ AND R. LIPOWSKY, *Stability of droplets and channels on homogeneous and structured surfaces*, Eur. Phys. J. E, **1** (2000), pp. 249–262.

188. P. LENZ, W. FENZL AND R. LIPOWSKY, *Wetting of ring-shaped surface domains*, Europhys. Lett., **53** (2001), pp. 618–624.

189. T.M. LIGGETT, *Interacting Particle Systems*, Springer, 1985.

190. T.M. LIGGETT, *Stochastic interacting systems: contact, voter and exclusion processes*, Springer, 1999.

191. P.L. LIONS AND A.-S. SZNITMAN, *Stochastic differential equations with reflecting boundary conditions*, Commun. Pure Appl. Math., **37** (1984), pp. 511–537.

192. R. LIPOWSKY, *Structured surfaces and morphological wetting transitions*, Interface Science **9** (2001), pp. 105–115.

193. R. LIPOWSKY, *Morphological wetting transitions at chemically structured surfaces*, Current Opinion in Colloid and Interface Science **6** (2001), pp. 40–48.

194. S. LU, *Hydrodynamic scaling limits with deterministic initial configurations*, Ann. Probab., **23** (1995), pp. 1831–1852.

195. S.L. LU AND H.-T. YAU, *Spectral gap and logarithmic Sobolev inequality for Kawasaki and Glauber dynamics*, Commun. Math. Phys., **156** (1993), pp. 399–433.

196. J. MAGNEN AND R. SÉNÉOR, *The infrared behavior of $(\nabla \varphi)^4_3$*, Ann. Phys., **152** (1984), pp. 136–202.

197. J.B. MARTIN, *Limiting shape for directed percolation models*, Ann. Probab., **32** (2004), pp. 2908–2937.

198. P. MASSART, *Concentration inequalities and model selection*, "Ecole d'Eté de Probabilités de Saint-Flour XXXIII-2003" (ed. Picard), Lect. Notes Math., Springer, 2005.

199. A. MESSAGER, S. MIRACLE-SOLÉ AND J. RUIZ, *Convexity properties of the surface tension and equilibrium crystals*, J. Statis. Phys., **67** (1992), pp. 449–470.

200. A.A. MOGUL'SKII, *Large deviations for trajectories of multi-dimensional random walks*, Theory Probab. Appl., **21** (1976), pp. 300–315.

201. S. MÜLLER, *Variational models for microstructure and phase transitions*, In: Calculus of variations and geometric evolution problems (Cetraro, 1996), Lect. Notes Math., **1713** (1999), Springer, pp. 85–210.

202. A. NADDAF AND T. SPENCER, *On homogenization and scaling limit of some gradient perturbations of a massless free field*, Commun. Math. Phys., **183** (1997), pp. 55–84.

203. T. NISHIKAWA, *Hydrodynamic limit for the Ginzburg-Landau $\nabla\phi$ interface model with a conservation law*, J. Math. Sci. Univ. Tokyo., **9** (2002), pp. 481–519.

204. T. NISHIKAWA, *Hydrodynamic limit for the Ginzburg-Landau $\nabla\phi$ interface model with boundary conditions*, Probab. Theory Relat. Fields, **127** (2003), pp. 205–227.

205. Y. NISHIURA, *Far-from-equilibrium dynamics*, Translations of Math. Monographs, **209**, Iwanami Series in Modern Math., Amer. Math. Soc., Providence, RI, 2002, xx+311 pp.

206. J. NORRIS, *Smoluchowski's coagulation equation: uniqueness, non-uniqueness and a hydrodynamic limit for the stochastic coalescent*, Ann. Appl. Prob., **9** (1999), pp. 78–109.

207. J. NORRIS, *Cluster coagulation*, Commun. Math. Phys., **209** (2000), pp. 407–435.

208. D. NUALART AND E. PARDOUX, *White noise driven quasilinear SPDEs with reflection*, Probab. Theory Relat. Fields, **93** (1992), pp. 77–89.

209. T. OHTA, *Instability of interfaces and pattern formation*, Frontiers of Physics, **10**, Kyôritsu, 1985, in Japanese.

210. T. OHTA, *Mathematical theory on interfacial dynamics*, Nihon-Hyôronsha, 1997, in Japanese.

211. T. OHTA, D. JASNOW AND K. KAWASAKI, *Universal scaling in the motion of random interfaces*, Physical Review Letters, **49** (1982), pp. 1223–1226.

212. C.-E. PFISTER AND Y. VELENIK, *Interface, surface tension and reentrant pinning transition in the 2D Ising model*, Commun. Math. Phys., **204** (1999), pp. 269–312.

213. A. PISZTORA, *Surface order large deviations for Ising, Potts and percolation models*, Probab. Theory Relat. Fields, **104** (1996), pp. 427–466.

214. B. PITTEL, *On a likely shape of the random Ferrers diagram*, Adv. Appl. Math., **18** (1997), pp. 432–488.

215. M. PRÄHOFER AND H. SPOHN, *Current fluctuations for the totally asymmetric simple exclusion process*, in "In and Out of Equilibrium (ed. Sidoravicius, Mambucaba, 2000)", pp. 185–204, Progr. Probab., **51**, Birkhäuser, 2002.

216. M. PRÄHOFER AND H. SPOHN, *Scale invariance of the PNG droplet and the Airy process*, J. Statist. Phys., **108** (2002), pp. 1071–1106.

217. J. QUASTEL, *Diffusion of color in the simple exclusion process*, Commun. Pure Appl. Math. **45**, 623–679 (1992).

218. M. REED AND B. SIMON, *Methods of Modern Mathematical Physics, vol. I: Functional Analysis, revised and enlarged edition; vol.II: Fourier Analysis, Self-Adjointness*, Academic Press, New York, 1980, 1975.
219. F. REZAKHANLOU, *Hydrodynamic limit for a system with finite range interactions*, Commun. Math. Phys., **129** (1990), pp. 445–480.
220. F. REZAKHANLOU, *Hydrodynamic limit for attractive particle systems on* $\mathbb{Z}^d$, Commun. Math. Phys., **140** (1991), pp. 417–448.
221. F. REZAKHANLOU, *Continuum limit for some growth models*, Stoch. Proc. Appl., **101** (2002), pp. 1–41.
222. F. REZAKHANLOU, *Continuum limit for some growth models II*, Ann. Probab., **29** (2001), pp. 1329–1372.
223. D. RUELLE, *Statistical Mechanics: Rigorous Results*, W. A. Benjamin, 1969.
224. Y. SAITO, *Statistical Physics of Crystal Growth*, World Sci., 1996.
225. H. SAKAGAWA, *The reversible measures of interacting diffusion system with plural conservation laws*, Markov Process. Related Fields, **7** (2001), pp. 289–300.
226. H. SAKAGAWA, *Entropic repulsion for a Gaussian lattice field with finite range interaction*, J. Math. Phys., **44** (2003), pp. 2939–2951.
227. H. SAKAGAWA, *Entropic repulsion for two dimensional multi-layered harmonic crystals*, J. Statist. Phys., **114** (2004), pp. 37–49.
228. H. SAKAGAWA, *Entropic repulsion for a high dimensional Gaussian lattice field between two walls*, preprint, 2004.
229. T. SEPPÄLÄINEN, *Existence of hydrodynamics for the totally asymmetric simple K-exclusion process*, Ann. Probab., **27** (1999), pp. 361–415.
230. S. SHEFFIELD, *Random surfaces: large deviations principles and gradient Gibbs measure classifications*, PhD thesis, Stanford Univ., 2003, arXiv:math.PR/0304049.
231. T. SHIGA AND A. SHIMIZU, *Infinite-dimensional stochastic differential equations and their applications*, J. Math. Kyoto Univ., **20** (1980), pp. 395–416.
232. I. SHIGEKAWA, *Stochastic Analysis*, Translations of Math. Monographs, **224**, Iwanami Series in Modern Math., Amer. Math. Soc., Providence, RI, 2004, 182 pp.
233. S. SHLOSMAN, *The Wulff construction in statistical mechanics and combinatorics*, Russian Math. Surveys, **56** (2001), pp. 709–738.
234. B. SIMON, *The $P(\phi)_2$ Euclidean (Quantum) Field Theory*, Princeton Univ. Press, 1974.
235. B. SIMON, *Functional Integration and Quantum Physics*, Academic Press, 1979.
236. YA.G. SINAI, *Distribution of some functionals of the integral of a random walk*, Theoret. Math. Phys., **90** (1992), pp. 219–241.
237. J. SJÖSTRAND, *Correlation asymptotics and Witten Laplacians*, St. Petersburg Math. J., **8** (1997), pp. 123–147.
238. F. SPITZER, *Principles of Random Walk*, Nostrand, 1964.
239. H. SPOHN, *Large Scale Dynamics of Interacting Particles*, Springer, 1991.
240. H. SPOHN, *Interface motion in models with stochastic dynamics*, J. Statis. Phys., **71** (1993), pp. 1081–1132.
241. A. STÖHR, *Über einige lineare partielle Differenzengleichungen mit konstanten Koeffizienten, I, II, III*, Math. Nachr., **3** (1950), pp. 208–242, pp. 295–315, pp. 330–357.
242. D.W. STROOCK AND S.R.S. VARADHAN, *Diffusion processes with boundary conditions*, Commun. Pure Appl. Math., **24** (1971), 147–225.

243. S. TAKANOBU AND S. WATANABE, *On the existence and uniqueness of diffusion processes with Wentzell's boundary conditions*, J. Math. Kyoto Univ., **28** (1988), pp. 71–80.

244. H. TANAKA, *Stochastic differential equations with reflecting boundary condition in convex regions*, Hiroshima Math. J., **9** (1979), pp. 163–177.

245. H. TANEMURA AND N. YOSHIDA, *Localization transition of d-friendly walkers*, Probab. Theory Relat. Fields., **125** (2003), pp. 593–608.

246. S.R.S. VARADHAN, *Large Deviations and Applications*, SIAM, 1984.

247. S.R.S. VARADHAN, *Nonlinear diffusion limit for a system with nearest neighbor interactions - II*, "Asymptotic problems in probability theory: stochastic models and diffusions on fractals" (eds. Elworthy and Ikeda), Longman, 1993, pp. 75–128.

248. S.R.S. VARADHAN, *Large deviations for the asymmetric simple exclusion process*, Proceedings of Shonan/Kyoto meetings "Stochastic Analysis on Large Scale Interacting Systems" (2002, eds. Funaki and Osada), Adv. Stud. Pure Math., **39**, Math. Soc. Japan, 2004, pp. 1–27.

249. Y. VELENIK, *Entropic repulsion of an interface in an external field*, Probab. Theory Relat. Fields, **129** (2004), pp. 83–112.

250. A.M. VERSHIK, *Statistical mechanics of combinatorial partitions, and their limit shapes*, Funct. Anal. Appl., **30** (1996), pp. 90–105.

251. A. VISINTIN, *Models of Phase Transitions*, Birkhäuser, 1996.

252. G.S. WEISS, *A free boundary problem for non-radial-symmetric quasi-linear elliptic equations*, Adv. Math. Sci. Appl., **5** (1995), pp. 497–555.

253. J. WLOKA, *Partial differential equations*, Cambridge Univ. Press, 1987.

254. G. WULFF, *Zur Frage der Geschwindigkeit des Wachsthums und der Auflösung der Krystallflächen*, Z. Krystallogr., **34** (1901), pp. 449–530.

255. K. YOSIDA, *Functional Analysis, 4th edition*, Springer, 1974.

256. L. ZAMBOTTI, *Fluctuations for a $\nabla\phi$ interface model with repulsion from a wall*, Probab. Theory Relat. Fields, **129** (2004), pp. 315–339.

257. M. ZHU, *Equilibrium fluctuations for one-dimensional Ginzburg-Landau lattice model*, Nagoya Math. J., **117** (1990), pp. 63–92.

258. M. ZHU, *The reversible measures of a conservative system with finite range interactions*, "Nonlinear Stochastic PDEs: Hydrodynamic Limit and Burgers' Turbulence" (eds. Funaki and Woyczynski), IMA volume **77** (1995), Univ. Minnesota, Springer, New York, pp. 53–64.

List of Participants

Lecturers

DEMBO Amir	Stanford Univ., USA
FUNAKI Tadahisa	Univ. Tokyo, Japan
MASSART Pascal	Univ. Paris-Sud, Orsay, F

Participants

AILLOT Pierre	Univ. Rennes 1, F
ATTOUCH Mohammed Kadi	Univ. Djillali Liabès, Sidi Bel Abbès, Algérie
AUDIBERT Jean-Yves	Univ. Pierre et Marie Curie, Paris, F
BAHADORAN Christophe	Univ. Blaise Pascal, Clermont-Ferrand, F
BEDNORZ Witold	Warsaw Univ., Poland
BELARBI Faiza	Univ. Djillali Liabès, Sidi Bel Abbès, Algérie
BEN AROUS Gérard	Courant Institute, New York, USA
BLACHE Fabrice	Univ. Blaise Pascal, Clermont-Ferrand, F
BLANCHARD Gilles	Univ. Paris-Sud, Orsay, F
BOIVIN Daniel	Univ. Brest, F
CHAFAI Djalil	Ecole Vétérinaire Toulouse, F
CHOUAF Benamar	Univ. Djillali Liabès, Sidi Bel Abbès, Algérie
DACHIAN Serguei	Univ. Blaise Pascal, Clermont-Ferrand, F
DELMOTTE Thierry	Univ. Paul Sabatier, Toulouse, F
DJELLOUT Hacène	Univ. Blaise Pascal, Clermont-Ferrand, F
DUROT Cécile	Univ. Paris-Sud, Orsay, F

FLORESCU Ionut	Purdue Univ., West Lafayette, USA
FONTBONA Joaquin	Univ. Pierre et Marie Curie, Paris, F
FOUGERES Pierre	Imperial College, London, UK
FROMONT Magalie	Univ. Paris-Sud, Orsay, F
GAIFFAS Stéphane	Univ. Pierre et Marie Curie, Paris, F
GUERIBALLAH Abdelkader	Univ. Djillali Liabès, Sidi Bel Abbès, Algérie
GIACOMIN Giambattista	Univ. Paris 7, F
GOLDSCHMIDT Christina	Univ. Cambridge, UK
GUSTO Gaelle	INRA, Jouy en Josas, F
HARIYA Yuu	Kyoto Univ., Japan
JIEN Yu-Juan	Purdue Univ., West Lafayette, USA
JOULIN Aldéric	Univ. La Rochelle, F
KLEIN Thierry	Univ. Versailles, F
KLUTCHNIKOFF Nicolas	Univ. Provence, Marseille, F
LEBARBIER Emilie	INRIA Rhône-Alpes, Saint-Ismier, F
LEVY-LEDUC Céline	Univ. Paris-Sud, Orsay, F
MAIDA Mylène	ENS Lyon, F
MALRIEU Florent	Univ. Rennes 1, F
MARTIN James	CNRS, Univ. Paris 7, F
MARTY Renaud	Univ. Paul Sabatier, Toulouse, F
MEREDITH Mark	Univ. Oxford, UK
MERLE Mathieu	ENS Paris, F
MOCIOALCA Oana	Purdue Univ., West Lafayette, USA
NISHIKAWA Takao	Univ. Tokyo, Japan
OBLOJ Jan	Univ. Pierre et Marie Curie, Paris, F
OSEKOWSKI Adam	Warsaw Univ., Poland
PAROUX Katy	Univ. Besançon, F
PASCU Mihai	Purdue Univ., West Lafayette, USA
PICARD Jean	Univ. Blaise Pascal, Clermont-Ferrand, F
REYNAUD-BOURET Patricia	Georgia Instit. Technology, Atlanta, USA
RIOS Ricardo	Univ. Central, Caracas, Venezuela
ROBERTO Cyril	Univ. Marne-la-Vallée, F
ROITERSHTEIN Alexander	Technion, Haifa, Israel

ROUX Daniel	Univ. Blaise Pascal, Clermont-Ferrand, F
ROZENHOLC Yves	Univ. Paris 7, F
SAINT-LOUBERT BIE Erwan	Univ. Blaise Pascal, Clermont-Ferrand, F
SCHAEFER Christin	Fraunhofer Institut FIRST, Berlin, D
STOLTZ Gilles	Univ. Paris-Sud, Orsay, F
TOUZILLIER Brice	Univ. Pierre et Marie Curie, Paris, F
TURNER Amanda	Univ. Cambridge, UK
VERT Régis	Univ. Paris-Sud, Orsay, F
VIENS Frederi	Purdue Univ., West Lafayette, USA
VIGON Vincent	INSA Rouen, F
YOR Marc	Univ. Pierre et Marie Curie, Paris, F
ZACHARUK Mariusz	Univ. Wrocław, Poland
ZEITOUNI Ofer	Univ. Minnesota, Minneapolis, USA
ZHANG Tao	Purdue Univ., West Lafayette, USA
ZWALD Laurent	Univ. Paris-Sud, Orsay, F

List of Short Lectures

Jean-Yves AUDIBERT   Aggregated estimators and empirical complexity for least squares regression

Christophe BAHADORAN   Convergence and local equilibrium for the one-dimensional asymmetric exclusion process

Fabrice BLACHE   Backward stochastic differential equations on manifolds

Gilles BLANCHARD   Some applications of model selection to statistical learning procedures

Serguei DACHIAN   Description of specifications by means of probability distributions in small volumes under condition of very weak positivity

Thierry DELMOTTE   How to use the stationarity of a reversible random environment to estimate derivatives of the annealed diffusion

Ionut FLORESCU   Pricing the implied volatility surface

Joaquin FONTBONA   Probabilistic interpretation and stochastic particle approximations of the 3-dimensional Navier-Stokes equation

Pierre FOUGERES   Curvature-dimension inequality and projections; some applications to Sobolev inequalities

Magalie FROMONT   Tests adaptatifs d'adéquation dans un
                            modèle de densité

Giambattista GIACOMIN   On random co-polymers and disordered
                            wetting models

Christina GOLDSCHMIDT   Critical random hypergraphs: a
                            stochastic process approach

Yuu HARIYA   Large time limiting laws for Brownian
                            motion perturbed by normalized
                            exponential weights (Part II)

Aldéric JOULIN   Isoperimetric and functional inequalities
                            in discrete settings: application to the
                            geometric distribution

Thierry KLEIN   Processus empirique et concentration
                            autour de la moyenne

Céline LEVY-LEDUC   Estimation de périodes de fonctions
                            périodiques bruitées et de forme
                            inconnue; applications à la vibrométrie
                            laser

James MARTIN   Particle systems, queues, and geodesics
                            in percolation

Renaud MARTY   Théorème limite pour une équation
                            différentielle à coefficient aléatoire à
                            mémoire longue

Oana MOCIOALCA   Additive summable processes and their
                            stochastic integral

Takao NISHIKAWA   Dynamic entropic repulsion for the
                            Ginzburg-Landau $\nabla\phi$ interface model

Jan OBLOJ   The Skorokhod embedding problem for
                            functionals of Brownian excursions

Katy PAROUX   Convergence locale d'un modèle booléen
                            de couronnes

Mihai N. PASCU   Maximum principles for Neumann /
                            mixed Dirichlet-Neumann eigenvalue
                            problem

Patricia REYNAUD-BOURET	Adaptive estimation of the Aalen multiplicative intensity by model selection
Cyril ROBERTO	Sobolev inequalities for probability measures on the real line
Alexander ROITERSHTEIN	Limit theorems for one-dimensional transient random walks in Markov environments
Gilles STOLTZ	Internal regret in on-line prediction of individual sequences and in on-line portfolio selection
Amanda TURNER	Convergence of Markov processes near hyperbolic fixed points
Vincent VIGON	Les abrupts et les érodés
Marc YOR	Large time limiting laws for Brownian motion perturbed by normalized exponential weights (Part I)
Ofer ZEITOUNI	Recursions and tightness

Lecture Notes in Mathematics

For information about earlier volumes
please contact your bookseller or Springer
LNM Online archive: springerlink.com

Vol. 1782: C.-H. Chu, A. T.-M. Lau, Harmonic Functions on Groups and Fourier Algebras (2002)

Vol. 1783: L. Grüne, Asymptotic Behavior of Dynamical and Control Systems under Perturbation and Discretization (2002)

Vol. 1784: L.H. Eliasson, S. B. Kuksin, S. Marmi, J.-C. Yoccoz, Dynamical Systems and Small Divisors. Cetraro, Italy 1998. Editors: S. Marmi, J.-C. Yoccoz (2002)

Vol. 1785: J. Arias de Reyna, Pointwise Convergence of Fourier Series (2002)

Vol. 1786: S. D. Cutkosky, Monomialization of Morphisms from 3-Folds to Surfaces (2002)

Vol. 1787: S. Caenepeel, G. Militaru, S. Zhu, Frobenius and Separable Functors for Generalized Module Categories and Nonlinear Equations (2002)

Vol. 1788: A. Vasil'ev, Moduli of Families of Curves for Conformal and Quasiconformal Mappings (2002)

Vol. 1789: Y. Sommerhäuser, Yetter-Drinfel'd Hopf algebras over groups of prime order (2002)

Vol. 1790: X. Zhan, Matrix Inequalities (2002)

Vol. 1791: M. Knebusch, D. Zhang, Manis Valuations and Prüfer Extensions I: A new Chapter in Commutative Algebra (2002)

Vol. 1792: D. D. Ang, R. Gorenflo, V. K. Le, D. D. Trong, Moment Theory and Some Inverse Problems in Potential Theory and Heat Conduction (2002)

Vol. 1793: J. Cortés Monforte, Geometric, Control and Numerical Aspects of Nonholonomic Systems (2002)

Vol. 1794: N. Pytheas Fogg, Substitution in Dynamics, Arithmetics and Combinatorics. Editors: V. Berthé, S. Ferenczi, C. Mauduit, A. Siegel (2002)

Vol. 1795: H. Li, Filtered-Graded Transfer in Using Noncommutative Gröbner Bases (2002)

Vol. 1796: J.M. Melenk, hp-Finite Element Methods for Singular Perturbations (2002)

Vol. 1797: B. Schmidt, Characters and Cyclotomic Fields in Finite Geometry (2002)

Vol. 1798: W.M. Oliva, Geometric Mechanics (2002)

Vol. 1799: H. Pajot, Analytic Capacity, Rectifiability, Menger Curvature and the Cauchy Integral (2002)

Vol. 1800: O. Gabber, L. Ramero, Almost Ring Theory (2003)

Vol. 1801: J. Azéma, M. Émery, M. Ledoux, M. Yor (Eds.), Séminaire de Probabilités XXXVI (2003)

Vol. 1802: V. Capasso, E. Merzbach, B.G. Ivanoff, M. Dozzi, R. Dalang, T. Mountford, Topics in Spatial Stochastic Processes. Martina Franca, Italy 2001. Editor: E. Merzbach (2003)

Vol. 1803: G. Dolzmann, Variational Methods for Crystalline Microstructure – Analysis and Computation (2003)

Vol. 1804: I. Cherednik, Ya. Markov, R. Howe, G. Lusztig, Iwahori-Hecke Algebras and their Representation Theory. Martina Franca, Italy 1999. Editors: V. Baldoni, D. Barbasch (2003)

Vol. 1805: F. Cao, Geometric Curve Evolution and Image Processing (2003)

Vol. 1806: H. Broer, I. Hoveijn. G. Lunther, G. Vegter, Bifurcations in Hamiltonian Systems. Computing Singularities by Gröbner Bases (2003)

Vol. 1807: V. D. Milman, G. Schechtman (Eds.), Geometric Aspects of Functional Analysis. Israel Seminar 2000-2002 (2003)

Vol. 1808: W. Schindler, Measures with Symmetry Properties (2003)

Vol. 1809: O. Steinbach, Stability Estimates for Hybrid Coupled Domain Decomposition Methods (2003)

Vol. 1810: J. Wengenroth, Derived Functors in Functional Analysis (2003)

Vol. 1811: J. Stevens, Deformations of Singularities (2003)

Vol. 1812: L. Ambrosio, K. Deckelnick, G. Dziuk, M. Mimura, V. A. Solonnikov, H. M. Soner, Mathematical Aspects of Evolving Interfaces. Madeira, Funchal, Portugal 2000. Editors: P. Colli, J. F. Rodrigues (2003)

Vol. 1813: L. Ambrosio, L. A. Caffarelli, Y. Brenier, G. Buttazzo, C. Villani, Optimal Transportation and its Applications. Martina Franca, Italy 2001. Editors: L. A. Caffarelli, S. Salsa (2003)

Vol. 1814: P. Bank, F. Baudoin, H. Föllmer, L.C.G. Rogers, M. Soner, N. Touzi, Paris-Princeton Lectures on Mathematical Finance 2002 (2003)

Vol. 1815: A. M. Vershik (Ed.), Asymptotic Combinatorics with Applications to Mathematical Physics. St. Petersburg, Russia 2001 (2003)

Vol. 1816: S. Albeverio, W. Schachermayer, M. Talagrand, Lectures on Probability Theory and Statistics. Ecole d'Eté de Probabilités de Saint-Flour XXX-2000. Editor: P. Bernard (2003)

Vol. 1817: E. Koelink, W. Van Assche(Eds.), Orthogonal Polynomials and Special Functions. Leuven 2002 (2003)

Vol. 1818: M. Bildhauer, Convex Variational Problems with Linear, nearly Linear and/or Anisotropic Growth Conditions (2003)

Vol. 1819: D. Masser, Yu. V. Nesterenko, H. P. Schlickewei, W. M. Schmidt, M. Waldschmidt, Diophantine Approximation. Cetraro, Italy 2000. Editors: F. Amoroso, U. Zannier (2003)

Vol. 1820: F. Hiai, H. Kosaki, Means of Hilbert Space Operators (2003)

Vol. 1821: S. Teufel, Adiabatic Perturbation Theory in Quantum Dynamics (2003)

Vol. 1822: S.-N. Chow, R. Conti, R. Johnson, J. Mallet-Paret, R. Nussbaum, Dynamical Systems. Cetraro, Italy 2000. Editors: J. W. Macki, P. Zecca (2003)

Vol. 1823: A. M. Anile, W. Allegretto, C. Ringhofer, Mathematical Problems in Semiconductor Physics. Cetraro, Italy 1998. Editor: A. M. Anile (2003)

Vol. 1824: J. A. Navarro González, J. B. Sancho de Salas, $\mathscr{C}^{\infty}$ – Differentiable Spaces (2003)

Vol. 1825: J. H. Bramble, A. Cohen, W. Dahmen, Multiscale Problems and Methods in Numerical Simulations, Martina Franca, Italy 2001. Editor: C. Canuto (2003)

Vol. 1826: K. Dohmen, Improved Bonferroni Inequalities via Abstract Tubes. Inequalities and Identities of Inclusion-Exclusion Type. VIII, 113 p, 2003.

Vol. 1827: K. M. Pilgrim, Combinations of Complex Dynamical Systems. IX, 118 p, 2003.

Vol. 1828: D. J. Green, Gröbner Bases and the Computation of Group Cohomology. XII, 138 p, 2003.

Vol. 1829: E. Altman, B. Gaujal, A. Hordijk, Discrete-Event Control of Stochastic Networks: Multimodularity and Regularity. XIV, 313 p, 2003.

Vol. 1830: M. I. Gil', Operator Functions and Localization of Spectra. XIV, 256 p, 2003.

Vol. 1831: A. Connes, J. Cuntz, E. Guentner, N. Higson, J. E. Kaminker, Noncommutative Geometry, Martina Franca, Italy 2002. Editors: S. Doplicher, L. Longo (2004)

Vol. 1832: J. Azéma, M. Émery, M. Ledoux, M. Yor (Eds.), Séminaire de Probabilités XXXVII (2003)

Vol. 1833: D.-Q. Jiang, M. Qian, M.-P. Qian, Mathematical Theory of Nonequilibrium Steady States. On the Frontier of Probability and Dynamical Systems. IX, 280 p, 2004.

Vol. 1834: Yo. Yomdin, G. Comte, Tame Geometry with Application in Smooth Analysis. VIII, 186 p, 2004.

Vol. 1835: O.T. Izhboldin, B. Kahn, N.A. Karpenko, A. Vishik, Geometric Methods in the Algebraic Theory of Quadratic Forms. Summer School, Lens, 2000. Editor: J.-P. Tignol (2004)

Vol. 1836: C. Năstăsescu, F. Van Oystaeyen, Methods of Graded Rings. XIII, 304 p, 2004.

Vol. 1837: S. Tavaré, O. Zeitouni, Lectures on Probability Theory and Statistics. Ecole d'Eté de Probabilités de Saint-Flour XXXI-2001. Editor: J. Picard (2004)

Vol. 1838: A.J. Ganesh, N.W. O'Connell, D.J. Wischik, Big Queues. XII, 254 p, 2004.

Vol. 1839: R. Gohm, Noncommutative Stationary Processes. VIII, 170 p, 2004.

Vol. 1840: B. Tsirelson, W. Werner, Lectures on Probability Theory and Statistics. Ecole d'Eté de Probabilités de Saint-Flour XXXII-2002. Editor: J. Picard (2004)

Vol. 1841: W. Reichel, Uniqueness Theorems for Variational Problems by the Method of Transformation Groups (2004)

Vol. 1842: T. Johnsen, A.L. Knutsen, K3 Projective Models in Scrolls (2004)

Vol. 1843: B. Jefferies, Spectral Properties of Noncommuting Operators (2004)

Vol. 1844: K.F. Siburg, The Principle of Least Action in Geometry and Dynamics (2004)

Vol. 1845: Min Ho Lee, Mixed Automorphic Forms, Torus Bundles, and Jacobi Forms (2004)

Vol. 1846: H. Ammari, H. Kang, Reconstruction of Small Inhomogeneities from Boundary Measurements (2004)

Vol. 1847: T.R. Bielecki, T. Björk, M. Jeanblanc, M. Rutkowski, J.A. Scheinkman, W. Xiong, Paris-Princeton Lectures on Mathematical Finance 2003 (2004)

Vol. 1848: M. Abate, J. E. Fornaess, X. Huang, J. P. Rosay, A. Tumanov, Real Methods in Complex and CR Geometry, Martina Franca, Italy 2002. Editors: D. Zaitsev, G. Zampieri (2004)

Vol. 1849: Martin L. Brown, Heegner Modules and Elliptic Curves (2004)

Vol. 1850: V. D. Milman, G. Schechtman (Eds.), Geometric Aspects of Functional Analysis. Israel Seminar 2002-2003 (2004)

Vol. 1851: O. Catoni, Statistical Learning Theory and Stochastic Optimization (2004)

Vol. 1852: A.S. Kechris, B.D. Miller, Topics in Orbit Equivalence (2004)

Vol. 1853: Ch. Favre, M. Jonsson, The Valuative Tree (2004)

Vol. 1854: O. Saeki, Topology of Singular Fibers of Differential Maps (2004)

Vol. 1855: G. Da Prato, P.C. Kunstmann, I. Lasiecka, A. Lunardi, R. Schnaubelt, L. Weis, Functional Analytic Methods for Evolution Equations. Editors: M. Iannelli, R. Nagel, S. Piazzera (2004)

Vol. 1856: K. Back, T.R. Bielecki, C. Hipp, S. Peng, W. Schachermayer, Stochastic Methods in Finance, Bressanone/Brixen, Italy, 2003. Editors: M. Fritelli, W. Runggaldier (2004)

Vol. 1857: M. Émery, M. Ledoux, M. Yor (Eds.), Séminaire de Probabilités XXXVIII (2005)

Vol. 1858: A.S. Cherny, H.-J. Engelbert, Singular Stochastic Differential Equations (2005)

Vol. 1859: E. Letellier, Fourier Transforms of Invariant Functions on Finite Reductive Lie Algebras (2005)

Vol. 1860: A. Borisyuk, G.B. Ermentrout, A. Friedman, D. Terman, Tutorials in Mathematical Biosciences I. Mathematical Neurosciences (2005)

Vol. 1861: G. Benettin, J. Henrard, S. Kuksin, Hamiltonian Dynamics – Theory and Applications, Cetraro, Italy, 1999. Editor: A. Giorgilli (2005)

Vol. 1862: B. Helffer, F. Nier, Hypoelliptic Estimates and Spectral Theory for Fokker-Planck Operators and Witten Laplacians (2005)

Vol. 1863: H. Fürh, Abstract Harmonic Analysis of Continuous Wavelet Transforms (2005)

Vol. 1864: K. Efstathiou, Metamorphoses of Hamiltonian Systems with Symmetries (2005)

Vol. 1865: D. Applebaum, B.V. R. Bhat, J. Kustermans, J. M. Lindsay, Quantum Independent Increment Processes I. From Classical Probability to Quantum Stochastic Calculus. Editors: M. Schürmann, U. Franz (2005)

Vol. 1866: O.E. Barndorff-Nielsen, U. Franz, R. Gohm, B. Kümmerer, S. Thorbjønsen, Quantum Independent Increment Processes II. Structure of Quantum Levy Processes, Classical Probability, and Physics. Editors: M. Schürmann, U. Franz, (2005)

Vol. 1867: J. Sneyd (Ed.), Tutorials in Mathematical Biosciences II. Mathematical Modeling of Calcium Dynamics and Signal Transduction. (2005)

Vol. 1868: J. Jorgenson, S. Lang, $Pos_n(R)$ and Eisenstein Sereies. (2005)

Vol. 1869: A. Dembo, T. Funaki, Lectures on Probability Theory and Statistics. Ecole d'Eté de Probabilités de Saint-Flour XXXIII-2003. Editor: J. Picard (2005)

Recent Reprints and New Editions

Vol. 1200: V. D. Milman, G. Schechtman (Eds.), Asymptotic Theory of Finite Dimensional Normed Spaces. 1986. – Corrected Second Printing (2001)

Vol. 1471: M. Courtieu, A.A. Panchishkin, Non-Archimedean L-Functions and Arithmetical Siegel Modular Forms. – Second Edition (2003)

Vol. 1618: G. Pisier, Similarity Problems and Completely Bounded Maps. 1995 – Second, Expanded Edition (2001)

Vol. 1629: J.D. Moore, Lectures on Seiberg-Witten Invariants. 1997 – Second Edition (2001)

Vol. 1638: P. Vanhaecke, Integrable Systems in the realm of Algebraic Geometry. 1996 – Second Edition (2001)

Vol. 1702: J. Ma, J. Yong, Forward-Backward Stochastic Differential Equations and their Applications. 1999. – Corrected 3rd printing (2005)

Printing and Binding: Strauss GmbH, Mörlenbach

4. Manuscripts should in general be submitted in English. Final manuscripts should contain at least 100 pages of mathematical text and should always include

   – a general table of contents;

   – an informative introduction, with adequate motivation and perhaps some historical remarks: it should be accessible to a reader not intimately familiar with the topic treated;

   – a global subject index: as a rule this is genuinely helpful for the reader.

   Lecture Notes volumes are, as a rule, printed digitally from the authors' files. We strongly recommend that all contributions in a volume be written in the same LaTeX version, preferably LaTeX2e. To ensure best results, authors are asked to use the LaTeX2e style files available from Springer's web-server at

   ftp://ftp.springer.de/pub/tex/latex/mathegl/mono.zip (for monographs) and
   ftp://ftp.springer.de/pub/tex/latex/mathegl/mult.zip (for summer schools/tutorials).

   Additional technical instructions, if necessary, are available on request from:

   lnm@springer-sbm.com.

5. Careful preparation of the manuscripts will help keep production time short besides ensuring satisfactory appearance of the finished book in print and online. After acceptance of the manuscript authors will be asked to prepare the final LaTeX source files (and also the corresponding dvi-, pdf- or zipped ps-file) together with the final printout made from these files. The LaTeX source files are essential for producing the full-text online version of the book. For the existing online volumes of LNM see:

   http://www.springerlink.com/openurl.asp?genre=journal&issn=0075-8434.

   The actual production of a Lecture Notes volume takes approximately 8 weeks.

6. Volume editors receive a total of 50 free copies of their volume to be shared with the authors, but no royalties. They and the authors are entitled to a discount of 33.3 % on the price of Springer books purchased for their personal use, if ordering directly from Springer.

7. Commitment to publish is made by letter of intent rather than by signing a formal contract. Springer-Verlag secures the copyright for each volume. Authors are free to reuse material contained in their LNM volumes in later publications: A brief written (or e-mail) request for formal permission is sufficient.

Addresses:

Professor J.-M. Morel, CMLA,
École Normale Supérieure de Cachan,
61 Avenue du Président Wilson, 94235 Cachan Cedex, France
E-mail: Jean-Michel.Morel@cmla.ens-cachan.fr

Professor F. Takens, Mathematisch Instituut,
Rijksuniversiteit Groningen, Postbus 800,
9700 AV Groningen, The Netherlands
E-mail: F.Takens@math.rug.nl

Professor B. Teissier, Institut Mathématique de Jussieu,
UMR 7586 du CNRS, Équipe "Géométrie et Dynamique",
175 rue du Chevaleret, 75013 Paris, France
E-mail: teissier@math.jussieu.fr

For the "Mathematical Biosciences Subseries" of LNM :
Professor P. K. Maini, Center for Mathematical Biology,
Mathematical Institute, 24-29 St Giles,
Oxford OX1 3LP, UK
E-mail : maini@maths.ox.ac.uk

Springer, Mathematics Editorial I, Tiergartenstr. 17,
69121 Heidelberg, Germany,
Tel.: +49 (6221) 487-8410
Fax: +49 (6221) 487-8355
E-mail: lnm@springer-sbm.com